Crop Loss Assessment and Pest Management

Edited by P. S. Teng

APS PRESS
The American Phytopathological Society
St. Paul, Minnesota

This book has been reproduced directly from typewritten
copy submitted in final form to APS Press
by the editor of the volume. No editing
or proofreading has been done by the Press.

Reference in this publication to a trademark, proprietary product,
or company name by personnel
of the U.S. Department of Agriculture or anyone else
is intended for explicit description only
and does not imply approval or recommendation
to the exclusion of others that may be suitable.

Material written by Federal employees as part of their jobs
is in the public domain.

Library of Congress Card Catalog Number: 87-070150
International Standard Book Number: 0-89054-079-9

Printed in the United States of America

The American Phytopathological Society
3340 Pilot Knob Road
St. Paul, Minnesota 55121, USA

TABLE OF CONTENTS

PREFACE

Pest management is an integral part of crop production activities, whether it is in a developing or a developed part of the world, and whether it is in the context of a subsistence multi-component agricultural system or a cash-cropped monocultural system. Much has been written on the rationale for pest management, in particular on the concept on "Integrated Pest Management" (IPM) and on its applications. In many respects, IPM represents the basis for problem-solving in pest management, leading ultimately to increased benefits to the farmer, the consumer and society as a whole. There is however, another aspect of pest management -- the problem-definition aspect -- which has been receiving more attention by scientists and administrators alike, because of the realization that limited resources require a judicious use of inputs to address a well-defined problem. This problem-definition aspect of pest management contains a collection of methods which has come to be known as **CROP LOSS ASSESSMENT**. This book contains a collection of papers by different authors which deal with topics commonly included in crop loss assessment.

The idea for this book grew out of an international training course on crop loss assessment as a means to improve crop production and pest management, held at the University of Minnesota in July 1984. The course was modeled after workshops held by the Plant Protection Service of the Food and Agriculture Organization, United Nations, in different regions of the world over the past decade. In this book, as in the course, an attempt has been made to gather together the topics that are identified with Crop Loss Assessment per se, but at the same time, include other topics that demonstrate the use of this technology in pest management. The aim of this book is to give practitioners and students of plant protection a solid introduction to the topic of crop loss assessment, with emphasis on the link between theory and application. It has been written for an audience that includes graduate and senior undergraduate students of plant protection, and for scientists working in developing countries who desire a "catch-up" text on the topic.

The first chapter by J.C. Zadoks, a pioneer in epidemiology, crop loss assessment and disease management, deals with the conceptual and historical basis for crop loss assessment. Following this, the topic of **MEASUREMENT** is addressed in Chapters 2, 3 and 4, focussing respectively on diseases, insects and the crop. The **EXPERIMENTAL** aspects of crop loss assessment are discussed in Chapters 5 through 10, in which authors address pest population dynamics, sampling, field data collection and methods for generating different pest intensity levels. Chapters 11 through 16 deal with the synthesis of data into meaningful **PEST-LOSS MODELS** which account for aspects of the environment and crop physiology. The use of statistical techniques such as regression and factor analysis are explored throughout these chapters. The above methodology in the context of **PEST MANAGEMENT** is next dealt with by the authors of Chapters 17 through 21, in which the systems approach, thresholds, predictive systems and economic decision making are discussed. Finally, selected **CASE STUDIES**, representing various uses of crop loss assessment technology, are presented by the authors of Chapters 22 through 25.

The selection of topics for this book was based on their relevance to practical situations. This theme is also reflected in the selection of the lecturers and subsequent authors, most of whom have considerable experience in applying crop loss assessment to cropping systems in developed and developing countries. I thank all the authors for their perseverance with this project and to the dedication they have shown in ensuring this volume goes to press. Thanks are also due to Bill Fry and reviewers of the chapters for their helpful comments, and to Claudine Pardo and Joseph O'Brien for their painstaking efforts in helping to get the manuscript into a final form. All remaining errors or omissions must remain mine.

Lastly, I crave indulgence in dedicating this book to three persons without whom crop loss assessment would have remained still an impossible art. They are - Dr. W. Clive James of the International Wheat and Maize Improvement Center (CIMMYT), Mexico, Dr. John F. Fulkerson, United States Department of Agriculture, Cooperative States Research Service, Washington D.C. and Dr. Luigi Chiarappa, formerly Chief of the Plant Protection Service, Food and Agriculture Organization, United Nations, Rome, Italy but now enjoying his well-deserved retirement in Davis, California.

Paul S. Teng
St. Paul, Minnesota.
January 1987

CHAPTER 1

RATIONALE AND CONCEPTS OF CROP LOSS ASSESSMENT FOR IMPROVING PEST MANAGEMENT AND CROP PROTECTION

J. C. Zadoks

Laboratory of Phytopathology,
Agricultural University, Binnenhaven 9,
6709 PD Wageningen, The Netherlands

Why is the interest in crop loss research so strong at present? Crop loss assessment has not always received the recognized interest as was once evident during the late 19th century. The area of crop loss research may historically be viewed as belonging to three periods -- the exploratory, emergency, and implementation. These periods differ in their origin of interest in crop loss research and the direction in which this originating interest led.

THE EXPLORATORY PERIOD

During the 19th century, when purchasing power was less equitably distributed than at present and transportation means were less powerful, crop failure through pest and diseases meant hunger and death. The potato murrain in 1845 and 1846, caused by Phytophthora infestans, is well documented. Over a million people died in Ireland (10), and hundreds and maybe thousands died in The Netherlands (6). The potato blight was one of the incidents that contributed to the beginning of phytopathology; however, the disastrous incident was not directly approached by scientists of that particular period. Though minor in comparison to the potato murrain of 1845, the famine in Europe, due to the rye crop failure and possibly caused by Puccinia striiformis (20), also failed to attract the attention of scientists.

Near the end of the 19th century, however, the interest in crop loss research grew as particular incidents such as the destruction of the California citrus groves by the San Jose scale (Quadraspidiotus pernicious) triggered scientific investigation. Scientists and administrators began to realize the threat due to international dispersal of harmful agents.

What is presently called disease and loss assessment started in Germany in 1880, when the German government was requested to establish a central plant disease and pest registration system. However, the government did not respond to the request and the German Agricultural Society assumed the task in 1890. The Prussian Ministry of Agriculture later replaced the Agricultural Society in 1895 (27).

Initiated by the Swedish phytopathologist, J. Ericksson, international consultations which addressed the area of crop loss research began at the International Congress of Agriculture and Forestry in Vienna, 1890. The International Agricultural Congress of 1903 in Rome initiated an International Phytopathological Committee headed by P. Sorauer. In 1905, the International Agriculture Institute founded in Rome, direct predecessor of the United Nations Food and Agriculture Organization (F.A.O.), organized an International

Phytopathological Conference in 1914. National and international statistical surveys were initiated; however, with the onset of World War I these efforts were abruptly discontinued.

This era may be termed the "exploratory period". Standard methods for disease and loss assessment were absent, transportation and communication were awkward, and objectives were not clearly identified. However, the period revealed the growing demand of governments, scientists, and practicians for information concerning crop loss.

THE EMERGENCY PERIOD

Under the stressed conditions created by World War I, agricultural research was forced to become efficient. Though labor and money were scarce, there existed a need for food production to be safeguarded.

The German Empire, large but isolated by the Allies, felt the impact of food scarcity during World War I. In 1916, the economic resilience of the country was seriously threatened by the decreased potato harvest (18). The famine caused by potato late blight, indeed, broke the resistance of the German soldier whose will to fight became paralyzed as his thoughts returned to his relatives at home who were tormented by hunger (13).

In the first World War, the United States led in field oriented, phytopathologic research (7). A Plant Disease Survey was initiated in 1917. G.R. Lyman, a noted phytopathologist, stated: "How can we expect practical men to be properly impressed with the importance of our work and to vote large sums of money for its support when in place of facts we have only vague guesses to give them and we do not take the trouble to make careful estimates" (14). This statement remains appropriate up till today.

The period between the World Wars revealed less stressful conditions. However, with the onset of the second World War, the acute interest in food production again emerged as evident with the establishment of the War Emergency Committee by the American Phytopathological Society (21). The "emergency period" consisted of two war time situations a quarter century apart.

Field oriented research was again accelerated (7, 15). Within the United Kingdom, war time food restrictions have played an incubator role in the formulation of crop loss research, including disease and pest assessment. W.C. Moore (16), in his introduction to the new British Journal of Plant Pathology, announced disease and weather surveys and crop loss estimates, thus launching the later famous Harpenden school of crop loss assessment (4, 9, 12, 19).

British entomologists led the way in pest assessment schemes (19). A nation wide network of field plots for the study of potato late blight forecasting was established, complete with crop loss assessment (11). J. Grainger (5), working in the United Kingdom, did much original crop phenology and crop loss work. These people established the basic methodology in pest, disease and loss assessment which, with modifications and improvements, is still used today. Their major aim was to establish national and regional crop loss figures that could support policy decisions on crop protection research. Their work was essentially nationally oriented.

This national orientation of crop loss assessment as a means to support governmental policy was the _Leitmotive_ of the Symposium of Crop Loss, organized by F.A.O. in Rome 1967. The Crop Loss Manual, edited by L. Chiarappa (1, 2), marks the period when basic crop loss methodology was developed.

THE IMPLEMENTATION PERIOD

Yield and loss figures can be used in a retrospective and in a prospective manner, and at the national and individual production level.

The retrospective method registers past events. Yield and loss figures used retrospectively at the national or regional levels can be useful for:
- Crop production inventories
- Crop insurance policies
- Farmer subsidies after a catastrophe
- Research policy
- Regulatory measures including eradication and quarantine
- Policy of harvesting, transportation, and storage
- Policy of pesticide supply
- Variety and gene deployment.

The prospective method utilizes information such as knowledge of crop development, actual situation of harmful agents, and weather. This prospective information leads to a loss expectation or prediction which is helpful in predicting:
- Timing of pesticide treatments
- Emergency measures for large scale control
- Logistics of pesticide supply
- Logistics of equipment supply
- Logistics of harvesting, transportation, and storage.

Thus far, the methodology discussed pertains to national or regional levels. Recently, however, crop loss considerations have been applied at the level of farm and individual field situations. However, the distinction between what constitutes retrospective methodology versus prospective methodology is not clearly defined on the individual production level as the farmer's reliance upon experience becomes the major guideline for action. Crop protection considerations have often played a determining role in the choice of crop, cultivar, and crop rotation. Tillage methods serve the purpose of crop protection. Plowing and harrowing are classical weed control methods.

The new element in recent crop loss science is the prospective application on the farm and, especially, at the field level. The knowledge of population dynamics and epidemiology has increased to a level where it is possible to make reasonable predictions of future loss early in the season, when there is time to take appropriate action. Threshold theory teaches that such action is economically rewarding only if the expected loss exceeds the expected costs of control (24). This is the basic principle of integrated control of pests and diseases (17).

Though the basic concept of integrated pest management is simple, its implementation is not. The vast amount of crop protection knowledge available today has not been focused on the implementation procedure so that much work remains to done. Several examples of successful implementation practices discussed in the present volume include EPIPRE (24) and SIRATAC (8). Examples from the U.S.A. can also be found in the "Success Stories" by CIPM (3).

I consider the introduction of crop loss methodology on the farm and in the field as a major step forward. Therefore, the recent period was called the "implementation period". Though the implementation period began about 1950, most of the work lies still ahead of us. The 1984 international training course on "Crop Loss Assessment" at St. Paul, Minnesota, and the present book, that resulted from it, have addressed many of these issues.

TERMINOLOGY

The F.A.O. Crop Loss Manual has caused some degree of unification of concepts and terms. It is good to refer to it as much as possible, especially to Supplement 3 (2). Still, a few simple terms might help. In another contribution to this volume (25), I define the concepts "injury", "damage", and "loss" in an elementary and logically consistent way.

In the same contribution I draw attention to the fact that crop yield and therewith crop loss may mean different things in different parts of the world. If we disregard the concept of "production level", our efforts will come to nothing. This warning is particularly appropriate for those areas of the world where the production level is too low to allow economical use of pesticides (but not necessarily of other pest control methods), and to those areas where the production level is too high, with economically unacceptable environmental pollution. The latter situation, overproduction, is triggering a political change which eventually, will lead to a deliberately chosen level of production that is lower than the maximum possible. In northwest Europe, some countries are moving in this direction. Sweden probably leads in this endeavor. Discussions in the Netherlands, where research on optimalization of production levels is ongoing (26), point in a similar direction (22).

Finally I address the question concerning the appropriate terminology for this area of study which integrates epidemiology, agronomy, and economics, and also deals with the science of crop loss and corollary problems. I propose to call it "blaptology", derived from the Greek word "blapto" which means "I damage". It sounds better than "croplossology" and agrees with terms like "phytopathology" and "entomology" or "epidemiology", "agronomy", and "economics", all of which have Greek roots.

LITERATURE CITED

1. Chiarappa, L., ed. 1971. Crop Loss Assessment Methods. F.A.O. manual on the for the evaluation and prevention of losses by pests, diseases and weeds. Commonw. Agric. Bur., Farnham Royal (UK): loose leafed.
2. Chiarappa, L., ed. 1981. Crop Loss Assessment Methods. Supplement 3. FAO/Commonw. Agric. Bur., Farnham Royal (UK): 123 pp.
3. CIPM. 1983. Success Stories. College Station, Texas: Consortium for Integrated Pest Management. 198 pp.
4. Grainger, J. 1967. Economic aspects of crop losses caused by diseases, pp. 55-98. F.A.O. Symposium on crop losses, Rome. 330 pp.
5. Hooijer, C. 1847. De groote nood des hongers in en bij den Boemelerwaard. Norman, Zaltbommel. 33 pp.
6. Horsfall, J.G. 1969. Relevance: are we smart outside? Phytopathology News 3:5-9.
7. Ives, P.M., Wilson, L.T., Cull, P.O., Palmer. W.A., Haywood, C., et al. 1984. Field use of SIRATAC: An Australian computer-based pest management system for cotton. Protection Ecology 6:1-21.
8. James, W.C. 1974. Assessment of plant diseases and losses. Annu. Rev. Phytopathol. 12:27-48.
9. King, J.E. 1980. Cereal survey methodology in England and Wales. In Crop Loss Assessment, ed. P.S. Teng, S.V. Krupa, pp. 124-133. Misc. Publ. no. 7, Univ. of Minn. Agric. Exp. Stn., St. Paul, MN. 327 pp.
10. Large, E.C. 1950. The Advance of the Fungi. 3rd ed. London: Jonathan Cape. 488 pp.
11. Large, E.C. 1953. Potato blight forecasting investigation in England and Wales, 1950-52. Plant Pathol. 2:1-15.
12. Large, E.C. 1966. Measuring plant disease. Annu. Rev. Phytopathol. 4:9-28.
13. Lohr von Wachendorf, F. 1954. Die grosse Plage. Herkul, Frankfort. 2te. Aufl. 595 pp.

14. Lyman, G.R. 1918. The relation of phytopathologists to plant disease survey work. Phytopathology 8:219-228.
15. Maan, G.C., Zadoks, J.C. 1977. Trends in research reported in the Netherlands Journal of Plant Pathology, 1895-1973. Neth. J. Plant Pathol. 83:91-95.
16. Moore, W.C. 1952. Introduction "Plant Pathology." Plant Pathol. 1:1.
17. Stern, V.M., Smith, R.F., van den Bosch, R., Hagen, K.S. 1959. The integrated control concept. Hilgardia 29:81-101.
18. Stoermer. 1918. Der Kartofferlbau. In Arbeitsziele der Deutschen Landwirtschaft nach dem Kriege, ed. E. von Braun, pp. 292-342. Berlin. 986 pp.
19. Strickland, A.H. 1953. Pest assessment work in England and Wales, 1946-1951. Plant Pathol. 2:78-79.
20. Treviranus, L.C. 1846. Der Spelzenbrand im Roggen. Bot. Z. 4:629-631.
21. War Emergency Committee. 1942. The War Emergency Committee of the American Phytopathological Society. Phytopathology 32: 917-919.
22. Weijden, W.J. van der, et al. 1984. Bouwstenen voor een geintegreerde landbouw. WWR voorstudies en achtergronden V44. Staatsuitgeverij, Den Haag. 196 pp.
23. Zadoks, J.C. 1984. A quarter century of disease warning, 1958-1983. Plant Dis. 68:352-355.
24. Zadoks, J.C. 1984. EPIPRE, a computer-based scheme for pest and disease control in wheat. In Cereal Production, ed. E.J. Gallagher, pp. 215-225. London: Butterworths. 354 pp.
25. Zadoks, J.C. 1985. The concept of thresholds: warning, action and damage thresholds. Chapter 18, this book.
26. Zadoks, J.C. 1986. The Dutch approach toward integrated agriculture. In press.
27. Zadoks, J.C., Koster, L.M. 1976. A historical survey of botanical epidemiology. A sketch of the development of ideas in ecological phytopathology. Meded. Landbouwhogeschool, Wageningen 76-12:56 pp.

CHAPTER 2

MEASUREMENT OF DISEASE AND PATHOGENS

R. E. Gaunt

Microbiology Department, Lincoln University College of Agriculture,
Christchurch, New Zealand

The measurement of disease and pathogen present is fundamental to crop loss studies and to many types of disease prediction and management systems. However, despite the interest, and the rapid development of good objective methods for disease diagnosis, disease measurement has been an art rather than a well defined science, and many methods today are still subjective and inaccurate. In addition to crop loss assessment, disease measurement is done by scientists specializing in epidemiology, plant breeding and chemical efficacy, leading to a diversification of methods to suit particular requirements. Recently, there has been progress in standardization whilst recognizing the specific requirements for each type of investigation. The subject has previously been reviewed by Moore (39), Chester (8), Large (32), Grainger (19), James (24), James & Teng (26) and Berger (5).

An analysis of objectives often defines criteria for the method of disease or pathogen measurement. Disease and pathogen measurements are used to quantify the development of epidemics in space and time (61), for analysis of the factors that affect disease development (e.g. temperature, surface wetness), as a basis for disease and yield loss prediction, and for the definition of disease action or threshold levels for management programs. These investigations may involve measurement before crop establishment or during crop growth to assess disease risk. Measurements for studies of the physiological basis of yield loss (Chapter 16) preferably should take some account of the effect of disease on the ability of the plant to grow and develop. Crop loss surveys require simple methods, preferably based on similar concepts to those used in the original studies. Measurements are also used to assess plant resistance to disease and chemical efficacy. In this chapter the criteria for measurement methods are examined for a range of disease types and objectives, and the main methods of measurement are described. The selection of methods is discussed, and it is shown that for many purposes it is possible to standardize the basic methods, although the application of the methods in terms of time, frequency and sample unit may vary.

CRITERIA FOR METHODS OF DISEASE MEASUREMENT

Growth Stage Keys

There is an increasing need for data on disease development to be transferable, so that comparisons may be made between seasons, locations and investigators. This imposes some constraints and requirements on the methodologies adopted. For example, a growth stage key is required to describe the development stage of the crop, and this preferably should also describe the amount of growth which has occurred at each development stage. This allows useful comparisons in situations where the time and length of growing season may be different. Growth stage keys are usually based on detailed phenological studies of plants, such as that described by Feekes (30) for cereals, and by Robinson

(48) for sunflower. Each of these keys lacks detail in parts of the growth cycle, and there is little provision for quantification of growth. The keys are essentially descriptive and therefore do not lend themselves readily to computer based data handling. Zadoks et al. (62) developed a detailed decimal code for cereals suited to computer analysis, which has been used as a model for the development of other detailed growth stage keys such as that for sunflower described by Siddiqui et al. (52). Some useful keys are to be found in the manual produced by MAFF (37), but others are readily devised with a knowledge of phenology.

Choice of Methods

It is necessary to consider the degree of accuracy, precision and repeatability required of any measurement method. The accuracy, or relation of the observed value to the true value, and the precision, or range of values for estimates by a single operator, are influenced by the level of subjectivity in the method and the calibration, whereas the repeatability is influenced largely by the level of standardization and training that is available. It is simple to assess the accuracy, precision and repeatability of a method (36, 51), but there are few reports in the literature of such tests. The performance characteristics of any method should be checked carefully before the adoption of that method for investigations.

The suitability of methods is determined by the type of disease-causing organism and the characteristic spatial pattern and location of the pathogen or disease. The type of organism and degree of biotrophy will influence the method of measurement of pathogen populations more than disease measurement. The techniques for counting viruses, bacteria, fungi and nematodes are very different and are also often very different to the methods of estimating insect populations (Chapter 3). The location of wind, splash and aphid dispersed organisms influences the sampling techniques more than measurement methods, because the spatial pattern (e.g., random or aggregated) determines the most efficient method of detecting populations, especially at low densities. Although the spatial pattern is the factor most affected, consideration should also be given to the size and shape of sample units (17). The location of carryover inoculum also influences measurement methodology, especially for soil and seed borne diseases. Sampling techniques are discussed later in this chapter and in Chapter 6.

Each type of investigation imposes constraints on the choice of measurement method. The analysis of epidemics is sometimes based on pathogen monitoring, as this is often simple, objective and intrinsically accurate. The logic for using pathogen monitoring is that the subsequent development of an epidemic is dependent on the amount of inoculum available for spread and infection. Empirical models of yield loss define the time and location for measurement of pathogen or disease, and any variable which may be simply correlated to yield is useful in this context. On the other hand, mechanistic models (Chapter 16) define time, location and unit measured, and are most likely based on variables which describe the effect of disease on the plant.

MEASUREMENT METHODS

Several terms have been used to define aspects of measurement, including prevalence, incidence, severity and intensity. Prevalence, the proportion of production units in which at least some disease or pathogen may be found, is used in survey work but rarely in other types of investigation. Incidence, the proportion of plants or plant parts infected, is used more widely for both pathogen and disease monitoring for many purposes. Severity, the amount of tissue damaged, is the most demanding and the most diverse form of disease measurement. Disease intensity is used as a general term for the amount of disease present, often based on incidence and/or severity.

There are a number of well developed techniques available for collecting and counting some pathogen propagules. Fungal spores, especially those which are airborne and are produced abundantly, can be trapped readily using such instruments as the volumetric spore trap designed by Hirst (21) and the jet spore trap described by Schwartzbach (54). Alternatively, spores may be collected from plant surfaces using small cyclone traps (58) or by washing plant surfaces to remove spores from lesions (20). Once collected, spore numbers can be counted either by the tedious examination of spores under a microscope, by the use of an automated counting instrument such as a Coulter Counter or indirectly by weight. The occurrence of races or non-viable spores complicates pathogen measurement, and to avoid this problem the use of bait plants is recommended, especially for negative prognosis of disease risk. Techniques for spore collection and measurement were reviewed by Johnson & Taylor (28) for investigations of host resistance, but the methods are equally applicable to epidemiology and crop loss studies.

It has been argued that cumulative spore counts are related to disease development because spore production integrates the effects of pathogen infection on host plants (27), but the major application of such methods has been the analysis of changes in pathogen populations with time. Components of an infection cycle can be analyzed in detail by such measurements (56). The measurements are accurate and cause minimum disturbance to the investigated site. Pustule and pycnidium numbers may be counted for rusts (6) and speckled leaf blotch (13), respectively, rather than spore numbers. Spore counting and related techniques are useful for pathogen monitoring for the identification of disease risk periods, when inoculum potential is high. These measurements are especially useful if it is also possible to monitor meteorological conditions suitable for germination and infection. Jeger (27) showed that cumulative spore counts were a good estimator of powdery mildew incidence in apples, but not of apple scab where disease progress was influenced markedly by the weather. Techniques for monitoring low population levels, especially of splash dispersed organisms, are less well developed. It is often not practical to monitor for bacterial pathogens directly, because of identification problems, although methods are being developed based on specific sera for special needs (18). Aphid trapping for vector borne viruses is a practical solution for these pathogens (see Chapter 3). Pathogen numbers have been shown in most cases not to be well correlated to disease severity and yield loss and these techniques are not suitable for either empirical or mechanistic models.

Pathogen monitoring is especially useful for pre-season analyses of risk in soil and seed borne diseases. Many specific tests have been designed for seed borne diseases, as described by the working sheets of the International Seed Testing Association, including direct examination, incubation and immunological tests. Several ELISA tests for bacteria and viruses are used for routine seed inspection (41). The estimation of inoculum density in the soil is not so well developed and is fraught with technical problems. Many of the pathogens are present at low densities, requiring large samples to detect the presence of the propagules. The distribution of the pathogen may be one of several distinct patterns and some knowledge of this is desirable before analysis, but this data is often difficult to obtain (53). A "most probable number" system was used in conjunction with a bioassay for _Aphanomyces_ _euteiches_ by Pfender et al. (44). The technique provided a good estimate of disease risk for pea crops and, though tedious, was an elegant solution to a difficult problem. This approach should prove to be useful in other situations; for example, nematode populations may be monitored both before and during crop growth by similar methods (14). The size of inoculum particles affects infection efficiency (60) and may need to be considered in monitoring systems.

Chemical analyses, especially of chitin and mannans, have been used as estimators of the amount of fungal pathogen in plant tissues. Ride & Drysdale (47) determined the amount of *Fusarium oxysporum* f sp *lycopersici* in infected tomato stems by measuring acetylglucosamine after suitable hydrolysis. The technique has been used in several investigations, especially for detailed studies of the infection process and host reaction. Whipps et al. (59) reviewed the potentially useful chemicals for a wide range of fungal pathogens and symbionts. However, the techniques are complicated and time consuming, and are not suitable for routine pathogen measurement. ELISA systems may be used quantitatively in the future to measure the amount of pathogen in host tissues.

Disease Measurement

It may be argued that in many situations disease measurement, based on symptoms, is more closely related to yield loss than measurement of pathogen populations. Measurements of pathogen propagules take no account of the ability of the propagules to cause disease, nor host reaction to successful infection by a pathogen. For these reasons, disease measurement has received most attention in studies of yield loss but until recently the techniques available have been subjective and prone to error. Nonetheless, disease measurement, by skilled operators who understand the limitations of the measurement systems, has been the cornerstone of our understanding of disease: yield loss relationships.

PREVALENCE AND INCIDENCE For survey purposes, crop loss on a regional basis may be predicted readily by relatively simple measurements of disease prevalence and incidence, provided that a good model is available which relates incidence to yield loss. For some diseases the model may be very simple, such as the smut diseases where an infected ear represents a total loss of production. In many horticultural crops where the cosmetic appearance of marketed produce is important, any infection on a plant part represents a total loss, although the rejected produce may have some residual value in, for example, the juice extraction industry. In these cases it is sufficient to record the incidence of disease on sampled plants, or plant parts, to estimate crop loss. The sampling method is critically important in this context (see Chapters 3 and 6), but measurement is extremely simple once disease is diagnosed.

SEVERITY The measurement of incidence is not sufficient, for many diseases, where there is a quantitative relationship between the degree of infection, or severity, and the amount of yield loss. Thus methods have been developed for estimation of disease severity, which include a variety of approaches and may be carried out on individual plant parts, whole plants or at the whole crop level. Most methods rely on the observable difference between healthy and infected tissues, quantified with varying degrees of accuracy and precision. Descriptive keys were once popular as a method of severity measurement, such as the potato late blight key first described by Moore (39) and revised by the British Mycological Society (2). This key categorized levels of disease on the basis of the degree of leaf area affected by disease, as summarized below:

Blight %	description
0	not seen in field
0.1	only a few plants affected; up to one or two spots in a 10.6m radius
1.0	up to ten spots per plant or general light infection
5.0	about 50 spots per plant or up to one leaflet in ten attacked
25	nearly every leaflet infected, but plants retain normal form; field may smell of blight but looks green although every plant affected.
50	every plant affected and about 50% leaf area destroyed; field looks green flecked with brown.

9

75	about 75% of leaf area destroyed by blight; field looks neither predominantly green nor brown.
95	only a few leaves left green, but stems green
100	all leaves dead, stems dying or dead.

This and similar keys have been useful in some contexts, but are generally too subjective and imprecise for detailed work, especially for diseases with poorly defined symptoms. Many keys are based on the whole plant rather than on individual plant parts, such as that developed by Saari & Prescott (50) for wheat foliar diseases. The approach has been useful for plant breeders who require a method suitable for rapid screening of thousands of breeding lines. The accuracy normally required by breeders is low relative to that required by epidemiologists and for investigations of crop loss.

STANDARD AREA DIAGRAMS   A logical development from the use of disease keys was the use of standard area diagrams as a guide to visual estimates of the area of plant parts occupied by disease. Initially developed for foliar diseases, the principle has been extended to other plant parts, including shoot buds (11), awns (10), ears and flowers, stems, fruit and roots (37). Some diagrams have been developed to describe the amount of disease on whole plants (16). The aim of a standard area diagram is to depict, realistically, several levels of severity taking account of lesion size, shape and distribution. The diagrams show typical disease symptoms for a range of severities, often up to 50% of total area, and may also show the proportional distribution of leaf area as shown in Figure 1. They are used to estimate severity, often with the provision for interpolation between the depicted levels of disease. Standard area diagrams have been developed for plants with compound leaves of varying complexity, as shown in Figure 2. In some diagrams the areas of chlorosis or necrosis associated with disease are included in the area affected, but in others only the areas actually infected are depicted. It has been suggested that for some purposes, particularly deriving physiologically meaningful models of the severity: yield loss relationship, it is logical to measure the remaining green leaf area, but this usually requires that healthy plants are available for comparison (33). One advantage of this approach for surveys is that multiple disease constraints may be assessed collectively, thus avoiding the problem of linear averaging of losses of several factors. Zadoks and Schein (63) described percent attack as the combined area of lesions, chlorosis directly associated with disease plus any other effects attributable to the pathogen. In some cases weighting may be applied to affected areas in particular locations, such as scald lesions at the bases of leaf laminas and root rot lesions on primary rather than secondary roots. There are many standard area diagrams in the literature for a wide range of crops and diseases such as those in the FAO (15) and MAFF (37) manuals. There is often a marked variation in the appearance of specific diseases in different locations and on different cultivars. Any diagram which depicts the observed symptoms is useful, whether or not it was designed initially for that purpose. James (22) and Allen et al. (1) described methods for the construction of standard area diagrams where there are none suitable in the literature, using graph paper, planimeters or computer based graphics.

It is claimed that visual estimates of severity with the aid of standard area diagrams by trained personnel are reliable, particularly when the measurements are made without prior knowledge of treatment. However, there is little evidence in the literature to support the assumption that measurements are both accurate and precise. Sherwood et al. (51) investigated these aspects for measurement of disease severity on orchard grass (Dactylis glomerata) leaves infected by Stagonospora arenaria. Visual estimates were compared with actual areas of leaf spots determined by weighing paper replicas of photographs. Nine out of ten operators overestimated the degree of infection to different extents and, perhaps more significantly, overestimation was density dependent. Thus no transformation of data was possible to remove operator error. It was also shown that a leaf with

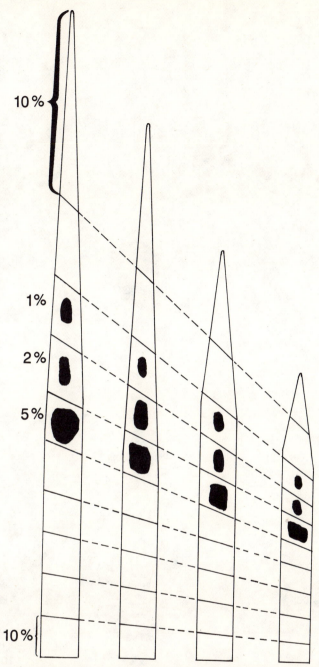

Figure 1. Example of a standard area diagram, with four leaf sizes and equivalent diseased areas (after Ref. 23).

a large number of small spots was scored higher than another leaf with fewer but larger spots, even though the actual area occupied by disease was identical. The accuracy of measurement was not high, and the high coefficients of variation indicated a low precision. The measurements were carried out by experienced people, and one can assume that inexperienced people would provide date which was less accurate and precise. Lindow and Webb (36) showed that visual measurements of _Ascochyta_ _pteridium_ severity on fern leaves were less accurate than estimates by planimeter, especially at intermediate levels of disease severity. There was no consistent error associated with the estimates, so correction for the error was not possible. A lack of accuracy may not be of concern in some situations, provided that the errors are consistent and identifiable, but a lack of precision will usuallly be a source of concern.

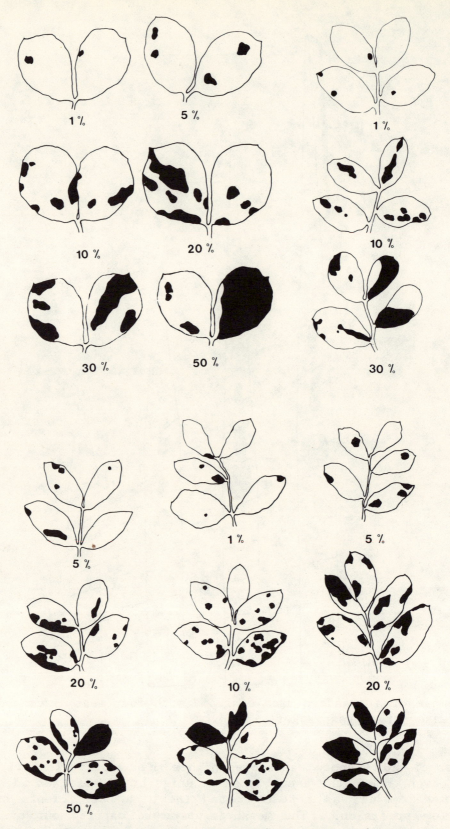

Figure 2. Example of a standard area diagram for compound leaves.

IMAGE ANALYSIS   The lack of objectivity in disease measurement has provided impetus to the search for better technologies, including the use of video image analysis by computer.   Image analysis systems have been developed on mainframe computers for a variety of purposes, including large scale analyses of cropping patterns, river flows and urban development.   Lindow & Webb (36) developed a system to measure disease severity on individual leaves based on an APPLE IIe microcomputer, using grey imagery and color filters to enhance contrast between green and necrotic tissues. There were close correlations between planimeter and computer estimates over a wide severity range for several necrotic diseases, and inaccuracies were consistent and could therefore be corrected by appropriate factors. Further development of the technique is required, both to improve the throughput of large sample numbers and to increase the discrimination between healthy and diseased tissues. Color based image analysis is an obvious further development, which may allow the measurement of biotrophic diseases where the visual impact on leaf color is less obvious than for necrotic diseases. Such systems are being tested (Teng, personal comm.) and one can be confident that this technology will be available, at least for detailed research investigations, in the near future. Some commercial units are already available for a variety of purposes, but these are not yet suited to routine analysis of disease severity.

PROPORTIONAL AND ACTUAL SCALES   James (22) suggested that the percentage scale should be used for disease measurements. The following reasons were given: the upper and lower limits of a percentage scale are uniquely defined; the percentage scale is flexible in that it can be conveniently divided and sub-divided; it is universally known and can be used equally for measuring proportion of plants infected (incidence) or the proportion of plant area damaged (severity); and it can be transformed easily for any subsequent epidemiological analysis. The merits of different scales and of arithmetic, logarithmic and other intervals were reviewed by James & Teng (26). Whilst recognizing the limitations in some cases of the percentage arithmetic scale, especially in epidemiology, this scale should be adopted as the basis of standardization. Disease categories or groups, favored by many plant breeders, are not generally suitable for epidemiology and crop loss studies.

Proportional scales, especially when mean values are calculated for different plant parts, introduce representational errors to disease measurement. Lim & Gaunt (33) pointed out that an arithmetic mean percentage of infection on cereal plants will either underestimate or overestimate disease if leaf size varies at different positions on the plant and the disease is not uniformly distributed on those leaves. Estimates of percentage severity also ignore the effect of disease on the size of new leaves, especially when disease is present from early growth stages. Step-wise regression of actual green leaf area on whole shoots was more closely related to barley yield and yield loss than percentage severity (34). Similar results were reported by Carver et al. (7). It is recommended that measurements of actual green leaf area should be considered for physiological investigations. At present the routine use of green leaf area measurements is limited by the time consuming nature of the methods, but it is likely that image analysis technology will change this situation, as discussed later.

INDIRECT MEASUREMENT   The methods of severity measurement discussed above are all based on the examination of individual leaves or plants directly. Severity may also be estimated by a variety of indirect techniques, including the use of incidence: severity relationships and remote sensing of light, or thermal, reflectance. James & Shih (25) showed that severity could be estimated at low to moderate levels by an estimate of incidence on plant parts. Rouse et al. (49) analyzed the incidence: severity relationships of wheat powdery mildew. The relationship changed with leaf position and time, and they concluded that the relationship was too variable as an indicator of severity unless an adjustment was made for these factors. The relationship between severity and mean incidence on all leaves was consistent, but it was not possible to measure variability in this relationship between sites or seasons. At Lincoln, we found (9) that the relationship between stripe rust incidence and severity was affected by leaf position, time, site and season, but that a relationship based on the mean incidence on the top three leaves was a stable predictor of severity over three seasons and several sites. The relationship was used in a disease management program based on sequential sampling.

Remote sensing may be used as a measure of both disease incidence and severity. Infected and healthy plant tissues have different reflectance characteristics, and most remote sensing techniques developed to date measure this difference. Direct reflectance measurements have mostly analyzed the differences in the near infrared wavelength region (700-950 nm). The difference in reflectance has also been detected by the use of photographic techniques, especially by the use of false color infrared film. With suitable calibration, film images can distinguish readily between different reflectance characteristics and can be examined by eye or by image analysis. Remote sensing may also be carried out by other techniques, such as infrared thermometry. Infected plants are frequently more water stressed than healthy plants, and stress usually results in a temperature increase of 2 C-5 C at the plant surface. These temperature differences may be detected by hand-held or other infrared thermometers, as described by Pinter et al. (45) working with sugar beet and cotton plants infected by soil borne root rotting pathogens. The differences between healthy and infected plants were detectable over a range of soil moisture conditions, and may present some solution to the difficult problems of measurement of these diseases. Reflectance measurement systems can be operated from ground rigs, small aircraft (including remote controlled model systems) and satellites, and there has been particular interest in the Earth Resources Technology Satellites and their equivalents for survey purposes. Nagarajan et al. (40) reported that wheat rust infections were detected by the Landsat-2 satellite using multispectral scanners.

Remote sensing has been most useful with those diseases where infection is associated with a complete yield loss. In these cases incidence only is measured, which avoids the more difficult problem of measuring severity by remote sensing. Losses in peanut fields were measured by remote sensing of the areas of diseased plants, on the assumption that these areas did not contribute to yield (46). More recently, color infrared photography was used to measure the mortality of oak trees and to identify oak wilt in forestry areas in central Texas in the USA (3). Quantitative relationships between severity and yield loss were described using a portable radiometer (42). Reflectance in several wavebands, automatically corrected for variations in incident radiation, was correlated with yield loss for several crop/disease systems and is a rapid method for measuring canopy green leaf area (43). Severity has also been measured by photographic methods as described by Manzer & Cooper (38) for detection of potato diseases. They used a portable videotape camera to record images on black and white film, using appropriate filters to enhance the differences. Such equipment may, with further development, allow routine measurements to be recorded and analyzed at a later date or transferred to image analysis systems. At present most of the technology is limited in application and expensive. The most likely immediate use will be confined to perennial crops and the surveillance of disease spread.

TIMING, FREQUENCY AND POSITION OF MEASUREMENTS

Decisions on when to measure disease or pathogen populations, and on which plant parts, are perhaps the most difficult in the design of surveys or disease-yield loss research programs. The objectives of any investigation must be analyzed critically before making any decision, and the choice is often a compromise between what is desirable and what is feasible. The type of measurement technique used is related to objectives, as described above, but the choice is often limited by available resources and the volume of data which has to be recorded. Teng (55) showed that an inefficient measurement system, with high variance, can increase overall costs of analysis by requiring a larger sample size to offset the innate variability. The following is a brief review of the suggested approaches for several types of investigation, including the analysis of epidemics, disease surveys and the effect of disease on plant growth and development.

Epidemic monitoring and analysis

Epidemics develop in both time and space, and an analysis requires that this be taken into account. If the objective is to monitor the progress of an epidemic to estimate when threshold levels are exceeded, a measurement of disease, or pathogen, incidence, on plant parts or in the environment at weekly intervals is satisfactory for many diseases. The analyses must be carried out more frequently if a combination of available inoculum and suitable weather is required for infection. Thus, for example, monitoring for the presence of spores of apple scab and meteorological measurements may be necessary every day for a disease control prediction program. A full epidemic analysis requires more detailed measurement, possibly every few hours, based for example on spore production on all green plant parts, the measurement of expanding lesions or the severity of infection on all plant parts. Epidemic analysis is often based on the infection rate which, on a rapidly growing plant, must be corrected for the appearance of new tissue and the senescence of existing, infected tissues (29). Bennett and Westcott (4) reported that the ranking of cultivars for resistance was affected by the growth stage at which assessment was carried out: similar difficulties are likely to occur in studies of epidemics and yield loss. The techniques described above for the measurement of spore numbers, fungal biomass in tissues and disease incidence and severity are all appropriate for some types of epidemic analysis.

Crop loss surveys

Disease surveys to estimate crop losses and establish research priorities are, by their very nature, carried out usually on a large scale, and thus many measurement methods are inappropriate for this purpose. However, it should be borne in mind that the adopted techniques should be related to the methods used to derive disease: yield loss relationships in any survey of crop losses on a regional basis (26). The measurement may be based on an empirical model provided that the survey is not conducted outside the environmental circumstances in which the initial study of the relationship was carried out. Methods involving the measurement of incidence and reflectance by remote sensing are especially suited to survey work, where there is a demand for a rapid technique which can be used by relatively unskilled personnel or by automated analyses of aerial photographs or similar material. The sampling method is critically important, as described in Chapter 6. The measurements often need to be taken only once or twice during the season, at a time related to a critical point model (Chapter 11).

Yield loss analysis

Measurements for analysis of the physiological basis of yield loss should be carried out at least once every two weeks, and should preferably be based on a host based character rather than a pathogen character. Measurement of percent, or preferably actual, green area on the whole plant is required as discussed elsewhere. It is only by a regular measurement schedule throughout the epidemic that the relationship between disease and yield loss can be analyzed (34). At the same time, more detailed measurements of plant growth and development are also required to provide a basis for physiological analysis (35; see Chapter 16).

LITERATURE CITED

1. Allen, S.J., Brown, J.F., Kochman, J.K. 1983. Production of inoculum and field assessment of Alternaria helianthi on sunflower. Plant Dis. 67:665-668.
2. Anon. 1947. The measurement of potato blight. Trans. Brit. Mycol. Soc. 31:140-141.
3. Appel, D.N., Maggio, R.C. 1984. Aerial survey for oak wilt incidence at these locations in Central Texas. Plant Dis. 68:661-664.
4. Bennett, F.G.A., Westcott, B. 1982. Field assessment of resistance to powdery mildew in mature wheat plants. Plant Pathol. 31:261-268.

5. Berger, R.D. 1980. Measuring disease intensity. In Crop Loss Assessment, pp. 28-31. Misc. Publ, 7. Agric. Exp. Stn. Univ. of Minn., St. Paul, Minn. 327 pp.

6. Buchenau, G.W. 1975. Relationship between yield loss and area under the wheat stem rust and leaf rust progress curves. Phytopathology 65:1317-1318.

7. Carver, T.L.W., Griffiths, E. 1982. Effects of barley mildew on green leaf area and grain yield in field and greenhouse experiments. Ann. Appl. Biol. 101:561-572.

8. Chester, K.S. 1950. Plant disease losses: their appraisal and interpretation. Plant Dis. Rep. Suppl. 193:189-362.

9. Cole, M.J. 1985. The development of a sequential sampling plan for management of stripe rust (Puccinia striiformis West.) in winter wheat (Triticum aestivum L.). Ph.D. thesis. Canterbury University, Lincoln College, New Zealand. 120 pp.

10. Cooke, B.M., Brokenshire, T. 1975. Assessment keys for halo spot disease on barley, caused by Selenophoma donacis. Trans. Br. Mycol. Soc. 65:318-321.

11. Dixon, G.R., Doodson, J.K. 1971. Assessment keys for some diseases of vegetable fodder and herbage crops. J. Natn. Inst. Agric. Bot. 12:299-307.

12. Dunleavy, J.M., Keck, J.W., Gobelman-Werner, K.S., Thompson, M.M. 1984. Prevalence of soybean downy mildew in Iowa. Plant Dis. 68:778-779.

13. Eyal, Z., Brown, M.B. 1976. A quantitative method for estimating density of Septoria tritici pycnidia on wheat leaves. Phytopathology 66:11-14.

14. Ferris, H. 1981. Dynamic action thresholds for diseases induced by nematodes. Annu. Rev. Phytopathol. 19:427-436.

15. Food, and Agricultural Organization of the United Nations. 1971. Crop Loss Assessment Methods. FAO manual on the evaluation and prevention of losses by pests, diseases and weeds. Commonw. Agric. Bur., Slough, England. 255 pp.

16. Fullerton, R.A. 1982. Assessment of leaf damage caused by northern leaf blight in maize. N.Z. J. Expt. Agric. 10:313-316.

17. Gilligan, C.A. 1982. Size and shape of sampling units for estimating incidence of sharp eyespot, Rhizoctonia cerialis, in plots of wheat. J. Agric. Sci., Camb. 99:461-464.

18. Gitaitis, R.D, Jones, J.B., Jaworski, C.A., Phatak, S.C. 1985. Incidence and development of Pseudomonas syringae pv. syringae on tomato transplants in Georgia. Plant Dis. 69:32-35.

19. Grainger, J. 1967. Methods for use in economic surveys of crop disease. In Background papers prepared for the FAO symposium on crop losses, pp. 49-74. Rome: FAO, UN.

20. Heagle, A.S., Moore, M.B. 1970. Some effects of moderate adult resistance to crown rust of oats. Phytopathology 60:461-466.

21. Hirst, J.M. 1952. An automatic volumetric spore trap. Ann. Appl. Biol. 39:257-265.

22. James, W.C. 1971a. An illustrated series of assessment keys for plant diseases, their preparation and usage. Can. Plant Dis. Surv. 51:39-65.

23. James, W.C. 1971b. A manual of disease assessment keys for plant diseases. Can. Dept. Agric. Pub. 1458. 80 pp.

24. James, W.C. 1974. Assessment of plant diseases and losses. Annu. Rev. Phytopathol. 12:27-48.

25. James, W.C., Shih, C.S. 1973. Relationship between incidence and severity of powdery mildew and leaf rust on winter wheat. Phytopathology 63:183-187.

26. James, W.C., Teng, P.S. 1979. The quantification of production constraints associated with plant diseases. In Applied Biology, Vol. IV, ed. T.H. Coaker, pp. 201-267. New York: Academic Press. 285 pp.

27. Jeger, M.J. 1985. Relating disease progress to cumulative numbers of trapped spores: apple powdery mildew and scab epidemics in sprayed and unsprayed orchard plots. Plant Pathol. 33:517-523.

28. Johnson, R., Taylor, A.J. 1976. Spore yield of pathogens in investigations of the race-specificity of host resistance. Annu. Rev. Phytopathol. 14:97-119.

29. Kushalappa, A.C., Ludwig, A. 1982. Calculation of apparent infection rate in plant diseases: development of a method to correct for host growth. Phytopathology 72:1373-1377.

30. Large, E.C. 1954. Growth stages in cereals: illustration of the Feekes scale. Plant Pathol. 3:128-129.
31. Large, E.C. 1955. Methods of disease-measurement and forecasting in Great Britain. Ann. Appl. Biol. 42:344-354.
32. Large, E.C. 1966. Measuring plant disease. Annu. Rev. Phytopathol. 4:9-28.
33. Lim, L.G., Gaunt. R.E. 1981. Leaf area as a factor in disease assessment. J. Agric. Sci., Camb. 97:481-483.
34. Lim, L.G., Gaunt, R.E. 1985a. The effect of powdery mildew (_Erysiphe graminis_ DC. f.sp. _hordei_) and leaf rust (_Puccinia hordei_ Otth.) on spring barley in New Zealand. I. Epidemic development, green leaf area and yield. Plant Pathol. 34:44-53.
35. Lim, L.G., Gaunt, R.E. 1985b. The effect of powdery mildew (_Erysiphe graminis_ DC. f.sp. _hordei_) and leaf rust (_Puccinia hordei_ Otth.) on spring barley in New Zealand. II. Apical development and yield potential. Plant Pathol. 34:54-60.
36. Lindow, S.E., Webb, R.R. 1983. Quantification of foliar plant disease symptoms by microcomputer-digitized video image analysis. Phytopathology 73:520-524.
37. Ministry of Agriculture, Fisheries and Food. 1976. Manual of plant growth stages and disease assessment keys. Plant Pathol. Lab., MAFF., Harpenden, England. 20 pp.
38. Manzer, F.E., Cooper, G.R. Use of portable videotaping for aerial infrared detection of potato diseases. Plant Dis. 66:665-667.
39. Moore, W.C. 1943. The measurement of plant diseases in the field. Trans. Brit. Mycol. Soc. 26:28-35.
40. Nagarajan, S., Seibold, G., Kranz, J., Saari, E.E., Joshi, L.M. 1984. Monitoring wheat rust epidemics with the Landsat-2 satellite. Phytopathology 74:585-587.
41. Nolan, P.A., Campbell, R.N. 1984. Squash mosaic virus detection in individual seeds and seed lots of cucurbits by enzyme-linked immunosorbent assay. Plant Dis. 68:971-975.
42. Nutter, F.W., Peterson, V.D. 1983. Yield loss determination in the barley-spot blotch system using visual and remotely sensed disease assessment schemes. Phytopathology 73:803.
43. Nutter, F.W., Littrell, R.H., Petterson, V.D. 1985. Use of a low.cost, multispectral radiometer to estimate yield loss in peanuts caused by late leaf spot (_Cercosporidium personatum_). Phytopathology 75:502.
44. Pfender, W.F., Rouse, D.I., Hagedorn, D.J. 1981. A "most probable number" method for estimating inoculum density of _Aphanomyces euteiches_ in naturally infested soil. Phytopathology 71:1169-1172.
45. Pinter, P.J., Stanghellini, M.E., Reginato, R.J., Idso, S.B., Jenkins, A.D., Jackson, R.D. 1979. Remote detection of biological stresses in plants with infrared thermometry. Science 205:585-587.
46. Powell N.L., Garren, K.H., Griffin, G.J., Porter, D.M. 1976. Estimating cylindrocladium black rot disease losses in peanut fields from aerial infrared imagery. Plant Dis. Rep. 60:1003-1007.
47. Ride, J.P., Drysdale, R.B. 1972. A rapid method for the chemical estimation of filamentous fungi in plant tissue. Physiol. Plant Pathol. 2:7-15.
48. Robinson, R.G. 1971. Sunflower phenology -- year, variety and date of planting effects on day and growing degree-day summations. Crop Science. 11:635-638.
49. Rouse, D.I., MacKenzie, D.R., Nelson, R.R., Elliott, V.J. 1981. Distribution of wheat powdery mildew incidence in field plots and relationship to disease severity. Phytopathology 71:1015-1020.
50. Saari, E.E., Prescott, J.M. 1975. A scale for appraising the foliar intensity of wheat diseases. Plant Dis. Rep. 59:377-380.
51. Sherwood, R.T., Berg, C.C., Hoover, M.R., Zeiders, K.E. 1983. Illusions in visual assessment of stagonospora leaf spot of orchardgrass. Phytopathology 73:173-177.
52. Siddiqui, M.Q., Brown, J.F., Allen, S.J. 1975. Growth stages of sunflower and intensity indices for white blister and rust. Plant Dis. Rep. 59:7-11.

53. Stanghellini, M.E., Stowell, L.J., Kronland, W.C., von Bretzel, P. 1983. Distribution of <u>Pythium</u> <u>aphanidermatum</u> in rhizosphere soil and factors affecting expression of the absolute inoculum potential. Phytopathology 73:1463-1466.

54. Schwartzbach, E. 1979. A high throughput jet trap for collecting mildew spores on living leaves. Phytopathol. Z. 94:165-171.

55. Teng, P.S. 1978a. Sample surveys for disease-loss estimation. In Epidemiology and Crop Loss Assessment, pp. 36-1/36-14. Proc. APPS workshop, Lincoln College, 1979.

56. Teng, P.S. 1978b. System modelling in plant disease management. Ph.D. thesis, Lincoln College, University of Canterbury, New Zealand, 398 pp.

57. Teng, P.S. 1983. Estimating and interpreting disease intensity and loss in commercial fields. Phytopathology 73:1587-1590.

58. Teng, P.S., Close, R.C. 1977. Mass efficiency of two urediniospore collectors. N.Z. J. Exp. Agric. 5:197-199.

59. Whipps, J.M., Haselwandter, K., McGee, E.M.M., Lewis, D.H. 1982. Use of biochemical markers to determine growth, development and biomass of fungi in infected tissues, with particular reference to antagonistic and mutualistic biotrophs. Trans. Br. Mycol. Soc. 79:385-400.

60. Wilkinson, H.T, Cook, R.J., Alldredge, J.R. 1985. Relation of inoculum size and concentration to infection of wheat roots by <u>Gaeumannomyces</u> <u>graminis</u> var, <u>tritici</u>. Phytopathology 75:98-103.

61. Zadoks, J.C. 1972. Methodology of epidemiological research. Annu. Rev. Phytopathol. 10:253-276.

62. Zadoks, J.C., Chang, T.T., Konzak, C.F. 1974. A decimal code for the growth stages of cereals. Weed Res. 14:415-421.

63. Zadoks, J.C., Schein, R.D. 1979. Epidemiology and plant disease management. Oxford, England: Oxford University Press. 427 pp.

CHAPTER 3

MEASUREMENT OF INSECT PEST POPULATIONS AND INJURY

P. T. Walker

Formerly with the Tropical Development and Research Institute, London.
10 Cambridge Road, Salisbury, Wiltshire
SP1 3BW England

Pest management depends on decisions, from the farmer to the national level. For rational decisions to be made, a first essential is a representative measure or estimate of the population density of the pest. If it is impossible to do this for reasons of difficulty, cost or time, the effect on the crop in terms of injury is not measured. The following principles are intended to apply to insect or mite pests, but could, in many cases, equally refer to the assessment of rodents or birds.

Pest and injury assessment is important, if not essential, for several aspects of pest management: understanding the biology of the pest requires quantitative measures of the time and duration of life-cycle stages, of distribution and migration; pests surveys, pest forecasts and studies of the effects of pesticides rely heavily on methods of assessment. In particular, pest management decisions based on economic benefit demand a quantitative measure of pest attack to relate to the increase in yield resulting from pest management. If such decisions are based on an economic action threshold, a measure of the pest density or amount of injury is an essential part of the decision-making system.

A PEST MANAGEMENT DECISION SYSTEM

A simple pest management decision system is shown in Figure 1, where information needed is linked in a circular, feed-back system to the amount of pest attack or damage present. The pest density (i) is surveyed or monitored, the relation of this to crop production or yield (y) is found by experiment and the techniques of crop loss assessment, described in Chapters 11-15. The relation may be modelled as some form of $y = f(i)$ and can be simple or complex. An economic action threshold (AT) of pest density or amount of damage is taken, depending on the control costs, crop value and other factors. If the field infestation is greater than this value, control is applied, and there is an effect on the infestation. If it is not greater, no control is applied. To formulate decision-making in this way allows the different phases of the process to be considered, the different crop, climate, pest, economic and social factors affecting them to be examined, and quantitative values put in to produce a predictive model of the system.

PEST AND DAMAGE ASSESSMENT METHODS

Methods should be standardized, so that they can be studied and tested, and different results compared. The principles were discussed by Walker (46). They should be quick, cheap and simple, and clearly described in a survey manual, with details of the method, the number and size of samples, the time in the pest life-cycle and crop cycle at which they are taken. A key to identify the pest and

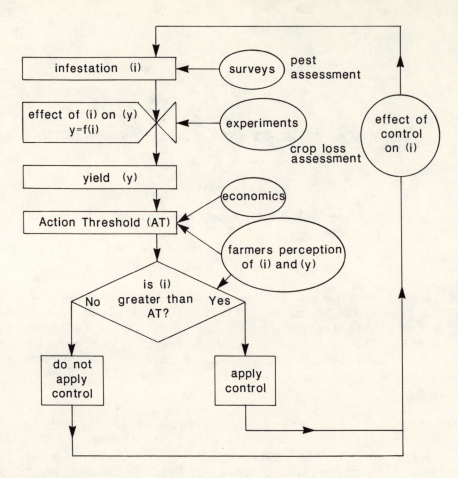

Figure 1. A pest management decision system, showing the place of pest assessment and crop loss assessment.

crop growth stage is important (see below). If damage is recorded, a standard area key is useful. An estimate of the accuracy, or nearness to the true value, and precision in relation to the mean is needed.

Reference Material

Many methods of assessment were described by Southwood (33), and good accounts are given by Bardner & Fletcher (3), Chiarappa (9), Bram (6) and in Youdeowei & Service (50) and Matthews (19). There is a general review by Strickland (36). Survey manuals have been produced by USDA (41), the Philippines, India and other countries, and methods given in publications on crop pests (22), on pest groups (43), in the works of International Agricultural Research Institutes such as IRRI, ICRISAT or CIAT, or by agrochemical companies (25).

Direct or Indirect Methods

Pests can be counted directly, or their effects on crops assessed indirectly as injury or damage. Direct counts can be considered in two ways: those which can be based on a standard unit, such as area of ground or weight of crop, or which can be converted to such a unit from the number of leaves, stems or plants per area or yield of crop. Alternatively pests are counted in the environment, for example in a light trap, giving no more than an estimate of the absolute population, which must be corrected for the conditions of trapping and for sampling error. In all cases, the method must be representative, or intended to give as true an estimate of the actual population as possible.

DIRECT COUNTS   There are a number of well-known methods:

Observation   The number of pests per leaf, fruit, root or stem or per area or volume of soil is counted. The absolute population is obtained by multiplying up to the desired population unit.

Cutting open fruits, seeds, stems, tubers to count the pests.

Beating, shaking, suction, knockdown with insecticide, is used to remove the pest population onto a sheet. The whole plant or a part, such as a sorghum head, can be put inside a container to remove the pest.

Sweep nets can give reliable and repeatable results if the net size and beat frequency are standardized (28).

Washing off pests: eggs, small insects or mites are removed with an egg solvent or detergent. They can be measured as a population -- related volume.

Brushing pests off leaves, sometimes mechanically, onto a sticky surface.

X-rays have been used for counting stem borers, seed pests, etc.

Crushing colored aphids or mites on absorbent or glossy paper, counting live weevil larvae in grains crushed on ninhydrin paper has been used, pest density being measured by a colorimeter.

The conditions of the method of assessment should be given clearly to make conversion to an absolute population density possible: number of leaves per tiller, tillers per plant, stems per plant, seedheads per row or rows per hectare.

COUNTS IN THE ENVIRONMENT   These include trapping in some way:

Chemical attraction by the food plant or a chemical constituent of it, or by attraction to chemicals emitted by the pest, pheromones or kairomones etc., are increasingly being used in pest assessment. They are associated with some means of trapping the pests; funnels, containers, water or sticky surfaces, some of which are colored.

There are many examples:

- Fruit or fruit extracts: esters, sugars, fermented materials, beer, hydrolysate, plant extracts, eugenol etc.
- Fish meal which acts like sorghum break-down products, to attract shoot fly.
- Crop sections: pine to attract pine weevils, banana or plantain to attract weevils, Cosmopolites, coconut to attract Oryctes beetles, etc.
- Pheromones: together with aggregation chemicals have now been identified for many groups: lepidoptera (7), coleoptera, hemiptera and recently diptera. A particular mixture of constituents is usually most attractive, some being repellent. Males are usually attracted, but females are in some groups. Trap design, which can range from funnels, bags, flat or triangular sticky surfaces or enclosures, trap size and position in relation to crop height, and the condition and rate of release of the attractant usually have considerable effect on trap efficiency (34,18). The great advantage of pheromone traps is selectivity, as well as simplicity, although those that only attract males are less informative about mating status than light traps. Their main use is to show when pests are present, in what numbers and for how long. The relation between the catch and crop damage is not always as close as might be expected.

Attraction by color is often used in water or sticky traps. Yellow (21) is often the most attractive to daytime leaf feeders (43), but red may attract some fruit pests and blue has been slightly more attractive to some leafhoppers. Standardization of the color to, for example, BS 0.001 is preferred. Baffles are often used to increase the amount of attraction.

Sticky traps: flat, cylindrical or triangular shaped surfaces are covered with sticky material (29), sometimes colored or with a chemical attractant. It is necessary to check that the surface does not become covered with insects or dust. The surface can sometimes be removed and sent for examination elsewhere. The catch is removed with solvent. Southwood (33) compared the catch of two aphid species by cylindrical and flat stick traps and a water trap.

Water traps: shallow traps containing water, a detergent or perhaps an anti-evaporant oil may be used together with color or chemical attraction. Protection from predators and frequent attention may be necessary. The optimum area of trap should be standardized. The catch in such traps depends on the shape, size, direction and height above the crop and the condition and rate of release of the attractant (18).

Light traps are based on a light source, either oil or electric bulb, discharge or fluorescent tube with associated control gear, and time switch, which may be daylight activated (26). A generator may be needed. There may be reflectors or electrified vanes to increase the catch, a roof to protect against rain or to limit the light direction, and a container for the catch, usually entered through a funnel. There may be a suction fan to draw in the catch. Baffle material and a knockdown chemical such as dichlorvos is often put in the container to prevent damage to the catch.

The number of insects caught depends on the intensity and wavelength of the light (44), UV or black light often being more attractive, on the height of the trap above the ground, the shape and design of trap, and the weather. The brightness of the light in relation to that of the surroundings, in particular, moonlight is important (5).

The advantage of light trapping is the ease with which data on the time of appearance and duration of adult flight, the proportion of different species, and the relative change in populations can be obtained. Hence the time and size of pest outbreaks can be predicted, or related to causal factors such as climate. Data on absolute populations are difficult to derive unless the relation between them and the catch, and any correction factors are known. The disadvantage over pheromone traps are the large, unsorted catch and the need for a power source and expensive fittings.

Suction traps are of two main types, fixed and portable. The Johnson-Taylor type is free-standing, has a metal sieve cone with standard diameter opening of 23 cm, through which air is blown by an electric fan. Other sized openings have been used. In some, cones are enclosed for protection, the air is drawn through, and the opening may be at different heights to sample different kinds of populations. A control mechanism is often used to segregate the catch into time periods. The cone should be earthed to prevent static charge affecting the catch.

The size of opening chosen, and hence the volume of air sampled, depends on the expected density of insects and the expected windspeed. The catch can be corrected for these factors (33), for the height of the opening above the ground (43), and for insect size (39).

A second type of suction trap is portable and directable and is used for sampling vertically downwards or sideways over a quadrat, or, for example, round the base of rice hills. They may be driven by solid or car batteries, as described by Arnold et al. (2), Carino et al. (8) and Cinch (10) or by a gasoline motor, for example Dietrick (12) and Thornhill (40).

Sampling soil or debris: A standard volume sample is usually taken by digging or with a core borer to a standard depth, depending on the insect population density and distribution, and the soil characteristics. As with other sampling methods, an optimum plan is usually developed after taking a pilot survey. Insects are then separated from the substrate by dry or wet methods. In dry methods, a hot water jacket or light bulb may be used to activate and drive insects out of the sample. In wet methods, such as that of Salt & Hollick (9), the processing may include pre-soaking, washing, agitation, air bubbling, differential wetting and flotation in solution, centrifuging and separation. Termites can be trapped by covering the soil with paper (20).

Insect emergence from the soil is monitored by emergence traps covering the ground, the insects being attracted to a collecting tube by daylight.

Pitfall traps: mobile insects on the soil surface can be collected in smoothsided containers sunk level with the surface. They need frequent attention and protection from predators and rain.

Mark, release and recapture methods: Populations of mobile pests in a limited environment can be estimated by marking, releasing and recapturing them, using the Lincoln Index:

$$\text{population} = \frac{\text{number marked released} \times \text{total caught}}{\text{number marked caught}}$$

There are several methods, described by Blower et al. (4), the choice depending on whether pests are removed or replaced and whether survival and migration are constant or variable. There must be efficient mixing, efficient trapping, and no effect of the marking on the pests. The insects caught in these traps can be marked, inside or out, by paint or dye, UV fluorescence, radioactivity, bacteria or genetic markers (13).

Factors which affect direct pest counts include the following:
-- The number of pests present, affected by changes in the food crop, the weather or the pests biology.
-- The stage, generation, sex, reproductive or nutritional state of the pests.
-- The activity of the pest, as affected by temperature, stage, or sex or nutritional state, governing response to trapping or sampling.
-- The conditions of the trap or sample: weather, position, condition of the attractant, etc. There may be a threshold level below which there is no activity or response.

INDIRECT METHODS OF ASSESSMENT   Estimates or counts of the effects of pests may be used to measure the amount of pest attack indirectly, in terms of injury, sometimes taken as the simple visible effect, and damage, which results in economic loss. The difference between incidence, the number of attacks, and intensity, the degree or extent of attack should be made.

The whole plant   The number or percentage of plants missing or showing signs of attack is often recorded as a simple, quick measure. For example, plants attacked by aphids, stemborers, root pests etc.

Leaves   The presence or area of leaf damage, holes, lesions etc. is measured by such methods as weighing, using a dot matrix grid, photography, air-flow, light measurement, electronic scanning or comparison with the undamaged leaf obtained from "length x breadth x constant" calculations.

Stems   The number or percentage of deadhearts, exit holes, length of tunnels or nodes or internodes attacked by borers are used to assess pest populations in cereals and woody plants (47). Lesions or stems cut by cutworms, sawfly (Cephus) or Mythimna may be counted.

Fruit and seeds   Holes and lesions in fruit, damaged ears, cobs or seeds of cereals, for example whiteheads in rice, damaged cocoa pods, coffee berries, cotton bolls, shown by flaring of the bracts, etc. are all measures of the pest population. The area of damage can be measured.

Roots   The length, dry weight or volume of fibrous roots attacked by soil pests can be measured or estimated from samples or sections. The area of damage to tuberous roots or stems such as potatoes or yams can be measured, or given a scale rating by comparison with a standard area diagram.

Amount of by-product   The quantity of excreta (stem borers) honeydew (aphids, brown plant hoppers) produced has been used to assess pest populations.

Rating scales   A score or rating for the degree of infestation or damage is often given, for speed or ease. Numerical scores can be summed and analyzed, but it should be remembered that they are discontinuous, finite and may be non-normally distributed, requiring transformation before analysis. A scale of, say, 1, 2, 3, may be arithmetically related to the actual count, for example 0.1-1, 2-5, 6-25 etc. Small scale divisions may be difficult to separate (9), while information will be lost if they are too big. Ratings always give less information than actual counts or percentages (11). Standard area keys are valuable for estimating areas of damage (Chapter 2).

Remote Sensing   The assessment of the amount of damage from photographs or transmitted images taken by a plane or satellite is increasing. Filters, film or sensitivity to particular wave-lengths, such as infrared or false color, which measures the water-status and reflectivity of the crop, are used (9, 15, 23, 45).

The techniques may be used to identify areas where rainfall or crop growth may be suitable for outbreaks of pests, such as locusts, as described by Hielkema (17). Clouds and lack of ground control may lead to difficulty. Radar can also be used for monitoring moth, locust or bird populations (27).

The Relation Between Indirect and Direct Methods

The relation between the counts in an indirect method, score or rating (x) and counts by a direct method, which give the pest population in terms of a recognizable base unit such as area of ground (n) should be determined:

$$n = f(x)$$

The relation depends on the units used, and may be linear, when one sorghum midge damages one sorghum seed (n = x), or non-linear, perhaps rising to a limit, as in the relation between percentage of stems attacked by stemborer and number of larvae per square metre. If the relation is known, the indirect method can be used to assess the actual population quickly, and to continue with studies on the loss of yield.

Pest Identification

In a pest or loss survey, a clear description, diagram or picture of the pest, stage, sex etc. to be sampled is essential. Examples are the pest identification charts published by the UK Ministry of Agriculture, or the Spodoptera exempta armyworm chart issued by the Armyworm survey in East Africa. For continued monitoring, survey handbooks are needed.

Crop Growth Charts

For accurate identification of the stage of crop development at which to sample, decimal growth stage charts are valuable (9). They give a numerical code to the morphological and physiological stages at which pest attack may be critical. A time scale is less useful because of the variation due to variety or climate. Stages have been published for wheat, rice, maize, beans, soybean, cotton, oilseed rape, potato, sorghum, sugarbeet, sugarcane and temperate fruit. Stages based on the growing point of cereals are particularly useful when examining the effect of pests on yield.

The Best Time to Sample or Method to Use

Pest assessment is most valuable at the time when pests have the maximum effect on yield. At this critical growth stage, the best measure of pest attack may be the first appearance of eggs or adults, the cumulative number of eggs, larvae or adults, that is the area under the population-time curve, or the number at maximum attack. The time or measure which gives the highest correlation coefficient (r) or coefficient of determination (r^2) in a yield regression will be most useful in forecasting the effect of infestation on yield. Several methods or times can be examined in a multifactorial regression. The techniques are given in Gomez & Gomez (14).

The Effect of Frequency Distribution of the Pest or Damage

The frequency distribution, that is the number of sampling units having different numbers of pests or amounts of damage, should be known. This is to enable development of an effective sampling design, so that a representative sample obtained can be analyzed in a valid way with accurate limits of error. A pilot survey is usually advisable. Random, regular, clumped or normal distributions may be found. If, for example, the distribution is negative binomial or contains many zeros, analysis of the raw data by parametric statistics will not be valid, and the limits of error and significance of means will not be correct.

The data (i) can be transformed before analysis, as discussed in the chapters on statistics and in Gomez & Gomez (14). Taylor's power law (38) can be used to indicate a suitable transformation: the log variances (s^2) of samples are plotted against their log means ($\bar{i}$), or the variances against the means on log:log paper. The slope (b) of the line is:

$$\frac{\log (\bar{s}^2) - \log a}{\log \bar{i}}$$

where (a) is the intercept on log (s^2) at log (i) = 0.

Counts are transformed to the power (1 - 1/2b). If this is zero, log transforms of (i) are used, if 0.5, the square root of (i), if -0.5, the reciprocal of (i).

Another method is to plot the range of sample values against their means. The transformation (log i + k) is used if a straight line results, where (k) is the intercept on the ($\bar{i}$) axis. If the line is curved downwards, a square root transformation is used.

Percentages are usually transformed by the arc-sine or angular transformation, particularly if there are values outside the range 20 - 80%.

Frequency distributions may change with time, with the stage of the pest, the crop stage or with the units used. Data need to be detransformed before conclusions are presented.

Sampling Theory

No firm advice on the number (Figure 2), size or shape (Figure 3) of samples can be given without some knowledge of the number of pests, the amount of damage, the frequency distribution of the data, the variation expected in the population, the heterogeneity in the field, and the precision needed, in terms of least significant differences or coefficient of variation. The latter will depend on the purpose of the sample. Time and cost are sure to be important. The subject

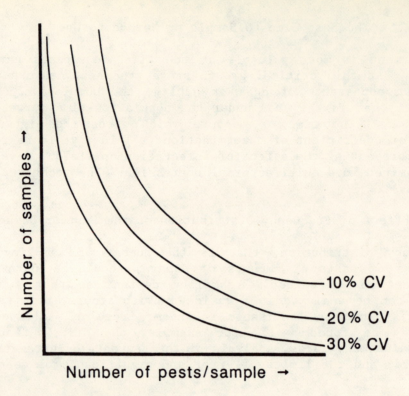

Figure 2. The effect of the number of samples and number of pests per sample on the coefficient of variation.

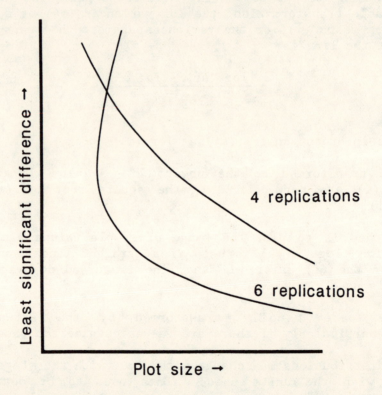

Figure 3. The effect of plot size on the least significant difference between means, with four and six replicates (after Smith 32)

is discussed later.

There are many examples and references: Simonet & Pienkowski (31), Smith (32), Steffey & Tollefson (35) are particular cases, Yates (49) and Church, in Chiarappa (9) are general treatments.

Sequential Sampling

The number of samples needed to make a decision varies with the population found. Upper and lower limits are drawn at the desired degree of precision, and sampling continued until the number of pests or amount of damage found is above or below these limits. Examples are given by Nishada & Torii (22), Shepard (30), Harcourt & Guppy (16) and Onsager (24).

Successive Sampling is a useful approach discussed by Abraham et al. (1).

Sampling in Particular Crops

Sampling methods and studies will be found for many crops, for example cotton, by Vaissayre (42) and Wilson (48), or rice by Gomez & Gomez (14), who give formulae for finding the percentage tiller infestation by stemborers from samples of infested hills:

Percentage infested tillers =

$$\frac{\text{number of infested tillers in } H_i}{\text{total tillers in } H_i} \times \frac{\text{number of } H_i}{\text{total hills}} \times 100$$

where H_i = infested hills

Sampling methods for locusts will be found in Symmons (37), while the general topic of pest sampling is discussed in Chapter 6.

CONCLUDING REMARKS

Standard methods of quantifying pest populations and their damage which are quick, easy, cheap and reproducible are an essential basis for integrated pest management. This is particularly so when data can be recorded quickly and easily in the field and processed by automated methods. Such methods themselves now need development and testing.

LITERATURE CITED

1. Abraham, T.P., Khosla, R.K., Kathuria, O.P. 1969. Some investigations on the use of successive sampling in pest and disease surveys. J. Ind. Soc. Agric. Stats. 21:43-57.
2. Arnold, A.J., Needham, P.H., Stevenson, J.H. 1954. A self-powered portable insect suction sampler and its uses to assess the effects of azinphos methyl and endosulfan on blossom beetle populations on oil seed rape. Ann. Appl. Biol. 75:229-233.
3. Bardner, R., Fletcher, K.E. 1974. Insect infestations and their effects on the growth and yield of field crops: a review. Bull. Entomol. Res. 64:141-160.
4. Blower, J.G., Cook, L.M., Bishop, J.A. 1981. Estimating the Size of Animal Populations. London: Allen & Unwin. 128 pp.
5. Bowden, J. 1982. An analysis of factors affecting catches of insects in light-traps. Bull. Entomol. Res. 72:535-556.
6. Bram, R.A. 1978. Surveillance and collection of arthropods of veterinary importance. USDA, ARS Handbook 518. 125 pp.
7. Campion, D.G., Nesbitt, B.E. 1981. Lepidopteran sex pheromones and pest management in developing countries. Trop. Pest Manage. 27:53-61.
8. Carino, F.O., Kenmore, P.E., Dyck, V.A. 1979. The farmcop suction sampler for hoppers and predators in flooded rice fields. Int. Rice Res. Newsletter. 4:21-22.

9. Chiarappa, L. 1971. Crop Loss Assessment Methods: FAO Manual on the evaluation and prevention of losses by pests, diseases and weeds. FAO and Commonw. Agric. Bureaux, Farnham Royal, UK. 255 pp.

10. Clinch, P.G. 1971. A battery operated vacuum device for collecting insects unharmed. N. Z. Entomol. 5:28-30.

11. Cothran, W.R., Summers, C.G. 1974. Visual economic thresholds and potential pesticide abuse: alfalfa weevils, an example. Environ. Entomol. 3:891-894.

12. Dietrick, E.J. 1961. An improved back-pack motor fan for suction sampling of insect populations. J. Econ. Entomol. 54:394-395.

13. Eddlestone, F.K., Setter, J., Schofield, P. 1984. Insect marking methods for dispersal and other ecological studies. Bibliography 4, Tropical Development and Research Institute, London. 63 pp.

14. Gomez, K.A., Gomez, A.A. 1984. Statistical Procedures for Agricultural Research. 2nd edition. London and New York: Wiley 680 pp.

15. Greaves, D.A., Hooper, A.J., Walpole, B.J. 1983. Identification of barley yellow dwarf virus and cereal aphid infestation in winter wheat by aerial photography. Plant Pathol. 32:159-172.

16. Harcourt, D.G., Guppy, J.C. 1976. A sequential decision plan for management of the alfalfa weevil, Hypera postica (Coleoptera: curculionidae). Can. Entomol. 108:551-555.

17. Hielkema, J.U. 1977. Application of landsat data in desert locust survey and control. AGP:LCC/77/11. Rome: FAO. 28 pp.

18. Lewis, T., Macauly, E.D.M. 1976. Towards rational design and elevation of pheromone. Pestic. Sci. 7:634-635.

19. Matthews, G.A. 1984. Pest Management. New York and London: Longman. 231 pp.

20. McMahen, E.A., Watson, J.A.L. 1977. The effect of separation by papering on caste ratios in Nasutitermes exitiosus (Hill) (Isoptera). J. Austr. Entomol.Soc. 16:455-457.

21. Moericke, V. 1980. Uber den Farbensinn der Pfirsichblattlaus Myzodes persicae.Z. Tierpsychol. 7:265-274.

22. Nishida, T., Torii, T. 1970. A handbook of field methods for research on rice stem borers and their natural enemies. Int. Biol. Programme Handbook 14. Oxford: Blackwell. 132. pp.

23. Olfert, O.O., Gage, S.H., Mukerji, M.K. 1980. Aerial photography for detection and assessment of grasshopper (Orthoptera: acrididae) damage to small grain crops in Saskatchewan. Can. Entomol. 112:559-566.

24. Onsager, J.A. 1976. The rationale of sequential sampling, with emphasis on its use in pest management. USDA/ARS/S. Carolina AES Technical Bull. 1526.Washington DC. 19 pp.

25. Puntener, W. (editor) 1981. Manual for Field Trails in Plant Protection. Documenta Ciba-Geigy, Basle. 205 pp.

26. Rabb, R.L., Kennedy, G.G. (editors) 1971. Movement of highly mobile insects. Proc. Conference, Raleigh, North Carolina. 456 pp.

27. Riley, J.R. 1979. Quantitative analysis of radar returns from insects. In Radar, insects population ecology and pest management, ed. C.R. Vaughan, W. Wolf, W. Classen, pp.131-158. Conf. Pub. 2070, NASA, Wallops Island, Virginia.

28. Ruesink, W.G., Haynes, D.L. 1973. Sweepnet sampling for the cereal leaf beetle, Oulema melanopus. Environ. Entomol. 2:161-172.

29. Ryan, L., Molyneux, D.H. 1981. Non-setting adhesives for insect traps. J. Insect Sci. & Appl. 1:349-355.

30. Shepard, M. 1973. A sequential sampling plan for treatment decisions on the cabbage looper. Environ. Entomol. 2:901-903.

31. Simonet, D.E., Pienkowski, R.L. 1979. Sampling plan development for potato leafhopper nymphs, Empoasca fabae (Homoptera: cicadellidae), on alfalfa. Can. Entomol. 111:481-486.

32. Smith, F.L. 1958. Effects of plot size, plot shape, and number of replications on the efficiency of bean yield trials. Hilgardia 28:43-63.

33. Southwood, T.R.E. 1978. Ecological methods: with particular reference to the study of insect populations. 2nd ed. London: Chapman and Hall. 524 pp.

34. Steck, W., Bailey, B.K. 1978. Pheromone traps for moths: evaluation of cone trap designs and design parameters. Environ. Entomol. 7:449-455.
35. Steffey, K.L., Tollefson, J.J. 1982. Spatial dispersion pattern of northern and southern corn rootworm adults in Iowa cornfields. Environ. Entomol. 11:283-286.
36. Strickland, A.H. 1961. Sampling crop pests and their hosts. Ann. Rev. Entomol. 6:201-220.
37. Symmons, P.M. 1981. Routine estimation of populations of the Australian plague locust, Chortoicetes terminifera. Tech. Rep. 2. Australian Plague Locust Commission, Barton, Canberra. 10 pp.
38. Taylor, L.R. 1961. Aggregation, variance and the mean. Nature 189:732-735.
39. Taylor, L.R. 1962. The absolute efficiency of insect suction traps. Ann. Appl. Biol. 50:405-421.
40. Thornhill, E.W. 1978. A motorised insect sampler. PANS. 24:205-207.
41. USDA. 1969. Survey methods for some economic insects. Agric. Res. Service, ARS-81-31. Hyattsville, USA. 137 pp.
42. Vaissayre, M. 1982. Methodes d'echantillonage des populations d'insectes nuisibles dans les cultures cotonnieres en Afrique. Entomophaga 27:25-29.
43. Van Emden, H.F. 1972. Aphid Technology. London: Academic Press. 344 pp.
44. Verheijen, F.J. 1960. Mechanisms of the trapping effect of artificial light sources upon animals. Arch. Neerl. Zool. 13:1-107.
45. Wallen, V.R., Jackson, H.R., MacDiarmid, S.W. 1976. Remote sensing of corn aphid infestation 1974 (Hemiptera: aphididae). Can. Entomol. 108:751-754.
46. Walker, P.T. 1980. Standardization of pest assessment and the FAO methods as applied to pests of cereals, olive and citrus. EPPO Bull. 10: 93-96.
47. Walker, P.T. 1981. The relation between infestation by lepidopterous stemborers and yield in maize: methods and results. EPPO Bull. 11:101-106.
48. Wilson, L.T. 1982. Development of an optimal monitoring program in cotton emphasis on spidermites and Heliothis spp. Entomophaga 27:45-50.
49. Yates, F.R.S. 1977. Sampling Methods for Censuses and Surveys. London: Griffin. 428 pp.
50. Youdeowei, A., Service, M.W. 1983. Pest and Vector Management in the Tropics: With Particular Reference to Insects, Ticks, Mites, and Snails. London and New York: Longman. 399 pp.

CHAPTER 4

MODELING OF CROP GROWTH AND
YIELD FOR LOSS ASSESSMENT

M. E. Pace and D. R. MacKenzie

Department of Plant Pathology and Crop Physiology
Room 302 Life Sciences Building, Louisiana State University,
Baton Rouge, LA 70803, U.S.A.

The concept of crop loss management is relatively new to agricultural research and it presents some exciting and unique perspectives and challenges. Crop loss management integrates pest management concepts with a crop's response to stress in a complicated interpretation of interacting biological and physical events. Clearly, one major imperative of the successful management of crop losses is the need for a better understanding of the response of a plant population in that environment irrespective of the pest-caused stress. Once this relationship can be established, then more comprehensive pest-caused crop stress relationships can be explored.

The preponderance of published yield loss research information has been expressed as a percentage of the related check or control plot. Typically the amount or intensity of a pest or disease is varied and the resulting yield of the plot is expressed as a percent of the "check". This tradition proves to be misleading and is subject to two hazards.

The first hazard was well demonstrated by a yield loss study reported by Reddy et al. (10) on bacterial leaf blight of rice. In an experiment designed to measure the relationship of bacterial leaf blight severity to rice yield, Reddy et al. compared two sets of data from different years for the same variety. When expressed as a percent of the zero-disease check plot Reddy et al. discovered a significant difference in the disease-severity-to-yield-loss slope as measured by the linear regression coefficient. The Y-axis intercepts were, of course, identical for the two years. This is because the intercepts were forced through "100 percent" yield. However, when the yields were expressed as metric tons per hectare no significant difference in the regression coefficients could be demonstrated. The difference between the two years was entirely explained by the yield response at zero disease -- the Y-axis intercept.

Yield potential, in the absence of the disease may vary between years. Intuitively one would expect this. By forcing a common intercept ("100 percent" yield) an artifact in the disease severity-yield loss relationship was created. Expressing yield as percent-of-the-check-plot is hazardous.

The second hazard comes from the difficulties encountered in using yield loss information when expressed as a percent. Extrapolation of percent loss to dollars, metric tons, or other units of measure becomes very difficult. A 10 percent yield loss in a potato field has little true meaning or use. One asks, "ten percent of what?" A metric ton per hectare estimate is much more useful.

We suspect that some scientists continue to express yield losses as a percent for convenience and to avoid the difficulties of estimating yield in the absence of stress. Sometimes, in small plots tests, the check plot can be maintained at zero disease. However, disease survey data for commercial production are oftentimes the only available information for yield loss estimates with no corresponding "zero disease" estimates.

For these reasons there is a growing interest in crop growth models to estimate better crop yield in the absence of stress. Crop growth models are useful for interpreting key environmental factors, physiological processes and other factors for computing crop growth over a season.

Significant progress has been made in crop growth modeling. The purpose of this chapter is to provide a conceptual framework of those accomplishments and to look at their application to crop yield loss assessment.

HISTORY OF CROP GROWTH MODELS

Crop growth modeling has a fairly recent history. Although considerable biological research lead up to contemporary crop growth modeling, it was the advent of large, high speed computers that made plant growth modeling practical. Complementary with computers were the contributions of systems science and several convenient computer languages useful for crop modeling.

A considerable portion of crop growth modeling has been done by agricultural engineers who were looking for areas to improve crop production. They, of course, borrowed very heavily from the plant physiological literature. Another very significant portion of crop modeling research has been conducted by the National Aeronautical and Space Administration (NASA) supporting research applications of remote sensing technology. Other clusters of modeling activity include integrated pest management research programs and some modest activities in agricultural economics.

Some of the existing crop growth models are adequate for applications in yield loss assessment. This varies by crop and by the problems to which the model is to be applied. In many cases, existing crop growth models are very inadequate. This represents a source of frustration for individuals undertaking yield loss research. Crop loss researchers require good crop growth models. Far more research needs to be done on crop growth modeling if yield loss research is to expand into areas of useful application.

Empirical Models

Crop growth and yield are often modeled as direct responses to weather, climate and environment. Among the more important factors in these models are solar radiation, water status, temperature and carbon dioxide. There are differences of opinion in the choice of these factors and the length of the time interval over which these factors are accumulated. Seif and Pederson (11) regressed wheat yields on rainfall over the growing season. Angus and others (1) used daylength, water status, temperature and radiation to regress growth rates for a grass crop using a time interval of seven days. Murata (9) regressed rice yield on radiation and temperature using phases of crop development instead of a fixed time interval.

Although regression models have been to some extent successful in predicting crop yields, problems may arise when the models are used outside the environmental ranges for which they were developed (6). Because of weather fluctuations, the time intervals become very important along with the development phase of the crop when establishing relationships between crop growth rates and environmental factors such as rainfall.

Instead of interpreting the effect of environmental factors directly on plant growth rates and yields, mechanistic models attempt to explain the effects of environment on lower level physiological processes. Among the most often considered processes are light interception and photosynthesis, root activity and nutrient uptake, transpiration, leaf area expansion, carbon and nutrient partitioning, growth of structural dry matter, growth and development of new organs and senescence. Mechanistic models differ in the number of processes considered and the degree of sublevel description of individual processes. Although one or more of the processes of special interest may be approached in mechanistic detail, the remaining processes are often treated empirically.

A number of partially mechanistic models for various crops have been published. Recent surveys of published empirical and mechanistic models have been presented by Baker (3), Legg (8) and France and Thornley (6).

The next section of this chapter will contain a discussion of one plant growth model in detail, to demonstrate how it was constructed, its underlying assumptions and its utility.

MODELING CORN GROWTH

The corn growth model of Splinter (12) used average daily light intensity, average daily temperature, and soil moisture readings to calculate the growth of corn during a crop season. We will use Splinter's model as an example of crop growth modeling as it was very well received at the time of publication. More recent models are, in many ways, only elaborations on the approach of Splinter.

Splinter (12) recognized that the photosynthate produced by a corn plant could be assigned to the functions of respiration and assimilation and to the growth of leaves and roots. Thus, by recognizing the partition of photosynthate into one of these four groups he could address the effects of selected environmental factors on the rates of utilization. Factors such as light intensity, temperature, and soil moisture would likely be determining factors in those various processes. The scientific literature certainly supports that assumption.

Splinter approached the corn modeling problem in the following way. The rate of photosynthetic accumulation and utilization can be described as instantaneous rates using calculus. For the reader unfamiliar with calculus the notation is quite simple. For example, the rate of photosynthate production can be expressed as: (dC/dt). This notation symbolized the instantaneous rate of carbohydrate production as a small increase in carbohydrate (dC) for a small increase in time (dt).

Similarly, instantaneous rates can be expressed for respiration (dR/dt), for root growth (dH/dt), and for leaf growth (dh/dt). The description of assimilation is a bit more complicated as it is assumed to be a function of the rate of respiration adjusted by two coefficients (b and c). These two coefficients were not known for corn at the time Splinter was preparing his growth model. He therefore used values for b and c that had been determined for tobacco grown in the North Carolina State University phytotron. It is quite common for crop growth modelers to utilize values for other species until more appropriate coefficients have been derived for the desired species. The instantaneous rate of assimilation was therefore given as $bc(dR/dt)$. The quantity b is the fraction of the respirational energy used in assimilation and is equal to 0.01. The quantity c represents the units of photosynthate assimilated for each unit of respiration and is equal to 3.0. The instantaneous rate of assimilation in the corn model thus becomes .03 times the instantaneous rate of respiration (or 3% of dR/dt).

The intended use of this model would be for determining the above ground growth of corn plants. To do this a rate equation is used to express the instantaneous rate of growth of leaves (dh/dt). This is given as the following equation:

$$dh/dt = dC/dt - dH/dt - dR/dt - bc(dR/dt) \qquad (1)$$

Splinter (12) then assumed that the instantaneous rate of photosynthate production (dC/dt) would be directly proportional to the light intensity and inversely proportional to the level of available carbohydrate in the leaves (h) and stems (H). Expressed symbolically this would be:

$$dC/dt \cong P/h \qquad (2)$$

where $P = 0.4996(I - 0.084)/1 + 0.912(I - 0.084)$. The quantity I is in langleys (cal/cm**2/sec).

The instantaneous rate of respiration (dR/dt) was assumed to be a function of temperature and the level of carbohydrate available for respiration. Again, using data developed for tobacco, Splinter chose:

$$dR/dt - (0.003857 + 0.003829T)h = gh \qquad (3)$$

where T is temperature in degrees Celsius, h is the level of carbohydrate available in the leaves and stems, and $g = 0.003857 + 0.003829T$. Equation 3 can then be used with the assimilation coefficients to derive the instantaneous rate of assimilation (calculated as 3% of instantaneous rate of respiration).

The instantaneous rate of carbohydrate translocation to the roots was assumed to be roughly equivalent to the instantaneous rate of respiration up to a limit of 25 C. Above 25 C, Splinter used the equation of Bohning et al. (4) as:

$$dH/dt = [0.1 - 0.003829(T-25)]h = qh \qquad (4)$$

Here $q = [0.1 - 0.003829 (T-212)]$. This equation comes from previous research conducted on tomato (4).

Splinter reported a linear relationship between the rate of photosynthesis (P') and soil moisture block electrical resistance which is an indication of soil moisture. For corn, Splinter gave the relationship as:

$$P' = P(0.135 - 0.000021p)/0.135 \qquad (5)$$

where p is soil moisture block resistance reading in ohms and P is defined as in Equation 2. Splinter then determined the amount of available carbohydrate in the leaves with the following equation:

$$dh/dt = dC/dt - dR/dt - bc(dR/dt) - dH/dt \qquad (6)$$

where dC/dt is the instantaneous rate of carbohydrate contributed through photosynthesis, dR/dt is the instantaneous rate of carbohydrate loss through respiration, bc(dR/dt) is the instantaneous rate of carbohydrate use in assimilation and dH/dt is the instantaneous rate of carbohydrate translocation to the roots. Inspection of this equation shows that each of these instantaneous rates is derived from separate equations. By substituting those equations into Equation 6 and using calculus to integrate from time t_o to t, the amount of photosynthate available for leaf and stem growth can be determined by using daily light intensity, average daily temperature, and measures of soil moisture (given as electrical resistance).

The mathematics becomes quite tedious and hence they are most suitable for a computer program which can incrementally step through a growing season one day at a time. Daily values for light intensity, temperature, and soil moisture can be interpreted over for a growing season to plot the expected "growth" of a corn crop.

Intuitively, one can recognize that the crop cannot be allowed to grow forever. Some limits must be placed on growth. Typically, corn essentially stops growing with the onset of flowering which is primarily a function of the accumulated degree days. Degree days are calculated from daily temperature averages minus a base ($10^{o}C$) and accumulated over the growing season. The limit of 450 degree days, beginning with the first field measurements, was chosen by Splinter to indicate the point at which flowering is initiated and growth ceases.

Even after flowering, a corn plant continues to accumulate dry matter. For this reason "photosynthesis" continues to 1400 degree days when kernel moisture is estimated to be 30 percent.

Splinter verified his corn growth model with actual field measurements of corn growth. The model performed very realistically.

YIELD LOSS ASSESSMENT BY SIMULATION

Once a crop growth model has been constructed and field tested, it can be used to provide answers about the influences of environmental variation on yield (e.g. planting date, nutrient levels, soil moisture and radiant energy). The flexibility of the model in its ability to predict yield changes depends on the variables input to the model, the state variables that the model calculates and the output of the model.

Most growth models calculate daily photosynthate accumulation and its partitioning from CO_2, radiant energy and available moisture levels plus the associated change in leaf photosynthetic area. These models are generally very reliable in predicting yield changes due to fluctuations in one or more of these variables. Any stress that directly reduces photosynthetic area (e.g. leaves) and subsequently, photosynthate production, can be used as a variable to calculate the loss to the crop growth model. Insects and plant pathogens are such stress factors. Including their impact on crop growth, however, is confounded by the pest organisms' reproductive rates, as well as the influence of the environment on these rates and on the organism per se. In addition, insects and plant pathogens often influence plant growth in ways other than directly reducing leaf area. Because of the complexity of the plant-environment pest system, the dynamics of each pest population is often a separate submodel, the output of which can then be used in calculating the effect of the single pest on other variables in the primary crop growth model.

Boote et al. (5) grouped insects and plant pathogens according to their general effects on crop growth. By delimiting pests in this manner, their influence on variables within the crop model can be more readily visualized and the respective pest submodels more accurately constructed. To add to the complexity, it is interesting to note that each pest has the potential to influence every other pest in either a positive, negative or neutral fashion. Boote's (5) seven classes of crop pests are:
1. stand reducers,
2. photosynthetic rate reducers,
3. leaf senescence accelerators,
4. light stealers,
5. assimilate sappers,
6. tissue consumers, and
7. turgor reducers.

The effects of another category, toxin production by insects and plant pathogens, are often extremely difficult to measure and to model.

A subroutine dealing with, for example, a light stealer such as sooty mold growing on aphid exudate on leaf surfaces might need to take into account the initial number of aphids present (or their rate of immigration), the rate of insect multiplication, the rate of exudate production per insect on the particular host, and the effects of environmental parameters on all of the above. In addition, the rate of fungal growth on the exudate and the respective amount of light reduction per unit leaf area accorded to that growth would also have to be calculated by one of the model's subroutines. Time interval calculations (e.g. daily) from this subroutine could then be used by the primary program (or another subroutine) to calculate the adjusted instantaneous rate of photosynthate production based on available light.

In order to incorporate the effects of the various pests into the model, Boote et al. (5) specified certain features. One of these was the incorporation of crop phenology to determine crop development rate and the expected time of physiological events. Another feature was the allocation of carbon, organogenesis, and fruit addition as affected by these events, which in turn may be affected by pests. Other features included the respiratory cost of maintaining and synthesizing plant and pest tissue, the effects of reduced photosynthetic area, the onset of leaf senescence, water relations, nutrient relations and physiological feedback mechanisms.

OVERVIEW

Critics of plant growth modeling oftentimes point out the need to include many variables in the plant growth model. The consideration of carbon dioxide level, soil fertility, variety differences, etc. might arguably improve Splinter's model. But that criticism is hardly fair. The model is reasonably accurate. And, because of its relative simplicity, it does have utilitarian appeal.

There are other plant growth models that operate with more complex physiological mechanisms. GLYCIM, a dynamic simulator of soybean growth (2) is an interesting example of a model in the category. In the GLYCIM model, carbon dioxide level and soil fertility are explicit variables. In some instances, when the mechanism of particular processes were unknown, the designers of GLYCIM chose to use plausible but hypothetical mechanisms instead of empirically derived equations.

Some modelers believe that empirical models are useful only for making predictions within the range of the experimental data used to generate the equations. Others dare to probe beyond experimental knowledge. The arguments rage. Highly detailed mechanistic models are not without critics. Many modelers believe that while complex computer algorithms have substantial predictive value, they are often as difficult to understand as are the biological systems that they are designed to simulate. Complicated models also require very fast computers with large amounts of memory. The advent of personal computers makes computer time and memory requirements for complex crop growth models a real concern. The designers of GLYCIM have estimated that it might take 2 hours for one "run" of their model on an 8087 math co-processor equipped IBM PC. This would surely limit many applications.

CONCLUDING REMARKS: FUTURE DIRECTIONS

Crop loss modeling has a strong need for much more effort in crop growth research. The focus of attention should include the development of specific submodels on such topics as leaf growth (area available for pest attack), stem growth (plant height and stress compensatory factors, etc.), fruit growth (related to yield) and root growth (disease relationships, water relationships, nutrients, etc.).

Many of the equations that will be used to develop needed submodels will undoubtedly be specific to the crop and perhaps even the crop cultivar. The key environmental factors that "drive" the model will need to be fully understood. There is a need to know the direct, indirect and feedback effects of stresses on the plant's growth and their biologically significant interactions.

The integration of the submodels will not be an easy task, but it is one on which progress is now being made. An example of this type of integration is demonstrated by the incorporation of parts of the rhizosphere simulator RHIZOS (7) into the GLYCIM model. However, the difficulties of submodel integration should not be accepted as anything more than an argument to initiate more such research. This is true because, without elegant crop growth and yield models, yield loss modeling and crop loss modeling will be severely limited in their applications.

LITERATURE CITED

1. Angus, J.F., Kornher, A., Torssell, B.W.R. 1980. A Systems Approach to Swedish Ley Production. Progress Report 1979/80. Report 85. Uppsala: Swedish University of Agricultural Sciences.
2. Acock, B., Reddy, V.R., Whisler, F.D., McKinion, J.M., Hodges, H.F., Boote, K.J. 1983. The soybean crop simulator, GLYCIM: model documentation 1982. 002 in the series "Response of vegetation to carbon dioxide." U.S. Dept. of Energy and U.S. Dept. of Agriculture. pp. 18-27.
3. Baker, D.N. 1980. Simulation for research and crop management. In World Soybean Research Conference II: Proceedings, ed. F.T. Corbin, pp. 533-546. Boulder, CO: Westview Press.
4. Bohning, R.H., Kendall, W.A., Linck, A.J. 19123. Effect of temperature and sucrose on growth and translocation in tomato. Am. J. Bot. 40:1120-1123.
5. Boote, K.J., Jones, J.W., Mishoe, J.W., Berger, R.D. 1983. Coupling pest to crop growth simulators to predict yield reductions. Phytopathology 73:1581-1587.
6. France, J., Thornley, J.H.M. 1984. Mathematical Models in Agriculture, pp. 161-174. Butterworth and Co.
7. Lambert, J.R., Baker, D.N. 1984. RHIZOS: a simulator of root and soil processes. S.C. Agric. Exp. Stn. Tech. Bull 1080. pp. 1-10.
8. Legg, B.J. 1981. Aerial environment and crop growth. In Mathematics and Plant Physiology, ed. D.A. Rose, D.A. Charles-Edwards, pp. 129-149. New york: Academic Press.
9. Murata, Y. 1975. Estimation and simulation of rice yield from climatic factors. Agric. Meteorol. 15:117-131.
10. Reddy, A.P.K., MacKenzie, D.R., Rouse, D.I., Rao, A.V. 1979. Relationship of bacterial leaf blight severity to grain yield of rice. Phytopathology 69:967-969.
11. Seif, J.E., Pederson, D.G. 1978. Effects of rainfall on the grain yield of spring wheat, with an application. Aust. Jour. Agric. Res. 29:1107-1115
12. Splinter, W.E. 1974. Modelling of plant growth for yield prediction. Agric. Meteorol. 14:243-253.

CHAPTER 5

DISEASE PROGRESS CURVES, THEIR MATHEMATICAL DESCRIPTION AND ANALYSIS TO FORMULATE PREDICTORS FOR LOSS EQUATIONS

Jerome Freedman and D. R. MacKenzie

Department of Plant Pathology and Crop Physiology,
Louisiana State University, Baton Rouge, LA 70803, U.S.A.

Plant epidemiology is the scientific discipline that addresses the increase of disease in populations over time. The study of epidemics and the application of that knowledge to plant pathological problems have created three branches of epidemiology: comparative, analytical and applied epidemiology (8). Comparative epidemiology includes those activities which describe the processes of infection and disease. It also maintains an ecological perspective of disease development. Analytical epidemiology is quite often empirical and quantitative in studying the effects of the environment and the genetics of the host and the pathogen. Statistical methodology is commonly used for model development and other applications. Applied epidemiology covers activities such as breeding for resistance, plant disease management, and other applications of epidemiological knowledge to the problems of plant protection.

Common to all three branches of epidemiology is the disease progress curve which describes the course of an epidemic and, from that, the mathematical description of the disease progress curve. The analysis of these curves is fundamental to all epidemiology. It is the topic of this Chapter.

THE ANALYSIS OF DISEASE PROGRESS CURVES

J.E. VanderPlank (11, 12, 13) was the first to systematically organize the analytical methods used for the study of disease progress curves. Primarily, the VanderPlankian approach to epidemiological analysis uses the logistic model (1, 11). That model recognizes that "disease begets more disease" in a "compound interest" relationship.

The logistic model also recognizes that infection resulting in disease results in the production of more propagules following a latent period. The rate at which the disease cycle progresses is one descriptor of an epidemic. In order to describe an epidemic's rate of development it is necessary to use some mathematics.

ALGEBRAIC DERIVATION OF THE LOGISTIC EQUATION

The starting point for this derivation is the germ theory. Simply stated, infection begets infection, and in order to get infection there must be some initial infection (I_o) (11).

We will represent the increase of disease in the first cycle as:

$$I = I_o + I_o r \tag{1}$$

In this equation r is a rate of infection increase that yields some new amount of infection (I). We can rewrite Equation 1 as:

$$I = I_o(1 + r) \qquad (2)$$

from which we can calculate the new amount of infection after one cycle. In order to make our rate increase more flexible we can divide r by u for reasons which will become apparent later.

$$I = I_o(1 + r/u) \qquad (3)$$

To represent the amount of infection at the second cycle we can expand Equation 3 to:

$$I = I_o(1 + r/u)(1 + r/u) \qquad (4)$$

This equation simply represents the fact that infection begets infection which begets infection. A convenient mathematical notation can be used to specify t intervals of cycles. This is given as:

$$I = I_o(1 + r/u)^{ut} \qquad (5)$$

Note that the intervals of the cycles t must be multiplied by u to compensate for our earlier decision to divide r by u. As u goes to infinity [$u \to \infty$]; some interesting relationships develop. We desire to have u go to infinity because we would like to have a continuous, or smooth line, expression. Infection is continuous in its development and the value of u should go to infinity in that case. This does open up some arguments. Conidiospore production and variable weather conditions are hardly continuous in real life, and this is one of the rigors of applying mathematics to biological phenomena.

Now we will change some notation to help out our algebra. Let's say that $r/u = y$. If this is the case, then $u = r/y$ and we can rewrite Equation 5 as:

$$I = I_o(1 + y)^{rt/y} \qquad (6)$$

which can then be rewritten algebraically as:

$$I = I_o[(1 + y)^{1/y}]^{rt} \qquad (7)$$

Now we will take a very close look at the relationship of $(1 + y)^{1/y}$. This expression has some very interesting properties that form the base of the natural log system. Our problem is this: evaluation of this relationship for a continuous expression requires that u go to infinity ($u \to \infty$). As u goes to infinity ($u \to \infty$), y goes to 0 ($y \to 0$). The expression $(1 + y)^{1/y}$ becomes a mathematical difficulty. Using the concept of limits we will choose a number so incredibly small that we will not violate the principle that division by zero is undefined. We will approach the limit as y goes to 0 [$\lim_{y \to 0} (1 + y)^{1/y}$].

y	$(1 + y)^{1/y}$
0.1	2.5934
0.01	2.7048
0.001	2.7169
0.0001	2.71814
0.000001	2.71828

It has been reported that a computer scientist, out of simple curiosity, evaluated $(1 + y)^{1/y}$ with y equal to a number so small that there were 28 zeroes following the decimal point. There is however, an easier way of doing it which does not require hours of computer time. Mathematicians simply say:

$$\lim_{y \to 0} (1 + y)^{1/y} \equiv e = 2.7182818... \qquad (8)$$

In this evaluation mathematicians define $\underline{e}$, the base of the natural log system. This identity ($\equiv$) greatly simplifies our expanded germ theory model. Equation 7 can now be written as Equation 9 because we want the limit as $y \to 0$ of $(1 + y)^{1/y}$. This is defined as $\underline{e}$. So:

$$I = I_0 e^{rt} \qquad (9)$$

which is the compound continuous interest equation. This is the equation from which the name for compound interest epidemics has been borrowed.

The compound continuous interest equation has some interesting characteristics that should be described. When:

$$I = I_0(2)$$

a population doubles. So when $e^{rt} = 2$ a population will double. And because $e^{0.693147} = 2$, we know that when $rt = 0.693147$, the population will double for $I = I_0 e^{rt}$ because $I = I_0(2)$.

To give an example, if $r = 0.03$ units per day then the doubling time would be calculated as $t = 0.693147/0.03 = 23$ days.

A gross approximation of doubling time is the "rule of seventy". Seventy divided by the percent rate of increase yields a doubling time in appropriate units. To use another example, 70 divided by 3% (the rate of increase as a percent) would give 70/3 or roughly 23 days.

THE COMPLICATIONS OF MULTIPLE INFECTION

Epidemics are limited by the availability of plant tissue for infection. As an epidemic progresses less plant tissue is available and an epidemic trails off to 100 percent disease. Plant tissue that is already diseased cannot again become diseased. These "second attempts" are referred to as multiple infections (11).

Early in the epidemic multiple infection is a very rare event. The probability of a multiple infection occurring is a function of the amount of healthy tissue and the amount of disease as proportions. That is, $x + h = 1.0$.

This says that the proportion of disease (x) plus the proportion of healthy tissue (h) is equal to 1.0. By rearrangement, $h = 1.0 - x$.

We stated earlier that infection (I) is not equal to disease (x) because of multiple infections. The increase of disease (x) is dependent upon the amount of available healthy tissue. We can now assume, as part of our mathematical development, that: $x = I(h)$.

This equation states that the amount of disease (x) will be equal to the amount of infection (I) times the proportion of healthy tissue available for infection (h). For example, if $h = 1$, (i.e. all tissues healthy) multiple infection would not be possible and $I(1) = x$. In this case, all "infections" (I) result in disease (x). If half of the tissue is diseased then the remaining half is healthy. In this case, $x = I(0.5)$. That is, one half of the infections are multiple infections that cannot result in new disease. This assumes that infection is at random and there are no interactions.

When there is no healthy tissue remaining, h = 0 and none of the infections can result in new disease. Mathematically stated, x = I(0.0). If x = I(h) and I = x/h and h = 1 - x, we can estimate I as I = x/(1 - x).

Simply stated, the amount of infection (including multiple infections) can be estimated by dividing the proportion of the disease (x) by the proportion of healthy tissue (1 - x). The function x/(1 - x) has been called the "multiple infection transformation".

For plant pathological purposes we are going to commit a violation of an algebraic rule, but we will compensate for the violation by the careful use of specific terms. We will estimate the amount of infection and allow us to use the compound continuous interest equation (Equation 9).

$$\text{If } I_o = x_o/(1 - x_o) \text{ and } I = x/(1 - x)$$

we can substitute these estimators into Equation 9 to yield:

$$x/(1 - x) = [x_o/(1 - x_o)]e^{rt} \qquad (10)$$

We pay a price for making this estimation; we must always refer to the rate of increase as the apparent infection rate. This is because we have not assessed the amount of infection (including multiple infections). We have estimated the amount of infection based on the proportion of disease and healthy tissue.

Figures 1 and 2 show graphically the relationship of the disease progress curve (sigmoid) and the effect of the multiple infection transformation on that curve. Converting x to x/(1 - x) ($\cong I$) yields the exponential increase curve described by the compound continuous interest formula.

However, to deal with the exponential curve and to calculate rates of increase are difficult tasks. The easy way around this problem is to use a logarithmic transformation. Remember, earlier we said $\underline{e}$ is the base of the natural log system, and this is what we will use.

We take the natural log of Equation 9 to give:

$$\ln I = \ln I_o + rt \qquad (11)$$

which is the equation for a straight line. Written in another form one would see it as y = a + bx. Figure 3 shows graphically these components.

We have chosen, however, to use the amount of disease (x) to estimate the amount of infection (I). Equation 10 can be converted to its linear equivalent taking the natural log of both sides to yield:

$$\ln[x/(1 - x)] = \ln[x_o/(1 - x_o)] + rt \qquad (12)$$

Again, Equation 12 is the equation for a straight line and these relationships are shown in Figure 4. Equation 12 can be reformed algebraically to yield the common form of the logistic model as applied by VanderPlank:

$$r = 1/t \; [\ln[x/(1 - x)] - \ln[x_o/(1 - x_o)]] \qquad (13)$$

Some important points must be stressed. When using the multiple infection transformation, doubling time calculations no longer work. With less than 5 percent disease, the doubling time estimate is fairly accurate because the effects of multiple infections are minimal. Doubling time calculations should not be used beyond 5 percent disease.

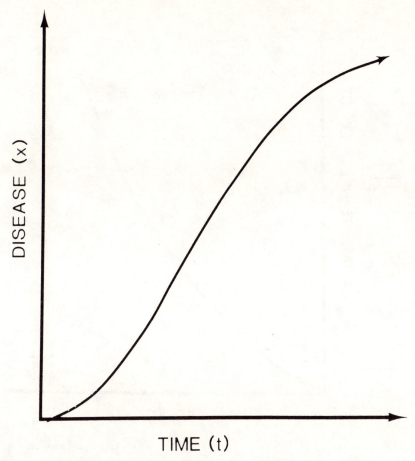

Figure 1. Sigmoid Disease Progress curve.

Below 5 percent disease it is also possible to use the amount of disease directly for an estimate of the amount of infection. This can be done without going through the multiple infection transformation. This estimate of the apparent infection rate has sometimes been called r_1 to symbolically distinguish it from the apparent infection rate which uses the multiple infection transformation (11).

The logistic model has had extensive use in the biological sciences since its description by Berkson in 1944 (1). Our correction term for $\ln[x/(1 - x)]$ is called the logit. Its application to plant epidemiology was made by VanderPlank (11). The logit, however, is by no means the exclusive domain of plant epidemiology.

PREDICTION WITH THE LOGISTIC MODEL

The logistic model can be transformed by algebraic manipulation to yield a prediction equation useful to epidemiology:

$$x = \frac{e^{rt + \ln[x_o/(1 - x_o)]}}{1 + e^{rt + \ln[x_o/(1 - x_o)]}} \tag{14}$$

The quantity x_o would be the present proportion of disease. The quantity r would be the anticipated apparent infection rate in units per day. The equation would be evaluated for t days to yield an expected disease proportion (x) at time t. Figure 5 shows the application of Equation 14 for predicting the course of an epidemic.

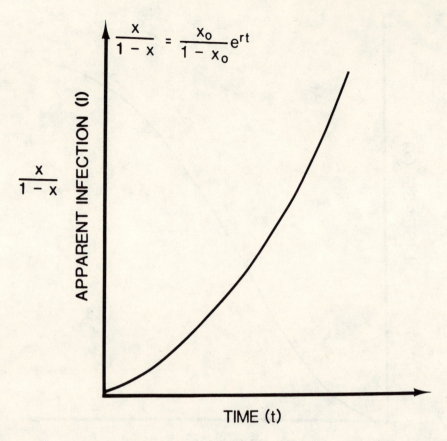

Figure 2. Multiple infection transformation of the disease progress curve: apparent infection rate.

APPLICATION TO YIELD LOSS EQUATION

One basic assertion of yield loss research is that expected yield (EY) is equal to attainable yield (AY) minus the yield loss attributed to various stresses. Because of certain limitations in our ability to predict yield loss, it has been preferred by some individuals to predict the terminal amount of disease (usually at harvest) and then interpret that predicted terminal disease severity yield (see Chapter 16).

In order to forecast expected yield (given present conditions) with a certain proportion of disease (x_o) and an expected increase of disease, Equation 14 can be substituted into our expected yield (EY) relationship to give:

$$EY = AY - b_1 \left[\frac{e^{rt} + \ln[x_o/(1 - x_o)]}{1 + e^{rt} + \ln[x_o/(1 - x_o)]} \right] \tag{15}$$

for $\underline{t}$ days and an expected apparent infection rate $\underline{r}$. The quantity b_1 would be the coefficient of stress.

One application of this method would be to vary the expected apparent infection rate to reflect different disease management strategies. For example, given a fungicide spray strategy of (1) do nothing, (2) spray once, (3) spray twice, an assessment could be made of their yield loss relationships versus the costs of the fungicide applications. The value of the resulting yield increments versus the cost of control strategy could be used in a cost/benefit analysis (5).

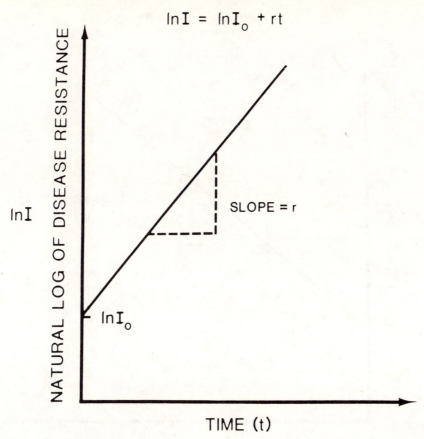

$$\ln I = \ln I_0 + rt$$

NATURAL LOG OF DISEASE RESISTANCE

$\ln I$

SLOPE = r

$\ln I_0$

TIME (t)

Figure 3. Straight-line (slope-intercept form) logarithmic transformation of the compound continuous interest equation.

One application of this method would be to vary the expected apparent infection rate to reflect different disease management strategies. For example, given a fungicide spray strategy of (1) do nothing, (2) spray once, (3) spray twice, an assessment could be made of their yield loss relationships versus the costs of the fungicide applications. The value of the resulting yield increments versus the cost of control strategy could be used in a cost/benefit analysis (5).

Epidemiologists have long conceded that not all days are equal in their contributions to an epidemic (11). Research is now being conducted at several institutions to substitute into specified models the contributions of weather to epidemic development. The attempt is to avoid the solar time scale of days which does not adequately reflect epidemiological relationships. An example developed in Chapter 1 is summarized here briefly. Severity values (SV's) are a means of expressing the duration of epidemiologically important temperature and relative humidity (RH) conditions. The temperature range during which the RH is greater than or equal to 90% is divided into three categories over the range from 7.2°C to 26.6°C. Each temperature category is related to the time period (in hours) during which the RH is greater than or equal to 90%. Each combination of time and temperature categories corresponds to a value from 0 (weather unlikely to contribute to a blight epidemic) to 4 (very highly conducive to a blight epidemic). The SV's are assigned daily and they represent the relationship between temperature and extended periods of high relative humidity on infection by Phytophthora infestans (Mont.) Debary.

Severity values can be used as the time scale for an epidemic's progress. This would have the effect of accounting for periods of time not conducive to an epidemic's development. Disease progress curves are rarely perfectly symmetrical, and this fact is sometimes an attribute of weather conditions during that disease

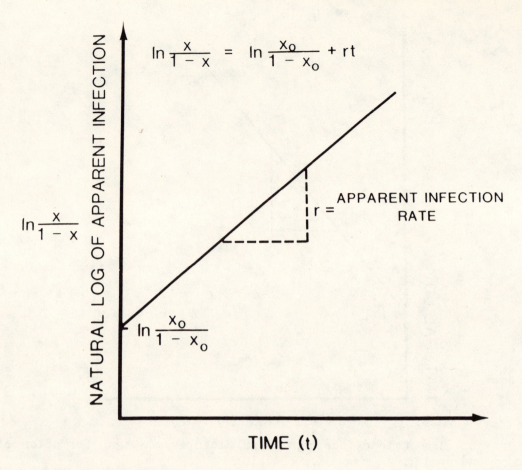

$$\ln \frac{x}{1-x} = \ln \frac{x_0}{1-x_0} + rt$$

Figure 4. Striaght-line (slope-intercept form) logrithmic transformation of the apparent infection rate expression.

development period. The use of severity values compensates for these differences. We see this as one contribution of disease forecasting to yield loss management. The use of severity values in conjunction with the BLITECAST system represents an application of expert systems in plant disease management. Expert systems are codified forms of problem-solving reasoning used by human experts, and have been applied to such activities as medical diagnosis and oil exploration (14). Though many of the non-agricultural applications of expert systems are computer-based, printed point-value systems have been developed and issued by extension personnel to growers as decision-making guides in spraying fungicides (9).

MONOMOLECULAR MODEL

The logistic model that is used for the analysis of epidemics is but one model available for the analysis of disease progress curves. We will briefly develop a second model, but we hasten to point out that there are many other models in the literature that are suitable for applications in plant epidemiology.

The monomolecular model is for use with "simple interest epidemics". It begins at the same point as the logistic model for its algebraic derivation - the germ theory. However, the significant difference is that multiple infections are taken to have a Poisson distribution. Also, as before, $x + h = 1.0$, and therefore $h = 1 - x$.

The Poisson distribution is given as:

$$h = (I^a/a!)e^{-I} \qquad (16)$$

44

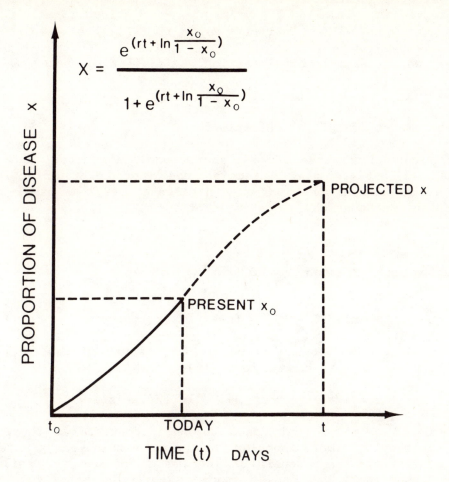

$$X = \frac{e^{(rt + \ln \frac{x_0}{1 - x_0})}}{1 + e^{(rt + \ln \frac{x_0}{1 - x_0})}}$$

Figure 5. Prediction of the course of an epidemic using the logistic model.

We have set up the relationship such that $a = 0$ for healthy plants and $a = 1$ for diseased plants. In the case of healthy plants, $0!$ is defined as 1 and $I^0 = 1$. These are mathematical facts.

So $I^0/0! = 1/1 = 1$. The Poisson distribution, in the case of healthy plants, is reduced to:

$$h = (1.0)\, e^{-I} \qquad (17)$$

Previously we defined $h = 1 - x$, and therefore we get:

$$1 - x = e^{-I} \qquad (18)$$

We must review some rules of exponents.

$$a^{-n}a^n = a^{-n+n} = a^0 = 1$$

If

$$a^{-n}a^n = 1,$$

Then

$$a^{-n}a^n/a^n = 1/a^n = a^{-n}$$

Given that rule we get:

$$1/(1 - x) = e^{I} \qquad (19)$$

45

We then take the natural log of both sides of Equation 19 to get:

$$\ln[1/(1 - x)] = I \qquad (20)$$

which is an estimator for multiple infection. By similar reasoning we get:

$$\ln[1/(1 - x_o)] = I_o \qquad (21)$$

to estimate the amount of initial infection.

The significant difference between the monomolecular model and the logistic model is that we assumed that infection does not beget infection. All new infection is a function of some increase in disease over time simply added to the initial infection. Mathematically stated, that is:

$$I = I_o + rt \qquad (22)$$

Substituting our Poisson estimators to convert the amount of expected infection we get:

$$\ln[1/(1 - x)] = \ln[1/(1 - x_o)] + rt \qquad (23)$$

which is known as the monomolecular model. That equation can be rearranged to solve for r which is the more common form (15):

$$r = 1/t[\ln[1/(1 - x)] - \ln[1/(1 - x_o)]] \qquad (24)$$

The monomolecular model would seem to be appropriate in those cases where disease does not result in more inoculum. Arguments persist among scientists as to the model's appropriate use. Generally, it has found use for soil-borne plant pathogens, although there are many instances where the model has proven to be inadequate for reasons not fully understood.

STATISTICAL REALITIES

One of the consequences of performing log transformations with these models is the effect they have on statistical estimations of error. When taking the natural log of the equations, as we have done, the error terms are compressed. Moreover, the statistical assumptions become quite complicated. For these and other reasons caution must be exercised when statistical procedures are used to test the significance of apparent infection rates or other estimators (10).

Further complications arise when attempting to distinguish the appropriateness of one model over another. These complications no doubt contribute to continuing scientific arguments and the lack of resolution as to which model is best used in specific disease situations. There are other hazards that should be recognized. These are primarily with the many assumptions of linear regression which are violated routinely by plant epidemiologists. Statistical consultation is advised.

Criticism aside, great progress has been made in the analysis of disease progress curves. The application of that knowledge to yield loss and crop loss forecasting has immense potential.

PREDICTORS FOR LOSS EQUATIONS

Several approaches have been used to relate disease progress curves to yield loss estimation. One application has been called the critical point method which uses the quantity of disease at a specific "critical point" in a crop's development to interpret the amount of yield loss. The ability to analyze an epidemic's progress has contributed greatly to exploring methods for predicting yield loss (2).

A related and significant extension of the use of disease progress curves for loss estimation is the multiple point crop loss model (5). The change in the amount of disease (dx) per unit time is weighted by partial regression coefficients. This method addresses the intuitively appreciated expectation that a quantity of disease occurring earlier in the season is not equal to a similar quantity that occurs later in the season.

In a series of articles, Burleigh et al. (2, 3, 4) applied stepwise multiple linear regression to estimate yield losses caused by wheat rusts. These researchers were able to identify the biological variables most useful in explaining the variation in leaf rust severity on 7, 14, 21 and 30 days after making the estimates. The coefficient for multiple determination (r^2) ranged from 0.745 for the 7 day forecast to 0.362 for the 30 day forecast, indicating that short-term forecasts could be made with a relatively high degree of accuracy (2). Studies on the yield loss caused by leaf rust lead to equations which explained 79% of the variability in loss from rust severities made at boot, early berry and early dough (3).

Other investigators have studied the area under the disease progress curve (AUDPC) for its relationship to yield loss. The AUDPC has proven in some cases to be a useful independent variable for predicting yield losses. As James (5) noted, the AUDPC approach may distinguish between two epidemics with different areas but with the same severity at a critical date, an advantage over the critical-point model. However, in applying AUDPC analysis in a potato late blight study (6), it was found to be unsuccessful for distinguishing between losses caused by early and late infections, as both these infections resulted in the same areas under curve.

CONCLUSIONS

The research objective of developing reliable prediction methods for yield loss and crop loss assessment seems realistic but not easily obtainable. One significant aspect of the progress that has been made to date has been in the area of the analysis of disease progress curves. In the past 20 years considerable knowledge has developed which describes the contribution of host resistance, pathogen variation, and the environment to the progress of an epidemic. These relationships are fundamental to our understanding of how to manage yield losses and crop losses.

Other areas of plant epidemiology need to be investigated to develop more reliable crop loss prediction systems. The prospects for the future are exciting for this area of plant protection research.

LITERATURE CITED

1. Berkson, J. 1944. Application of the logistic function to bio-assay. J. Amer. Stat. Assoc. 39:357-365.
2. Burleigh, J.R., Eversmeyer, M.G., Roelfs, A.P. 1972a. Development of linear equations for predicting wheat leaf rust. Phytopathology 62:947-953.
3. Burleigh, J.R., Roelfs, A.P., Eversmeyer, M.G. 1972b. Estimating damage to wheat caused by Puccinia recondita tritici. Phytopathology 62:944-946.
4. Eversmeyer, M.G., Burleigh, J.R., Roelfs, A.P. 1973. Equations for predicting wheat stem rust development. Phytopathology 63:348-351.
5. James, W.C. 1974. Assessment of plant diseases and losses. Annual Rev. of Phytopathology 12:27-48.

6. James, W.C., Shih, C.S., Hodgson, W.A., Callbeck, L.C. 1972. The quantitative relationship between late blight of potato and loss in tuber yield. Phytopathology 62:92-96.

7. Krause, R.A., Massie, L.B., Hyre, R.A. 1975. Blitecast: a computerized forecast of potato late blight. Plant Dis. Rep. 59:95-98.

8. MacKenzie, D.R., Elliot, V.J., Kidney, B.A., King, D.E., Royer, M.H., Theberge, R.L. 1983. Application of modern approaches to the study of epidemiology of diseases caused by Phytophthora. In It's Biology, Taxonomy, Ecology, and Pathology, ed. D.C. Erwin, S. Bartnicki-Garcia, P.G. Tsao, Chpt. 23. St.Paul, MN: American Phytopathological Society. 392 pp.

9. Teng, P.S. 1981. Validation of computer models of plant disease epidemics: a review of philosophy and methodology. Z. Pflanzenkr. Pflanzenschutz 88:49-63.

10. Tekrony, D.M., Stuckey, R.E., Egli, D.B., Tomes, L. 1985. Effectiveness of a point system for scheduling foliar fungicides in soybean seed fields. Plant Dis. 69:962-965.

11. VanderPlank, J.E. 1963. Plant Diseases: Epidemics and Control. New York: Academic Press. 349 pp.

12. VanderPlank, J.E. 1968. Disease Resistance in Plants. New York: Academic Press. 206 pp.

13. VanderPlank, J.E. 1975. Principles of Plant Infection. New York: Academic Press. 216 pp.

14. Weiss, S.M., Kulikowski, C.A. 1984. A Practical Guide to Designing Expert Systems. Totowa, New Jersey: Rowman and Allanheld. 174 pp.

15. Zadoks, J.C., Schein, R.D. 1979. Epidemiology and Plant Disease Management. New York: Oxford University Press. 427 pp.

CHAPTER 6

SAMPLING THEORY AND PROTOCOL FOR INSECTS*

P. M. Ives and R. D. Moon

Department of Entomology, University of Minnesota,
St. Paul, MN 55108, U.S.A.

The goal of sampling is usually either to estimate the value of one or more parameters for a population, or to decide in which of two (or more) possible categories (with respect to the value of a parameter) the population belongs. One major utility of sampling theory is to provide methods for calculating from sample data the estimators and associated confidence intervals. Confidence intervals quantify precision and confidence associated with a sample estimate, or with a decision about the relationship of a population attribute to a reference value. Commonly used methods involve random variables that are assumed to be normally or binomially distributed (9, 39). The data for deriving sample estimates or categorizing a population are generated by a sampling plan, a designed protocol for selecting and examining sample units in a meaningful and statistically acceptable manner.

A second major utility of sampling theory is to make sampling more efficient. One goal is to achieve required precision at a minimum cost. Another goal may be to predict achievable precision given a chosen sampling plan and a limited budget (9). In either case, the theory employs the frequency distribution of the estimator rather than what is known about the distribution of the underlying random variable. The estimator's distribution would be realized through repeated random sampling of the same population with the same method. With sample sizes in common practice, the Central Limit Theorem gives assurance that estimators of means and totals are approximately normally distributed. The estimator's distribution is defined by the sample mean, $\bar{x}$-and the associated standard error of the mean, $SE_{\bar{x}}$. The latter is computed from the sample variance, s^2, and sample size, n.

When sampling, we are concerned with minimizing bias or systematic errors/ departures of an estimator from its underlying parameter. Minimizing these maximizes accuracy. Systematic sampling or the deliberate selection of "representative" or "typical" samples is prone to yield biased results. The precision associated with a chosen estimator is a direct function of the number of sample units examined. The relationship between sample size and precision will be discussed later in this chapter.

The bulk of this chapter is devoted to the design of sampling plans that are commonly employed for field study of agricultural pests, primarily insects. Estimation of parameters will be covered first, followed by decisions about categorizing populations. The reliability of an estimator or decision depends directly on the validity of the sampling plan used to generate the sample data.

Decision	H_0 True	H_0 False
Accept H_0	Correct decision Confidence of test Proby. = $1 - \alpha$	Incorrect decision Type II error Proby. = β
	←------------------Proby. = OC------------------→	
Reject H_0	Incorrect decision Type I error Proby. = α	Correct decision Power of test Proby. = $1 - \beta$
	←------------------Proby. = 1 - OC------------------→	

Table 1. The relationships of Type I and Type II errors, the Power and the Confidence of a hypothesis test, and the Operating Characteristic (OC) curve of a test, to acceptance and rejection of a Null Hypothesis H_0.

SOME BASIC CONSIDERATIONS

Insect densities are most commonly characterized by μ, mean absolute or relative density per N possible sample units or arbitrary subdivisions of the population. Populations can also be characterized by $N\mu$, the total in the population occupying all units. Finally, populations can be characterized by p, the proportion of all individuals with an attribute of interest. None of these parameters are ever known perfectly. Rather, their values are inferred or estimated from data from sample units taken at a chosen time from the population.

Because increasing the number of sample units also increases the cost of sampling a trade-off exists between precision and cost. Different optimal solutions depend on whether the goal of sampling is estimation or decision.

Sampling plans for studying insect populations have the same elements as those for studying any other population. The elements may be grouped under the following general headings:
(1) Purpose of sampling
(2) Definition of target population and sample population
(3) Definition of sample unit, sample frame and methods of enumeration
(4) Preliminary sampling
(5) Precision, cost and number of sample units
(6) Schedule of sampling.
There are many variations of random sampling -- the simplest being unrestricted random sampling (URS). The elements of a sampling plan will be discussed individually in the context of unrestricted random sampling for the purpose of estimating parameters. Sampling designs more elaborate than URS (e.g. stratified RS, nested RS, combinations of these) for the purpose of estimating parameters will also be discussed in a later section.

Hypothesis testing involves choosing acceptable levels of probability of reaching an erroneous conclusion (Table 1). Rejecting the null hypothesis when in fact it is true is referred to as the Type I error, with a probability α. The probability of accepting the null hypothesis when it is false, a Type II error, is β. For most hypotheses concerned with estimating a parameter, only the Type I error, α, is of major concern. In this case, the error is one of failing to

50

bracket the parameter of interest within the estimator's confidence interval. Some hypothesis tests concerned with categorizing populations require that both α and β be chosen.

It is important that recording of data, its subsequent transformation and analysis, be planned integrally with planning of the sampling design. Also, although strict adherence of the sampling plan to the principles of sampling theory would result in maximum reliability of the estimator or decision, in practice, some trade-off of reliability for feasibility is often necessary. These two topics will be discussed in the final sections of the chapter.

ESTIMATING POPULATION PARAMETERS

Unrestricted Random Sampling (URS)

This is the simplest form of probability sampling, in which every possible set of n sample units has an equal chance of being selected. Expressed another way, every unit has an equal chance of appearing in the sample. URS is also the basis of other sample designs that will be described later.

Purpose of Sampling and Definition of Sampling Universe

The first step in developing any sampling plan is to define or specify clearly the purpose of sampling. Insects are sampled for a variety of reasons (Chapter 3), and several major aspects of sample design differ markedly according to purpose. Precisely, what is to be accomplished? A specific population parameter (e.g. number of nymphs of a chosen species) might be of interest, and sampling would be carried out to derive an estimate of the parameter. Alternatively, a pest management decision might only require that the population be classified reliably as needing one treatment or another.

For example, an alfalfa field might be sampled to assess weevil density. If studying efficacy of a new insecticide, then a precise measure of density in treated and untreated fields would be desired. Costs would be limited by the researcher's budget, and an appropriate plan would maximize precision given that budget. An appropriate plan might involve intensive effort to collect and process a large number of plant terminals, clipped quadrats, or other sample units.

In another circumstance, the same field might be one of many under management by an IPM supervisor. In this case, one would only want to know if density in the field at sampling time were above or below some action threshold density. Here, matters of scale would dominate over matters of precision. The researcher's sampling plan would likely be inappropriate for the alfalfa manager, and vice versa.

Definition of sampling purpose leads necessarily to definition of the universe to be sampled. The sampling universe is the set of all habitat units within which it is possible with a given sampling method to measure, count, or score as present or absent, members of the population of interest. In the above examples, different parts of the alfalfa (e.g. terminals, canopy, stubble) would be accessible to sampling for weevils depending on the sampling methods used. The universe would comprise the accessible part of the alfalfa, over the whole field. By contrast, in a study of the spread of an introduced parasite of the alfalfa weevil, the universe might comprise all alfalfa weevils (hosts) in all alfalfa fields within the range of distribution of the weevil. Alternatively it might comprise only weevils in alfalfa fields in that part of the range of the weevil that fell within the borders of the state funding the study.

It is of utmost importance that purpose be understood in advance of all other work, and the sampler would be well advised to write and edit a statement of purpose that clearly states the objective to be accomplished.

Target Population and Sample Population

The second step is to define the target population, i.e. the population about which we want to gain information and reach conclusions. Clear definition is essential so there is no doubt whether an item is or is not in the population. For example, Jansson & Smilowitz (19) found that populations of green peach aphids (GPA) increased more rapidly on lower than middle and upper leaves of potato plants. Rates of GPA population growth on middle and upper leaves did not differ significantly among different N fertilizer treatments, whereas on lower leaves they differed considerably among treatments. Had Jansson & Smilowitz sampled from only one level of leaves, their results would have been biased, and opposite results would occur depending on the layer of leaves that were sampled. Had they sampled from all 3 levels but pooled the samples, the treatment effects would undoubtedly have been weaker. Thus for the purposes of Jansson & Smilowitz's study, it was appropriate to define three separate target populations, and sample from each of them.

The composition of the sample population depends on the methods used, and choice of sample unit. The sample population should ideally be the same as the target population, but sometimes it is either more efficient or only possible to sample from a sub-set of the target population. For example Wilson et al. (49) found that sampling only the mainstem node leaves (MSNL) of cotton plants for Heliothis zea eggs took about 14% of the time required to sample the whole plant. About half the eggs were on MSNL, but they were predominantly near the terminal. Thus sampling only the upper 30% of MSNL yielded more than 70% of the MSNL eggs, i.e. more than 35% of all eggs on the whole plant, while taking less than 4% of the time needed to sample the whole plant. For similar reasons many workers have adopted procedures that sample only that portion of the habitat where the insect is most prevalent or most conveniently sampled. Sampling may be restricted to plant terminals (4, 16), one surface only of a subset of leaves (51), or single or few leaves from selected parts of the plant (3, 19, 57). In all cases, the sampler must be prepared to assume that densities in that portion of the habitat sampled are related linearly to densities elsewhere.

It is important to ensure that the sub-set of the population actually sampled represents the target population, whenever possible. Effective species-specific pheromone traps save enormous amounts of labor compared with light traps, as a tool for monitoring lepidopteran pests. However, pheromone traps catch only males. If the sample population is not representative (e.g., when information about the total adult population is wanted from pheromone trapping), then the type and amount of bias should be estimated, and its importance relative to the purpose of sampling assessed, to decide whether another method would be better. Unlike Jansson and Smilowitz, Bechinski & Stoltz (3) were interested in a single target population, mites on the whole canopy, as their goal was to develop commercially feasible sampling plans for snapbeans. They compared three separate sample populations (leaves from the upper, middle and bottom third of the bean canopy) to assess which was most representative of the target population. In their commercial sampling plan then, sample units (single trifoliate leaves) were taken only from the bottom (most representative) third of the canopy.

Care needs to be taken for situations in which the relationship between the sample population and the target population changes through time -- and methods need to be adjusted accordingly. For example, Schotzko and O'Keeffe (35, 36) found both seasonal and diel variation in sweepnet effectiveness for sampling Lygus hesperus in lentils. Effectiveness of sampling adults changed between early and late season, and effectiveness of sampling nymphs was strongly affected by time of day, as well as by relative humidity (RH), temperature and light. Thus in this case the relationship between target and sample populations when the target population comprised adults, differed from when it comprised nymphs. Pooling life-stages would have masked the patterns in temporal variation, and produced misleading results.

The proportion of the target population that is in the universe may differ with season, time of day or weather conditions. When such variation can be recognised, it will be necessary either to correct the population estimates with predictive relationships defined elsewhere, or change sampling methods. For seasonal or time of day effects, the temporal dimension of the universe should be defined so that such variation is avoided. For example, an appropriate universe for Schotzko and O'Keeffe's (35, 36) _Lygus_ nymphs would comprise that part of the lentil canopy accessible to a sweepnet, during the afternoon, when temperature, RH and light intensity were within critical limits. Sampling outside this universe would be deliberately avoided.

Methods of Measurement, Sample Unit and Sampling Frame

The third step involves choosing these three highly interdependent components. Walker (Chapter 3) and Southwood (39) present many alternative methods of measurement. Choice of methods and sample unit may need to be a two or more stage process, with the initial choice based on previous data from the same or related systems. It is advisable to specify the assumptions involved to be sure they are compatible with the purpose of sampling and testable by preliminary sampling.

SAMPLE UNIT   It is important to consider how the sample unit relates to the purpose of sampling, to the biology of the species and to its habitat. Considerations important for choosing a sampling unit have been extensively discussed elsewhere (9, 15, 28, 39) and will be summarized here:
  (i) In URS, all units must have an equal chance of selection.
 (ii) The unit must be stable and continuously available to the insects. If the sample unit is not stable, e.g. a growing crop, then its changes should be easily and continuously measured.
(iii) The proportion of the insect population using the sample unit must remain constant, at least within specified time periods. When the proportion changes between time periods (e.g. at different stages of plant growth, different stages of the season, or times with different weather conditions) then comparisons across time periods are only valid if the change in proportion is predictable and can be cancelled out by a correction factor. For absolute sampling the actual value of the proportion needs to be known; for relative sampling, it does not.
 (iv) It is essential to ensure that any element in the population belongs to one and only one sample unit, i.e. the sample unit must be easily delineated in the field. Extra precautions must be taken to deal with mobile organisms that can move between sample units within one sampling period.
  (v) For mobile organisms, the sample unit should be larger than the space occupied by the moving organism. The unit's size should provide a reasonable balance between sample variance and processing cost.

Optimizing unit size with respect to the density and spatial distribution of organisms leads to more efficient sampling and more powerful and robust statistical conclusions. Green (15) gives some guidelines on this:
(a) The ratio of area or volume of the organism being sampled to the area or volume of the sample unit should be negligible: no greater than 0.05. Likewise for methods that result in avoidance movement by the organism, the area or volume of avoidance movement should be no greater than 0.05 of the sample unit area or volume.
(b) When using presence/absence binomial sampling, efficiency will be maximized by choosing a sample unit size that results in roughly 50% occupancy at the most likely densities.

(c) When preliminary sampling or data from prior studies show that the organism is not distributed randomly, then a new sample unit should be chosen, with area either much larger or much smaller than the scale of the spatial pattern. Green recounts Goodall's (11, 12) advice: when the larger sample unit is chosen, the minimum area is the area of a square with side equal to the distance at which the variance between samples ceases to be a function of their spatial separation. Other factors, e.g. time and labor costs, will set the upper limit to the size of this sample unit. When a sample unit smaller than the scale of the spatial pattern is chosen, these same guidelines give the maximum unit area, and other factors, such as (a) or (b) above, set the minimum.

(d) If the organism has a hierarchical spatial pattern, with patches on one scale distributed within patches on a larger scale, more than one sample unit size may be acceptable and feasible. Sample unit sizes that coincide with any of the patch sizes should be avoided as they maximize among replicate variation.

SAMPLING FRAME   The sampling frame is a list of all possible sample units in the universe. The frame becomes defined when the universe and sample units have been defined. For example in the case of sweep or vacuum sampling, the sampling frame is a list of mutually exclusive sweeping or vacuuming locations, each to a depth in the canopy where insects are likely to be collected. Except when the crop canopy is shallow and simple, the sampling frames will differ between sweep sampling and vacuum sampling. A sampling frame is useful for numbering possible units and selecting a sample using a random number table or other randomization process.

Preliminary Sampling

   This is important to gain information on
 (i) the validity of assumptions about methods, appropriate sample unit, etc.
 (ii) the spatial distribution or dispersion of the organism and the sampling distribution that results with the chosen sample unit.
(iii) The costs of logistical activities in the plan.
Depending on the relationship between sample unit size and the scale of the spatial pattern, if any, the sampling distribution may either mask or reflect the dispersion pattern. For example, the frequency distribution of insects per sample unit may be random when a small sample unit is used, and contagious when a large sample unit is used, reflecting an aggregated spatial distribution at a larger scale. Green's guidelines (c) and (d) above are designed to cancel out the effect of spatial pattern on sampling distribution.

   Green (15) gives several examples showing how time spent on preliminary sampling is a wise investment. He advises nested preliminary sampling, to determine the optimal sample unit size and the spatial distribution pattern on different scales. Unless specifically designed to look for temporal variations in sampling efficiency, preliminary sampling probably will not reveal them.

Dispersion Patterns

   Knowledge of the dispersion pattern of the individuals in a population is necessary for devising optimal sampling plans. In theory, the dispersion pattern may fall anywhere along a continuum from regular (with variance, s^2 < the mean, $\bar{x}$), through random ($s^2 = \bar{x}$, described by the Poisson distribution), to increasingly aggregated ($s^2 > \bar{x}$).   In practice, however, many biological populations have aggregated spatial distributions.

   Various discrete probability distributions (e.g. Thomas, Double Poisson, Neyman Type A, Polya-Aeppli, Negative Binomial, and Discrete Lognormal) have been used to statistically describe such population patterns, both as a guide to appropriate transformation and also as an attempt to infer biological mechanisms from observed distributions. As emphasized by Southwood (39), Taylor (45), Campbell & Noe (7) and others, however, such inferences are rarely justified and usually misleading.

The Negative Binomial (NBD), which is the most commonly fitted of these distributions, is described by 2 parameters: the mean, μ, and the aggregation parameter, k. The variance of the NBD is $\mu + \mu^2/k$ (2). Southwood (39) gives three methods of calculating k, each of which is efficient over limited ranges of density and degree of aggregation. Although many sets of biological population data fit negative binomial distributions, the NBD may be generated by 15 or more different theoretical processes (7), making biological inferences from the distribution extremely arbitrary. Other probability distributions, such as Neyman's A, B, or C, are based on more singular models, but will not adequately describe a population's dispersion unless all underlying biological assumptions are met.

Even with the more valid uses of distributions as guides to choosing transformations and optimizing sampling, there are problems. The same set of sample data may not differ significantly from (and so may be said to "fit") several conflicting probability distributions, with no satisfactory way of determining which is more appropriate. Taylor et al. (43) argue that the value of a fitted frequency distribution model for stabilizing statistics via transformations is diminished if the model changes at different population densities or sample sizes. The degree of aggregation of biological populations often changes with changing mean density. Taylor (45) cited data of McGuire et al., (26) for European corn borer, and Brown & Cameron (6) for gypsy moth, that in each case successively fitted a series of different probability distributions as density increased.

An alternative, more empirical method of describing the spatial pattern of a population is by an index of aggregation. The variance to mean ratio ($s^2/\bar{x}$) is the simplest such index. The inverse of k of the N.B.D. has also been widely considered as such an index, and frequently, negative binomial distributions have been fitted to several data sets for the same species, with the goal of deriving a single "typical" or "common" value of k (39). But Taylor et al. (43) showed that while a few species are known to have a common k at different densities, this is not widely true. Taylor et al. (44) argue that an ecological aggregation parameter should be consistent over a range of taxa, population densities, and scales of space and time. Otherwise, subsequent manipulation of equations using the parameter may obscure the initial premises and extrapolate arguments beyond their range of validity, with misleading results. They show that the relationship of 1/k to μ may take several different forms, and conclude that k is an unstable parameter whose relationship with aggregation is doubtful and arises solely through its relationship with the variance.

Other indices that are less dependent on the data fitting a particular distribution have also been derived. Lloyd (24), describing the influence on an individual of other individuals in the same quadrat, developed the "mean crowding" index, $\overset{*}{x}$, where

$$\overset{*}{x} = \bar{x} + (s^2/\bar{x} - 1) \tag{1}$$

From this he derived a Patchiness Index ($\overset{*}{x}/\bar{x}$) for the population. This index, Green's (13) Coefficient of Dispersion (($s^2/\bar{x} - 1)/(\Sigma\bar{x} - 1)$), and Morisita's (27) I_δ, together with $s^2/\bar{x}$ and 1/k, are the most widely used indices that can be derived from a single set of data. Smith-Gill (37) standardized Morisita's I_δ to render it more independent of mean density, apparently with success (30).

However the two most widely used methods of describing aggregation are regressions that reflect the change in aggregation with density: Taylor's (42)

Power Law, and Iwao's (17) Patchiness Regression. Taylor (42) confirmed (by analysis of multiple data sets) and promoted the generality of the power relationship between mean and variance used first by Bliss (5). Taylor (42, 43, 44, 45) has shown that for over 100 species, the sample variance is related to the mean density by

$$s^2 = a\bar{x}^b \tag{2}$$

where a and b are empirical constants. Taylor's Power Law is simply an empirical description, with no specific underlying biological mechanism. Nevertheless it holds over a continuous series of distributions from regular, through random to highly aggregated. It does seem that b is species (and to some extent habitat) specific, but a is a property of the sample unit being used. Values of a and b may be calculated from log/log plots of s^2 vs. $\bar{x}$; Wilson et al. (53) caution, however, that at low density values both a and b may be biased by over-estimation of s^2.

Iwao (17) found that in a wide variety of situations, Lloyd's mean crowding index, $\overset{*}{x}$, is linearly related to mean density, μ, estimated by $\bar{x}$. Iwao used the notation α for the intercept, and β for the slope, but as we have already defined α and β as the probabilities of Type I and Type II errors, respectively, here we will use different notation for the terms in Iwao's Patchiness Regression (IPR). Parallels with other descriptors of dispersion will be clearer if we replace Iwao's α with (a' - 1) and his β with (b' + 1), i.e.

$$\overset{*}{x} = (a' - 1) + (b' + 1)\bar{x} \tag{3}.$$

Unlike Taylor's Power Law, Iwao's $\overset{*}{x}$ - $\bar{x}$ regression does have an associated biological hypothesis; with a' reflecting the average number of individuals in a colony, and (b' + 1) reflecting the distribution of colonies. In the case of the Poisson distribution, where individuals are distributed randomly, a' = 1, and b' = 0. For an aggregated distribution of colonies, b' > 0, and for randomly distributed colonies, a' > 1, b' = 0.

Taylor's Power Law (TPL), Iwao's Patchiness Regression (IPR), the Negative Binomial distribution (NBD), and the Poisson distribution (PD), are all used to describe variance-mean relationships (43):

$$\text{TPL:} \quad \ln(s^2) = \ln a + b \ln \bar{x} \tag{4}$$

$$\text{IPR:} \quad s^2 = a'\bar{x} + b'\bar{x}^2 \tag{5}$$

$$\text{NBD:} \quad s^2 = \bar{x} + \bar{x}^2/k \tag{6}$$

$$\text{PD:} \quad s^2 = \bar{x} \tag{7}$$

Precision, Cost and Number of Samples

SAMPLING A FIXED NUMBER OF UNITS   The number of units, $\hat{n}$, that it will take to achieve a desired level of precision in estimating a parameter depends on the sampling distribution of the population. In general, however, the Central Limit Theorem (CLT) enables calculation of $\hat{n}$ if the mean, variance, and relationship between them are known. For example, if the parameter is μ, to be estimated by $\bar{x}$, then the CLT states that the probability that μ lies within the confidence interval (CI)

$$\bar{x} \pm z_{\alpha/2} \, \sigma / \sqrt{(n)} \text{ is approx. } (1 - \alpha) \tag{8}$$

Descriptor of sampling distribution	Relative (D)	Absolute (d)
Binomial ≠	$\hat{n} = \left(\dfrac{z_{\alpha/2}}{D}\right)^2 \dfrac{\hat{q}}{\hat{p}}$	$n = \left(\dfrac{z_{\alpha/2}}{d}\right)^2 \hat{p}\,\hat{q}$
Poisson ≠	$\hat{n} = \left(\dfrac{z_{\alpha/2}}{D}\right)^2 \dfrac{1}{\bar{x}}$	$\hat{n} = \left(\dfrac{z_{\alpha/2}}{d}\right)^2 \bar{x}$
Normal ≠	$\hat{n} = \left(\dfrac{z_{\alpha/2}}{D}\right)^2 \dfrac{s^2}{\bar{x}^2}$	$\hat{n} = \left(\dfrac{z_{\alpha/2}}{d}\right)^2 s^2$
Negative Binomial ≠	$\hat{n} = \dfrac{(z_{\alpha/2})^2 \left(\dfrac{1}{\bar{x}} + \dfrac{1}{k}\right)}{D^2}$	$\hat{n} = \left(\dfrac{z_{\alpha/2}}{d}\right)^2 \left(\dfrac{k\bar{x} + \bar{x}^2}{k}\right)$
Taylor's power law (empirical)	$\hat{n} = \left(\dfrac{z_{\alpha/2}}{D}\right)^2 a\,\bar{x}^{\,b-2}$	$\hat{n} = \left(\dfrac{z_{\alpha/2}}{d}\right)^2 a\,\bar{x}^{\,b}$
Iwao's patchiness regression	$\hat{n} = \left(\dfrac{z_{\alpha/2}}{D}\right)^2 \left(\dfrac{a'}{\bar{x}} + b'\right)$	$\hat{n} = \left(\dfrac{z_{\alpha/2}}{d}\right)^2 \left(a'\bar{x} + b'\bar{x}^2\right)$

≠ Karandinos (1976) presented these equations in terms of the parameters μ, σ^2, & p. They are presented here in terms of the estimators $\bar{x}$, s^2 & $\hat{p}$, because usually that is all that is available.

Table 2. Sample size required to achieve desired relative precision D or absolute precision d, as a function of the underlying distribution of a population.

where α = chosen acceptable Type I error rate,
  $z_{\alpha/2}$ = value of standard normal deviate at upper $\alpha/2$ point, which is best replaced by Student's t when n is likely to be <30. If the latter is chosen, an assumed n must be used to calculate df's for t.
  σ = standard deviation of the population (estimated by s, the standard deviation of the sample).

Hence for a desired precision of $\mu = \pm D\bar{x}$, with $(1 - \alpha)$ probability, we get

$$D\bar{x} = z_{\alpha/2}\,(s/\sqrt{(n)}) \tag{9}$$

which, rearranged, gives

$$\hat{n} = (z_{\alpha/2}/D)^2\,(s^2/\bar{x}^2) \tag{10}.$$

Karandinos (20) derived equations to calculate $\hat{n}$ for the general case shown, and the Poisson, Negative Binomial, and Binomial distributions, for desired precision being either a proportion of the mean, D, or a fixed positive number, d (Table 2). While the theory is sound, practice requires
(i) a preliminary estimate of the mean, and
(ii) an estimate of sample variance, or at least, reasonable knowledge of the relationship between the mean and the variance.
Wilson and Room (55), symbolising $(z_{\alpha/2}/D)^2$ by C, and incorporating Taylor's Power Law, converted the sample size equations for the general distribution, the Poisson and Negative Binomial distributions into the common form

$$\hat{n} = C\,a\bar{x}^{(b-2)} \tag{11}.$$

This equation has the advantage of satisfying (ii), and so of allowing n to be estimated for a range of densities for any species whose Taylor coefficients have been or can be adequately estimated. However it is still limited by (i). Thus $\hat{n}$ from Karandinos' equations may sometimes turn out to be either much larger, or much smaller, than actually needed to achieve the desired CI and α.

SEQUENTIAL SAMPLING TO ACHIEVE AN ESTIMATE WITH FIXED PRECISION   When sample units can be processed as they are selected, sequential methods allow the size of each sample to vary according to the results obtained. Sequential sampling methods were developed to enable workers to categorize or classify a population. A quite different method of sequential sampling may also be used to estimate a parameter with a fixed level of precision (21, 22).

   Kuno's (21) method involves successive comparisons of cumulative total counts (T'_n) with critical totals that indicate a target precision has been reached. Kuno allows for two different sorts of fixed precision -- one in which a fixed value of the standard error (which we will call d') is chosen, and the other in which a fixed value is chosen for the ratio of the standard error to the mean (which we will call D'). The 100(1 - α) % confidence interval for the mean (estimated as T'_n/n) is given respectively by

$$CI = T'_n/n \pm z_{\alpha/2}\, d' \quad \text{and} \quad CI = T'_n/n \pm z_{\alpha/2}D'T'_n/n \qquad (12)$$

Thus there is a 68% probability (α = 0.32 when $z_{\alpha/2}$ = 1) that the mean lies in the range $T'_n/n \mp d'$. It seems more useful, therefore, not to consider just d' or D', but to present Kuno's equations in terms that more explicitly relate achievable precision to α, as well as n. Karandinos' d and D do just that. Kuno's d' and D' are very simply related to Karandinos' (Table 2), i.e.:

$$d'(Kuno) = d/z_{\alpha/2}(Karandinos) \qquad (13)$$
and

$$D'(Kuno)=D/z_{\alpha/2}(Karandinos) \qquad (14)$$

Hence D' is also related to Wilson & Room's C (Eqn.9), i.e.

$$D'= \sqrt{(1/C)} \text{ or } D'= C^{-1/2} \qquad (15)$$

The criterion to judge at any given point whether sampling must continue or can stop, is a line that shows, for a pre-chosen level of precision, the theoretical relationship between cumulative count (T'_n) and number of sample units (n). Sampling can stop when the observed cumulative total crosses the stop line. For each sort of fixed precision Kuno gives equations for this "stop line" for both a general case, and for specific types of sampling distributions. He also graphed the stop lines for a variety of sampling distributions, with and without the finite population corrections. The general case uses the mean-variance relationship derived from Iwao's patchiness regression (Eqn. 5). With the exception of the binomial distribution, the specific cases (Table 3) are worked out for chosen values of the coefficients a' and b' from Eqn. 5. Our use of $z_{\alpha/2}$ pre-supposes a large sample size; when sampling stops early, the confidence interval may be recalculated using t_α instead of $z_{\alpha/2}$. Unless exceptional precision is needed, use of $z_{\alpha/2}$ will suffice unless n << 30.

   Kuno (22) showed that estimates of μ obtained by this method have a bias, B, which is positive when D' is used and negative when using d'. He concluded that when using D', the relative bias, γ (where γ = B/[D'$\bar{x}$]), is usually fairly slight compared with the probable range of sampling variation (e.g. γ <10% of 95% C.I. so long as D' $\leq$ 0.3, and γ < 2.5% of C.I. when D' = 0.1). Thus in most circumstances it can be safely overlooked, but if a very precise estimate is required, then one may correct for the bias using Kuno's equations. When using d', the relative bias,

Population Distribution	Coeff's of $\overset{*}{x}-\bar{x}$ regression		Chosen measure of precision	
			$D = z_{\alpha/2}D'$	$d = z_{\alpha/2}d'$
General	a'	b'	$\dfrac{a'}{(D/z)^2 - (b'/n)}$	–
Generalised Poisson	≥ 1	0	$a'/(D/z)^2$	$\left(\dfrac{d}{z}\right)^2 \dfrac{n^2}{a'}$
e.g. Poisson	1	0	$1/(D/z)^2$	$\left(\dfrac{d}{z}\right)^2 n^2$
Generalised aggregated	≥ 1	>0	–	$\left(\dfrac{n}{2b'}\right)\left(\sqrt{a'^2 + 4nb'\left(\dfrac{d}{z}\right)^2} - a'\right)$
e.g. Negative Binomial with common k	1	$1/k$	$\dfrac{1}{(D/z)^2 - (1/nk)}$	–
Binomial	1	-1	$\dfrac{1}{(D/z)^2 + (1/n)}$	$\left(\dfrac{n}{2}\right)\left(1 \overset{+}{\underset{-}{}} \sqrt{1 - 4n\left(\dfrac{d}{z}\right)^2}\right)$

Population distribution	Chosen measure of precision	
	$D = z_{\alpha/2}D'$	$d = z_{\alpha/2}d'$
General	$\left(1 - \dfrac{n}{N}\right)\dfrac{a'}{(D/z)^2 - (1 - n/N)(b'/n)}$	–
Generalised Poisson	$\left(1 - \dfrac{n}{N}\right)[a'/(D/z)^2]$	$\left(\dfrac{N}{N-n}\right)\left(\dfrac{d}{z}\right)^2 \dfrac{n^2}{a'}$
Generalised aggregated	–	$\left(\dfrac{n}{2b'}\right)\left(\sqrt{a'^2 + (4nN/(N-n))b'\left(\dfrac{d}{z}\right)^2} - a'\right)$
Binomial	$\dfrac{(N - n/N)}{(D/z)^2 + 1/n - 1/N}$	$\left(\dfrac{n}{2}\right)\left(1 \overset{+}{\underset{-}{}} \sqrt{1 - [4\left(\dfrac{d}{z}\right)^2 Nn][N - n]}\right)$

* For clarity of reading $z_{\alpha/2}$ is represented in the equations simply by z.

Table 3. Equations for T_n, the "stop line" cumulative number of counts from n units sampled with Kuno's method of sequential sampling to estimate parameters with fixed precision. Stop Lines are functions of the chosen measure of precision (D or d), distributional characteristics of the population, and the sampling fraction (n/N).*

gamma ($\gamma = B/d'$) increases with increasing d' or with decreasing $\bar{x}$. So when $\bar{x}$ is very small Kuno advises setting a minimum number of sample units to be taken, regardless of the stop line, with the sequential method only being used if the stop line hasn't yet been crossed when the minimum sample has been taken. Again, he gives equations for calculating an unbiased estimator, if necessary.

When a proportion p is being estimated (by $\hat{p}$), however, the relative bias γ equals $D'/(1 - p)$, which tends to D' as p tends to 0.0, but becomes very large as p tends to 1.0. Thus this method is inappropriate for estimating proportions when p is near 1.0, and Kuno suggests using inverse sampling instead, i.e. using a horizontal line, $T_n = 1/(D')^2$ as the alternative boundary. In this case an unbiased estimator of p, $\hat{p}'$, is obtained from

$$\hat{p}' = \hat{p} - [(1 - \hat{p})/n_0] \tag{16}$$

where n_0 is the actual sample size.

Green (14) derived related sequential sampling lines by using Kuno's definition of D', but using Taylor's Power Law as the variance-to-mean relationship, rather than Iwao's $\overset{*}{x} - \bar{x}$ regression. Green's equations are also more usefully expressed in terms of Karandinos' D and $z_{\alpha/2}$, i.e.:

$$(D/z_{\alpha/2})^2 = an^{(1-b)}T_n^{(b-2)} \tag{17}$$

and

$$\ln(T_n) = \frac{\ln[(D/z_{\alpha/2})^2(1/a)]}{b-2} + \frac{(b-1)\ln(n)}{(b-2)} \tag{18}$$

unless b = 2, when $\ln(n) = \ln(a) - 2\ln(D/z_{\alpha/2})$. \hspace{1cm} (19)

The stop line, when $\ln(T_n)$ is graphed against $\ln(n)$, is a straight line: horizontal for b = 1, changing from a negative slope, for 1 < b < 2, through vertical for b = 2, to positive slope for b > 2, as the degree of aggregation of the population increases. For b = 1, (generalized Poisson distributions; with a = 1 as well, for the Poisson itself), sampling would continue until a fixed number of individuals:

$$\ln(T_n) = \ln[(z_{\alpha/2}/D)^2 a] \tag{20}$$

had been found, regardless of the number of samples it took to find them. For b = 2, (logarithmic series distribution) the number of units sampled is independent of the mean density and given by equation (19). In between, the negative slope would give an upper limit to the number of units sampled, even if no individuals had been found.

Green used his method with 24 sets of field data compiled by Taylor (42) to test the rule of thumb that in many cases it will be sufficient to examine

$$\hat{n} = (z_{\alpha/2}/D)^2, = (D')^{-2} \text{ units} \tag{21}.$$

He concluded that the rule is conservative, so such sample sizes will usually give a CI less than or equal to that for the chosen D and $z_{\alpha/2}$. However, a lower CI than necessary means unnecessary sampling. When labor costs for sampling are high, the trade-off of simplicity of calculation with more sampling effort will not necessarily be advantageous.

DOUBLE SAMPLING   Because there are practical limitations on the feasibility of following the strict guidelines for taking sequential samples, and also to allow for situations when it is not feasible to process samples and estimate parameters during sampling, Kuno (22) suggested that double sampling will sometimes be more practical. He therefore compared the biases in estimation by double sampling with those discussed above, and concluded that for estimation of μ, whether using D' or d', the biases were almost identical when using double sampling to when using his sequential method. Double sampling also gave equally biased estimates of p as did the sequential method, as p approached 1.0, in the case of binomial sampling.

Thus double sampling is an equally rigorous alternative to Kuno's sequential sampling, when it is not feasible to apply the latter, so long as the size of the first sample (n_1) is fairly large, but less than n_0, the expected sample size for the assigned precision. For populations for which b'> 0, Kuno suggests that n_1 calculated as follows

$$\hat{n}_1 = b'/(d')^2 \qquad\qquad (22)$$

will usually obey these guidelines. When b' $\leq$ 0, there exists no such convenient way of choosing an appropriate size for the first sample.

Schedule of Sampling

The final component of any sampling plan is scheduling when and how often to sample, including when to begin and end. These decisions need to be based on:
 (i) the purpose of sampling: the question(s) to be answered, and the definition of the sampling universe,
(ii) the interplay between the biology of the organism and the sampling methods and sample unit being used.

When defining the target population, sample population and universe, it is important to bear in mind that all sampling universes have temporal as well as spatial dimensions, i.e. a population exists in a specific area for a specific period of time. If the universe exists only as long as it takes to sample it, then it is essential to collect units in random order so that any time pattern present during sampling is distributed randomly among units, and not confounded with other effects. In this situation, the population mean µ, or proportion p with an attribute, being estimated applies only to the period of effort.

When the universe exists longer than it takes to sample, temporal variation may occur in the structure of habitat units comprising the sampling universe, in the behavior of the organisms being sampled, or in efficiency of the sampling method. It is important to consider which if any of these could introduce bias. Unless prior studies have shown that they do not vary in time, either preliminary sampling should be undertaken to assess temporal variation, or else sampling should be stratified through time.

Although restrictive redefinition of the universe may be necessary, in some cases it has been possible to calculate correction factors by which to adjust catch rates, so that sampling need not be restricted to quite such narrow limits. For example, Morton et al. (29) found a marked negative relationship between wind speed and hourly catch rate within the same night for _Heliothis_ spp. moths in blacklight traps. They developed equations enabling the catch rate to be appropriately adjusted upward when windspeed exceeds 0.2 m/s. Likewise, Cherry et al. (8) found that the efficiency of sweepnet sampling for potato leafhopper adults in alfalfa was inversely related to windspeed and directly, but less strongly, related to temperature. They calculated a sweepnet conversion factor to correct the catch for these effects. However, care must be taken not to extrapolate the use of such correction factors beyond the range of conditions used to derive them.

When the population parameter of interest fluctuates or oscillates measurably through time, then infrequent sampling will give a very misleading impression of population changes, peaks and troughs. When the trends are of interest, the population should be the sampled at least twice during each expected or average interval between the population peak and trough.

In some circumstances, other factors related to the purpose of sampling may dictate an upper limit to the sampling interval. For example, the SIRATAC pest management system for Australian cotton (Chapter 25), advises that arthropods be sampled three times a week. Sampling at intervals longer than three days can result in _Heliothis_ eggs being laid, hatching, and larvae feeding before the eggs have been detected. Most newly hatched _Heliothis_ larvae are sheltered in a square or terminal by the time they are detected, and hence invulnerable to contact insecticides. In the event of undetected heavy oviposition, considerable damage could result before the larvae were again vulnerable to insecticides.

The problem of when to begin (and sometimes also when to end) a sampling program is most acute when there is a critical window of time within which the insect must be sampled -- whether it is a short-lived, univoltine species, or, for example the first generation of the season is of particular interest. Obviously it is a waste of effort to sample when the insects are not present, so that for insects whose timing varies much, using the earliest recorded calendar date for a start date may involve much unnecessary cost. However, use of an average date may result in missing much of the critical sampling period in some years.

Several alternatives to calendar dates are possible, though not all may be available in any particular case. One may set up an artificial cohort in the field, or have the location of a known natural cohort marked, so it can be examined either directly or with emergence traps. Alternatively, one may seek to determine whether there is any correspondence between the population phenomenon of interest and other, more visible, biological indicators. Finally, when there is sufficient data on the temperature relationship and sometimes other relationships (e.g. diapause termination or induction) of the insect of interest, phenological models may be developed that enable prediction of the timing of the critical stage. The most commonly used phenological model is the degree-day model for insect development (33, 41, 52).

STRATIFIED AND NESTED SAMPLING DESIGNS

Two additional forms of probability sampling are stratified and nested (2-or multi-stage) designs. As alternatives to URS, they have features that are attractive in field situations. Stratified sampling often greatly improves precision, allows one to control location and cost of sampling effort, and lets one make use of knowledge about location of organisms in the universe. Nested designs make it possible to obtain probability samples from universes so large that defining the frame is not feasible. Discussion of other, more complex designs can be found in Cochran (9).

Stratified sampling

The procedure for stratified random sampling (SRS) is conceptually simple. The universe is divided into two or more strata of known size, each stratum is sampled separately using URS, and then data from all strata are assembled to obtain estimators for the universe.

One goal of stratification is to create sub-universes that are uniform internally, i.e., to minimize variation within strata and maximize variation among strata. Stratum boundaries are best based on the variable being sampled. For example, edge effects are gradients in insect density from high at field margins to low within. An appropriate approach would be to define one stratum containing marginal plants, and the other as containing central ones. Densities within each stratum would be less variable than across the entire plot. Where the sampled variable is not immediately visible, one could stratify using a covariate such as plant height, density, color, injury, etc., that is thought to predict the sampled variable.

Stratum boundaries can be defined before sampling begins, or they can be defined after the universe is inspected. Stratum definitions can change from one sampling occasion to the next. The strata need not be of equal size and sample units in a given stratum need not be contiguous. However defined, every unit must occur in one stratum or another, and the sampler must know how many units exist in each stratum. Alternatively, relative stratum sizes must be known. Cochran (9) advises that 2-6 strata are sufficient; further subdivision yields diminishing returns.

Notation and formulae below for estimators of density are derived from Cochran (9, Chapters 5 & 6). N possible units in the universe are divided into N_1, N_2, ..., N_h, ... N_H for the H strata, respectively, and n_h are actually examined. Within the hth stratum, one computes mean density ($\bar{x}_h$) and variance (S_h^2) as usual. The combined mean is a weighted sum of stratum means,

$$\bar{x}_{st} = \sum_{h}^{H} W_h \bar{x}_h \tag{23}$$

where weights, W_h are N_h/N. (To simplify the equations below, indices above and below the summation sign, Σ, will be omitted, but should remain inferred from the subscripts on symbols for the variables being summed.) The estimated mean has

$$SE_{st} = \sqrt{\Sigma \frac{W_h^2 S_h^2 (1-f_h)}{n_h}} \tag{24}$$

where $f_h = n_h/N_h$, and $(1 - f_h)$ is the finite population correction factor. The latter is usually omitted if all fh are < 0.05.

The effective degrees of freedom (df_e) can never be less than the smallest number of units examined in a single stratum, i.e. the smallest n_h, minus 1, nor greater than the total number of units examined, minus 1, and are given by

$$df_e = \frac{(\Sigma \ g_h s_h^2)^2}{\Sigma \left(\frac{g_h^2 s_h^4}{n_h-1} \right)} \tag{25}$$

where $g_h = N_h(N_h-n_h)/n_h$. A valid CI for the mean could be constructed using student's t with df_e if sampling from a normal distribution. Otherwise, transformation of the data would be in order. Equations for the estimators of p and SE are obtained respectively, by substituting $\hat{p}$ for $\bar{x}$ in Eqn. 23, and $(\hat{p}\hat{q})$ for S^2 in Eqns. 24 and 25.

An important remaining issue concerns allocation of effort among strata. If no prior data are available, then proportional allocation is a good choice, i.e., allot the total n to strata in proportion to each stratum's relative size or weight. If preliminary data are available, then an optimum allocation can be defined that will have the smallest SE per unit sampling cost. Optimum allocation calls for relatively more units in a stratum if it is more variable, if it is larger (N_h larger), or if its per unit cost, C_h, is lower. The optimal fraction of n in the hth stratum is

$$\hat{n}_h = \frac{\hat{n} \ W_h S_h / \sqrt{C_h}}{\Sigma \ (W_h S_h / \sqrt{C_h})} \tag{26}$$

The distribution of effort among strata is the same, whether limited by budget or sampling to achieve desired precision and confidence. If budget is fixed, then

$$C_{tot} = C_o + \Sigma \, (C_h \hat{n}_h) \qquad (27)$$

where C^o is usually an overhead cost such as travel, lab space, etc., that is independent of n. Affordable sample size is then

$$\hat{n} = (C_{tot} - C_o) \; \frac{\Sigma \, N_h S_h / \sqrt{C_h}}{\Sigma \; N_h S_h \sqrt{C_h}} \qquad (28)$$

The $\hat{n}$ units are distributed among strata according to the fractions from Eqn. 26. Notice that Eqns. 26-28 simplify if the C_h are equal among strata. Again, if sampling to estimate a proportion, substitute $\sqrt{(\hat{p}\hat{q})}$ for S_h in Eqns. 26-28.

If sampling to achieve desired precision, then one must first derive a desired SE_{st} from the appropriate equation for $\hat{n}$ in Table 2. $\hat{n}$ is then

$$\hat{n} = \frac{t_{\alpha/2}}{d}^2 \cdot \Sigma \, [(W_h S_h)^2 / w_h] \qquad (29)$$

where

$$w_h = \frac{W_h S_h / \sqrt{C_h}}{\Sigma W_h S_h / \sqrt{C_h}} \qquad (30)$$

If $\hat{n}/N > 0.05$, then adjust $\hat{n}$ downward to

$$\hat{n}' = \frac{\hat{n}}{1 + \left(\frac{t_{\alpha/2}}{d}\right)^2 \cdot \frac{1}{N} \cdot \Sigma \, W_h S_h^2} \qquad (31)$$

These $\hat{n}$ or $\hat{n}'$ units would then be distributed among strata according to the fractions from Eqn. 26.

In situations where certain strata are quite variable, or quite cheap to sample, the optimal allocation may be to examine 1 or more strata entirely. This event will be indicated where a computed $\hat{n}$ for a given stratum is greater than the corresponding N in that stratum. When this occurs, the remaining sample units, ($\hat{n} - N_h$) must be reallocated among the remaining strata.

Nested Sampling

Both URS and SRS require that the sampler be able to list all sample units in the universe. Many universes are so large, geographically and in number of sample units, that it is not practical to locate a truly random sample using a mechanical method. Fortunately, most sampling universes can be viewed as being organized in a natural or artificial hierarchy. Examples viewed from bottom to top in a hierarchy include twigs on branches on trees in a forest, and soil in cm^3 from cores in plots in large fields.

Nested sampling (NS), also known as <u>subsampling</u> and <u>multistage</u> sampling, exploits such a hierarchy. Using URS, a sample of first level (primary) units is drawn from a frame of possible primary units at the top level within the universe. Chosen primary units are then subsampled to locate a set of secondary units. Secondary units can be further subdivided into tertiary units, and so forth. This process of subsampling down the hierarchy is repeated until the final sample units have been selected. At each level, a list of possible sub-samples is manageably short, and one can control the subsample size or probability that a unit in the next lower division is selected. The latter permits an optimal mix of effort at each level in the hierarchy to balance sampling expense against achievable precision.

Estimators and formulae for effort allocation in a 3-stage design are presented below. Equations for the simpler 2-stage case can be obtained by eliminating reference to the 3rd, lowest level. Our discussion applies only to cases where units within each level are of equal size and are all selected with URS. Cochran's (9) notation and presentation is adopted here.

There are N, M, and K possible units at the primary, secondary and tertiary levels, respectively, of which n, m and k of them have been selected. Sampling fractions are $f_1 = n/N$, $f_2 = m/M$ and $f_3 = k/K$. The sample mean is simply

$$\bar{\bar{\bar{x}}} = \frac{\sum\limits_{}^{n} \sum\limits_{}^{m} \sum\limits_{}^{k} x_{iju}}{nmk} \qquad (32)$$

where x is the variable of interest. When sampling for a proportion, each sample unit's x would be coded as 1 if the attribute is present or 0 if otherwise.

Variance can be partitioned into components from each level in the hierarchy, i.e., variation among means of primary units, variation among means for secondary units within chosen primaries, and variation among means for tertiary units within chosen secondaries. These variances can be computed as a nested ANOVA (38) or as follows:

$$S_1^2 = \frac{\sum(\bar{\bar{x}}_i - \bar{\bar{\bar{x}}})^2}{n - 1} \qquad (33)$$

$$S_2^2 = \frac{\sum\sum(\bar{x}_{ij} - \bar{\bar{x}}_i)}{n(m - 1)} \qquad (34)$$

$$S_3^2 = \frac{\sum\sum\sum(x_{ij} - \bar{x}_{ij})^2}{nm(k - 1)} \qquad (35)$$

Standard error for $\bar{\bar{\bar{x}}}$ is

$$SE_{\bar{\bar{\bar{x}}}} = \left(\frac{(1 - f_i)}{n} S_1^2 + \frac{f_1(1 - f_2)}{nm} S_2^2 + \frac{f_1 f_2(1 - f_3)}{nmk} S_3^2 \right)^{1/2} \qquad (36)$$

if n, m, and u are all large, or x is normally distributed at all levels in the hierarchy. This $SE_{\bar{\bar{\bar{x}}}}$ has n.m(k-1) degrees of freedom. Equation (36) simplifies when the sampling fractions are all negligible.

It is worth noting here that improved precision (smaller $SE_{\bar{\bar{\bar{x}}}}$) can be obtained by bulking and mixing units from a level that is a particularly high source of variation. For example, a common 3-stage design involves soil cores nested within plots that are in turn nested within larger fields. High variances among cores within the same plots are often attributed to aggregation by the insects, pathogens, seeds, or other elements within the scale of the cores. The effect of mixing cores from within plots before further sub-sampling is to diminish the variance attributable to the level at which mixing is done. The result can be a highly improved precision, reduced cost, or both.

If preliminary estimates of s_1^2, s_2^2 and s_3^2 are available, then one can estimate the combination $\hat{n}$, $\hat{m}$ and $\hat{k}$ that will be most efficient. To do so, one needs to know the costs of sampling each primary unit, C_1, each secondary unit, C_2, and tertiary unit, C_3. Whether sampling with a fixed budget or to achieve desired precision, the optimal allocation is the same.

Provided that the limiting conditions $s_1^2 > s_2^2/M$ and $s_2^2 > s_3^2/K$ are both true, then for every primary unit, the optimal number of secondary units is

$$\hat{m} = \sqrt{C_1/C_2} \cdot \frac{\sqrt{s_2^2 - s_3^2/K}}{\sqrt{s_1^2 - s_2^2/M}} \qquad (37)$$

and the optimal number of tertiary units should be

$$\hat{k} = \sqrt{c_2/c_3} \cdot \frac{s_3}{\sqrt{s_2^2 - s_3^2/K}} \qquad (38)$$

each rounded up to the nearest integer. If the limiting conditions are unfulfilled or $\hat{m} > M$ or $\hat{k} > K$, then the sampler is being told to collapse the hierarchy; i.e., to examine all K if $\hat{k} > K$ or to examine all M if $\hat{m} > M$.

The remaining decision concerns the number of primary units to examine, which in turn governs the total effort. If the budget is fixed, then total sampling cost might be approximated by

$$C_{tot} = C_o + nC_1 + nmC_2 + nmkC_3 \qquad (39)$$

where C_o is an overhead cost independent of n, m, and k. The number of primary units is

$$\hat{n} = (C_{tot} - C_o)/(C_1 + mC_2 + \hat{m}\hat{k}C_3) \qquad (40)$$

Alternatively, if the budget is not limiting, then one can proceed by first deriving a desired $SE_{\bar{\bar{x}}}$, and then substituting $\hat{m}$ and $\hat{k}$ into Eqn. (36) and solving for the required n.

DECIDING

Sequential Sampling

When the goal of sampling is to categorize the population on the basis of some particular attribute, the value of the attribute needs to be estimated precisely only when it is close to the reference value that divides the categories. Sequential sampling methods take advantage of this, allowing a low number of sample units to be taken when the mean is far from the reference value.

Although first developed for quality control of munitions (46, 47), sequential sampling has gained favor with applied entomologists (1, 3, 32, 40, 48, 53, 54, 56) because of its potential to reduce sampling labor (costs) without jeopardizing the reliability of pest management decisions. In pest management, whenever possible we need to minimize the risk of deciding wrongly that pest abundance warrants control measures, as well as the more obvious risk of a wrong decision that pests are below the treatment threshold when they are actually causing economic damage. Sequential sampling allows the pest manager to specify acceptable Type I and Type II error rates. In practice, some modification of strict sequential sampling methods is usually unavoidable because of logistical limits. Such compromises will be discussed later; the theoretically correct methods will be presented here.

The Sequential Probability Ratio Test (SPRT)

The original, standard sequential sampling methods (47) are founded on the SPRT, a procedure for sequentially testing the null hypothesis (H_0) that the population being sampled occurs in a particular category with respect to an attribute of interest. Sample units chosen at random are examined in sequence, providing progressively more information each time the test is applied, until there is sufficient information to reach a decision with pre-chosen chances of errors. When the method is used by entomologists, usually a particular species or lifestage of an insect is of interest. The attribute of interest may be the actual numbers in or on the sample unit, or it may simply be the presence or absence of the insect.

Although sequential sampling may be used to decide in which of several alternative categories a population belongs, the simplest case, that of two categories, will be used to describe the sequential decision process. The H_0 is tested against an alternate hypothesis (H_a) that the population falls into the other category. As each sample unit is examined, the cumulative number (either of insects, or of infested sample units) is compared, either graphically or in a table, with two so-called decision values -- the acceptance value and the rejection value. Graphically, these values form parallel straight lines dividing an indecision zone from an acceptance region on one side and a rejection region on the other. When the cumulative number of insects (or infested sample units) crosses from the indecision zone to either the acceptance or rejection region, sampling can stop, as the hypothesis can then be accepted or rejected as the case may be, with the pre-chosen level of reliability.

In tabular form, the sequential decision values form two columns of numbers with which the cumulative total is successively compared. Again, once the cumulative total either falls below the lower number, or exceeds the upper number, sampling may stop, as a decision about categorizing the population can be reached.

Choice or design of an appropriate sequential sampling plan based on the SPRT requires a knowledge of the sampling distribution of the organism. Error levels α and β, and not one but two threshold values of the parameter of interest (μ_1 and μ_0, for the Poisson and Negative Binomially distributed cases, and p_1 and p_0, for the positive binomial case, in Table 4), must be chosen in order to calculate the decision lines. In pest management, usually the upper threshold is the action or economic threshold (whichever is available), and the lower threshold is some arbitrary proportion of it. However, μ_0 and μ_1 may also be chosen to represent acceptable confidence limits around the action threshold.

The general equation for the decision lines is:

$$CT_n = (Num\ I\ /\ CD) + (Num\ S\ /\ CD) \cdot n \qquad (41)$$

where n = no. of units examined

CT_n = cumulative total of insects or infested sample units

Num I = $\ln [\beta/(1 - \alpha)]$ for the intercept of the acceptance line

Num I = $\ln [(1 - \beta)/\alpha]$ " "   "   " " rejection "

Num S = the numerator of the slope equation,

and CD = a common denominator for all terms. Equations for both Num S and CD depend on the sampling distribution of the population (Table 4).

Distribution	Num S	CD	Conditions
Poisson	$\mu_1 - \mu_0$	$\ln(\mu_1 - \mu_0)$	$\mu_0 < \mu_1$
Binomial	$\ln\left(\dfrac{1 - P_0}{1 - P_1}\right)$	$\ln\left[\left(\dfrac{P_1}{P_0}\right)\left(\dfrac{1 - P_0}{1 - P_1}\right)\right]$	$0 < P_0 < P_1 < 1.0$
Negative binomial	$k.\ln\left[\dfrac{(1+\mu_1/k)}{(1+\mu_0/k)}\right]$	$\ln\left[\dfrac{(\mu_1/k)(1+\mu_0/k)}{(\mu_0/k)(1+\mu_1/k)}\right]$	$\mu_0 < \mu_1$

Table 4. Partial coefficients Num S and CD (see Eqn. 41 in text) of decision lines for sequential lines for sequential sampling based on the SPRT, as functions of the sampling distribution of the population, 47, 48).

The Operating Characteristic (OC) curve (Table 1) of a sequential sampling plan enables the sampler to examine for each likely parameter value (e.g. level of density or proportion infested), the probability that the population will be assigned to a particular category when the decision is reached (32, 47), and hence the probability of a correct decision. This information, plus information from the Average Sample Number (ASN) curve (32, 47), will enable the trade - off between reliability and the likely costs of the sampling to be optimized by adjusting α, β, and the category limits (upper and lower thresholds). The average number of sample units required at different densities is directly related to the distance between the upper and lower thresholds, and the chosen α and β. The closer together the thresholds are, or the lower α and β are, the higher the average amount of sampling required to reach a decision. Waters (48) warns that the confidence levels expressed by the values chosen for α and β, which need not necessarily be equal, "must be set with a practical eye to the feasibility of obtaining the specified reliability and the use to which the information will be put".

Other Sequential Sampling Methods to Categorize Populations

For many species the probability distribution that best describes the sampling distribution of the organism changes as a function of density, or at best, even if the form of the distribution remains the same (e.g. negative binomial) the value of parameters such as k is not constant with changing density. Yet the assumption of a common k is necessary if the NBD is to be used as the underlying model for generating sequential sampling plans based on the SPRT. Entomologists have approached this problem in various ways. Some have chosen to consider departures from a common k at extremes of the density range irrelevant, and have developed plans for populations with densities in the vicinity of μ_0 and μ_1 only, using a common $\hat{k}$. Other workers have chosen to use binomial sampling when feasible, while others again have developed independent sequential sampling plans not based on the SPRT (3, 18, 53).

Iwao's (18) method is based on the regression of mean crowding on mean density discussed above (Eqn. 3) to describe the mean-variance relationship of the population. He then calculated the confidence interval for a total based on the Central Limit Theorem as follows

$$\text{C.I. of } T_n = n\bar{x} \ I \ t \ n.SE_{\bar{x}}. \tag{42}$$

where t is Student's t statistic. So, incorporating the mean-variance relationship (Eqn. (5)) gives:

$$SE_{\bar{x}} = \sqrt{[(1/n)(a'\bar{x} + b'\bar{x}^2)]} \tag{43}$$

Thus C.I. of $T_n = n\bar{x} \pm t\sqrt{[n(a'\bar{x} + b'\bar{x}^2)]}$ (44).

Iwao's upper decision line is thus defined by the upper critical value of this confidence interval, for each sample size and mean density; likewise the lower decision line is given by the lower limit of this confidence interval. Recalling Kuno's sequential sampling for deriving an estimate with a fixed precision, and replacing t by $z_{\alpha/2}$, it can be seen that Iwao's upper and lower critical values can be expressed in terms of Kuno's D' or d' and $z_{\alpha/2}$, and hence in terms of Karandinos' D or d, i.e.:

$$T_n \text{ upper} = T_n(1 + D) \quad \text{and} \quad T_n \text{ lower} = T_n(1 - D) \tag{45}$$

$$T_n \text{ upper} = T_n + nd \quad \text{and} \quad T_n \text{ lower} = T_n - nd \tag{46}$$

These equations show that as n increases, so also will the confidence interval on T_n.

Iwao's method differs fundamentally from the SPRT in several respects. The SPRT uses two thresholds μ_0 and μ_1 ($\mu_0 < \mu_1$), and tests between the H_0 that $\mu \leq \mu_0$, and an alternative H_a that $\mu \geq \mu_1$, allowing the user to set both Type I and Type II errors. Iwao's method uses a single threshold or critical density (μ_c) only, and tests the H_0 that μ (estimated above by $\bar{x}$) = μ_c, against the H_a that $\mu \neq \mu_c$, considering the Type I error only. The method also differs from strict SPRT sequential sampling, in how an upper limit n_{max} is set to the number of sample units taken, when a decision cannot be reached within a reasonable number of units. The limit is calculated from the equation for the number of sample units necessary to estimate μ with a prechosen fixed precision, in Table 2, incorporating the IPR mean-variance relationship and assuming μ exactly equals μ_c:

$$n_{max} = (t^2/D^2)(a'\mu_c + b'\mu_c^2) \tag{47}$$

The notation in Iwao's paper (avoided here) is unfortunate in that, following the precedent of his original papers on the mean crowding-mean density regression, he uses α and β for the intercept and slope terms of the regression line. It must be emphasized that these are totally distinct from α and β representing Type I and Type II error, used in generating sequential sampling plans based on the SPRT, and this must be borne in mind when Iwao describes the interactions between the values of α, β, μ_c, the upper and lower limits (decision values), and the value of (Student's) t for a chosen Type I error, which he refers to as confidence probability.

Nyrop and Simmons (31) critically examined Iwao's methods and conclusions, using both statistical theory and numerical simulation of a realistic sampling situation. They concluded that the actual error rates that arise from using Iwao's sequential rules are much larger than the nominal ones presented by Iwao, although not necessarily to an unacceptable extent. They caution that before using the method, the actual error rates should be estimated via simulation, to determine whether they are acceptable. Nyrop and Simmons also refer to work (34, 10) in which actual error rates have been calculated for a number of sequential t tests.

Wilson et al. (53, 54, 56) and Bechinski and Stoltz (3) also have developed sequential sampling plans not based on the SPRT. It appears that some of Nyrop and Simmons' criticisms of Iwao's methods may also apply to Bechinski & Stoltz's and Wilson et al.'s methods, but it is not clear whether those that do are the more

69

serious ones.

In each case these authors were using binomial or presence-absence sampling, which also can save time and therefore sampling costs (50), and has the added advantage when it is feasible, that it helps circumvent the problem of the underlying population not having a constant distribution, e.g. that k is a function of the mean. Their plans are attractive if action thresholds correspond with densities at which the proportion of infested units is below ca. 0.5; as p tends to 1.0, the CI around $\bar{x}$ becomes very wide.

ANALYSIS

The final steps in any research project are to analyze sample data and draw conclusions regarding original questions. Analysis is best planned in advance of any actual sampling. Toward that end, it could be a useful exercise to imagine the planned study, step by step, and write down fictitious data in the form they will be collected. They should then be analyzed as intended. These activities will provide the researcher with a final opportunity to think through details of the sampling protocol, to devise data sheets, and to become familiar with the intended computational methods and computer software, if available. Once real data are in hand, the remaining work should include an examination of how well the data conform with assumptions underlying the chosen statistical procedure.

There are three assumptions common to the parametric two-sample comparisons, ANOVA, and linear regression. They are (1) that errors ($x_i - \bar{x}$) among observations (sample units and/or experimental units) are independent of each other, (2) that the errors are normally distributed about all means, and (3) that error variances are constant among means. Where treatments have been applied to groups of experimental units, then it must also be assumed (4) that treatment effects are additive. Less powerful non-parametric tests only require that assumption (1) be met. Sokal and Rohlf (38) provide a particularly readable treatment of tests for each of these assumptions. Green (15) concludes that t- and F-tests are reasonably robust, even when the assumptions are violated. False conclusions are usually in the direction of higher than nominal Type I errors, i.e., falsely concluding significance when in fact the results could be due to chance alone.

Errors will be independent if probability sampling is used, and if treatments are randomly assigned to experimental units. Independence can be checked with a nonparametric test of runs above and below the median if sample size is greater than ca. 50, or by inspection of residuals about treatment means. Lack of independence will bias estimates of treatment effects and cause actual Type I errors to depart in either direction from tabular ones. Since lack of independence cannot be overcome after data are collected, it is imperative that one use randomization when assigning treatments to experimental units and when choosing sample units from a frame. Systematic sampling or the deliberate selection of "representative" or "typical" sample units can introduce biases that no statistical tests can detect or adjust.

Violations of the three remaining assumptions lead to increased Type I error rates and biased estimates of means and treatment effects. Fortunately, all can be remedied by transformation to another scale before analysis.

Tests for normality in errors can be graphical. A more rigorous approach is to examine the frequency distributions for skewness and kurtosis. Finally, one can compare observed and expected frequencies with a Chi-square test, providing data are sufficient. Most commonly, distributions of biological data are skewed to the right. A log(x), log(x + 1), or power transformation with a power less than unity will shrink the right tail sufficiently. Taylor (45) asserts that a suitable power is 1 - b/2, where b is obtained from regression of log sample variances on log sample means. Poisson variates (b = 1) would require a square root transformation, or if counts are extremely low, $\sqrt{(x + 1/2)}$. Taylor's b = 2 indicates a logarithmic transformation is needed.

Bartlett's Chi-square test will reveal unequal variances. Log or power transformations of x, x + 1/2, or x + 1 will reduce heterogeneity where variances increase with means. Otherwise, treatment groups having either unusually small or large variances can be omitted from further analysis. Tukey's test will assess the additivity of treatment effects. Log or power transformations will often render the effects additive. Both Bartlett's and Tukey's procedures are outlined in Sokal and Rohlf (38).

PRACTICE

Costs of Adhering to Sampling Theory and Practical Compromises

When sampling a fixed number of units in a field situation, once the random locations for the units have been determined, they may be visited in any convenient order (23), without invalidating the assumption that errors are independent. For sequential sampling methods, this is unfortunately not true. A fundamental assumption of sequential sampling is that each of the sequentially examined sample units is chosen at random from the frame. This is rarely feasible in a field sampling situation because too much time is necessary to move randomly between locations. It is more expedient and hence more usual to select units in a regular pattern, such as some form of transect.

Three approaches have been used to increase the randomness of such sequential samples with relatively small increases in logistic costs. Most commonly, samplers set an arbitrary minimum number of sample units to avoid reaching premature decisions. Even with this modification, sampling may still be very systematic. A second approach is to stratify and sequentially sample within each stratum. Thirdly, Kuno (21) suggested taking a random series of small systematic sub-samples (e.g. 10 units), and comparing the cumulative total with the stop lines only at the end of each sub-sample. This procedure will result in some unnecessary units being sampled; a careful assessment of the trade-offs is necessary to choose the size of the systematic sub-samples.

Arbitrary upper limits on the number of units sampled, even though still in the indecision zone, are usually set, to limit absolute costs. Wald (47) and Nyrop & Simmons (31) discuss how this affects error rates in the case of the SPRT and Iwao's sequential sampling method, respectively, but their discussions are likely to be relevant when upper limits are set with other sequential sampling methods also.

Sampling to Achieve Multiple Objectives

Crop pest managers often require abundance information on several different pest populations or their damage symptoms. Some measure of crop development is also often needed, such as average plant height, percent of plants that have entered the reproductive stage, or numbers of fruits or other reproductive structures. Attempts to simultaneously sample organisms from such different groups must involve compromises.

The advantages and disadvantages of simultaneous sampling need to be carefully considered. In some cases, it may be more efficient to sample one group separately, and less frequently than the others; for example when the attribute of interest changes less rapidly for that group than the others. For instance, in the SIRATAC cotton pest management program (Chapter 25), fruit -- squares and bolls -- are sampled weekly, whereas insects are sampled three times a week. Main and Proctor (25) give a useful example of field tests of various sampling plans, and a flexible sampling design for surveying for presence of several different diseases at once.

The cost of sampling with a given reliability has three components: the cost of walking from one sample unit to the next, the cost to examine each unit, and the number of units that will give the desired reliability (56). When many organisms must be sampled simultaneously using separate sequential sampling plans for each, walking time cannot be reduced below that for the organism that requires most units to cross a decision line. If walking contributes most of the sampling cost, having even one organism that requires sampling to continue when it has stopped for all others may negate the overall advantage of sequential sampling. Processing time may differ among organisms, and for some of them it may contribute significantly to total cost. In this case, we are interested in the likelihood that decision lines will be crossed early for costly organisms. Unless outweighed by walking costs, this likelihood will determine the relative advantage of sequential sampling over fixed sampling. For organisms that are rare but have a low action threshold, a large number of empty units will be required to cross the lower stop line.

Sometimes aspects of the biology of the organisms will necessitate a modification of the sampling plan. For example, many crop plants grow bigger, and produce more fruit per plant when at low than high stand density, thus partially compensating on a per unit area basis, for the reduced stand. However, the degree of this compensation varies, especially when variations in stand density reflect variations in field conditions, e.g. water-holding capacity of the soil. Consequently results could be misleading if the stand density where fruit are sampled is not representative of the field in general. In theory this could be corrected by sampling a very large number of plants; in practice this is not feasible. A compromise is to reject any sample units in which stand density differs by more than a prechosen amount, e.g. 20%, from the mean stand density calculated from a less frequent large sample (16).

CONCLUDING REMARKS

Crop loss assessment and pest management are quantitative activities that require effective sampling plans to answer questions about field populations. Effective plans will be relevant to the question(s) being asked, they will be valid from a statistical perspective, and they will probably require much labor and expense. When the latter resources are limiting, the simple designs described in this chapter can be used to get the best information from the resources expended.

* Paper No. 15,104, Minnesota Agricultural Experiment Station.

LITERATURE CITED

1. Allen, J., Gonzalez, D., Gokhale, D.V. 1972. Sequential sampling plans for the bollworm, Heliothis zea. Environ. Entomol. 1:771-780.
2. Anscombe, F.J. 1949. The statistical analysis of insect counts based on the negative binomial distribution. Biometrics 5:165-173.
3. Bechinski, E.J., Stoltz, R.L. 1985. Presence-absence sequential decision plans for Tetranychus urticae (Acari: Tetranychidae) in garden-seed beans, Phaseolus vulgaris. J. Econ. Entomol. 78:1475-1480.
4. Beeden, P. 1974. Bollworm oviposition on cotton in Malawi. Cotton Grow. Rev. 51:52-61.
5. Bliss, C.I. 1941. Statistical problems in estimating populations of Japanese beetle larvae. J. Econ. Entomol. 34:221-232.
6. Brown, M.W., Cameron, E.A. 1982. Spatial distribution of adults of Ooencyrtus kuvanae (Hymenoptera: Encyrtidae), an egg parasite of Lymantria dispar (Lepidoptera: Lymantriidae). Can. Entomol. 114:1109-1120.

7. Campbell, C.L., Noe, J.P. 1985. The spatial analysis of soilborne pathogens and root diseases. Ann. Rev. Phytopathol. 23:129-148.

8. Cherry, R.H., Wood, K.A., Ruesink, W.G. 1977. Emergence trap and sweep net sampling for adults of the potato leafhopper from alfalfa. J. Econ. Entomol. 70:279-282.

9. Cochran, W.G. 1977. Sampling Techniques. 3rd ed. New York: Wiley & Sons, 428 pp.

10. Fowler, G.W., O'Regan, W.G. 1974. One sided truncated sequential t-test: application to natural resource sampling. U.S. Dep. Agric. For. Serv. Res. Pap. PSW-100.

11. Goodall, D.W. 1961. Objective methods for the classification of vegetation. IV. Pattern and minimal area. Aust. J. Bot. 9:162-196.

12. Goodall, D.W. 1973. Numerical methods of classification. In Ordination and classification of communities, Part V, Handbook of vegetation science, ed. R.H. Whittaker, pp. 575-618. The Hague: W. Junk.

13. Green, R.H. 1966. Measurement of non-randomness in spatial distributions. Res. Popul. Ecol. 8:1-7.

14. Green, R.H. 1970. On fixed precision level sequential sampling. Res. Popul. Ecol. 12:249-251.

15. Green, R.H. 1979. Sampling design and statistical methods for environmental biologists. New York: Wiley-Interscience, 257 pp.

16. Hearn, A.B., Ives, P.M., Room, P.M., Thomson, N.J., Wilson, L.T. 1981. Computer-based cotton pest management in Australia. Field Crops Res. 4:321-332.

17. Iwao, S. 1968. Use of the regression of mean crowding on mean density for estimating sample size and the transformation of data for the analysis of variance. Res. Popul. Ecol. 10:210-214.

18. Iwao, S. 1975. A new method of sequential sampling to classify populations relative to a critical density. Res. Popul. Ecol. 16:281-288.

19. Jansson, R.K., Smilowitz, Z. 1986. Influence of nitrogen on population parameters of potato insects: abundance, population growth, and within- plant distribution of green peach aphid, _Myzus persicae_ (Homoptera:Aphididae). Environ. Entomol. 15:49-55.

20. Karandinos, M.G. 1976. Optimum sample size and comments on some published formulae. Entomol. Soc. Amer. Bull. 22:417-421.

21. Kuno, E. 1969. A new method of sequential sampling to obtain the population estimates with a fixed level of precision. Res. Popul. Ecol. 11:127-136.

22. Kuno, E. 1972. Some notes on population estimation by sequential sampling. Res. Popul. Ecol. 14:58-73.

23. Legg, D.E., Yeargan, K.V. 1985. Method for random sampling insect populations. FORUM: J. Econ. Entomol. 78:1003-1008.

24. Lloyd, M. 1960. Mean crowding. J.Anim. Ecol. 36:1-30.

25. Main, C.E., Proctor, C.H. 1980. Developing optimal strategies for disease-loss sample survey. In Crop loss assessment, Proc. of E.C. Stakman Commem. Symp. pp. 118-123. Misc. Publ. #7, Agric. Expt. Stn., Univ. Minn.

26. McGuire, J.U., Brindley, T.A., Bancroft, T.A. 1957. The distribution of European corn borer larvae _Pyrausta nubilalis_ (Hbn.) in field corn. Biometrics 13:65-78.

27. Morisita, M. 1962. The I_δ index, a measure of dispersion of individuals. Res. Popul. Ecol. 4:1-7.

28. Morris, R.F. 1955. The development of sampling techniques for forest insect defoliators, with particular reference to the spruce budworm. Can. J. Zool. 33:225-294.

29. Morton, R., Tuart, L.D., Wardaugh, K.G. 1981. The analysis and standardisation of light-trap catches of _Heliothis armiger_ (Hubner) and _H. punctiger_ Wallengren (Lepidoptera: Noctiduidae). Bull. Entomol. Res. 71:207-225.

30. Myers, J.H. 1978. Selecting a measure of dispersion. FORUM: Environ. Entomol. 7:619-621.

31. Nyrop, J.P., Simmons, G.A. 1984. Errors incurred when using Iwao's sequential decision rule in insect sampling. FORUM: Environ. Entomol. 13:1459-1465.

32. Onsager, J.A. 1976. The rationale of sequential sampling, with emphasis on its use in pest management. U.S. Dep. Agric. Tech. Bull. No.1526. 19pp.

33. Pruess, K.P. 1983. Day-degree methods for pest management. Environ. Entomol. 12:613-619.

34. Rushton, S. 1950. On a sequential t-test. Biometrica 37:326-333.

35. Schotzko, D.J., O'Keeffe, L.E. 1986a. Comparison of sweepnet, D-vac, and absolute sampling for Lygus hesperus (Heteroptera:Miridae) in lentils. J. Econ. Entomol. 79:224-228.

36. Schotzko, D.J., O'Keeffe, L.E. 1986b. Evaluation of diel variation of sweepnet effectiveness in lentils for sampling Lygus hesperus (Heteroptera:Miridae). J. Econ. Entomol. 79:447-451.

37. Smith-Gill, S.J. 1975. Cytophysiological basis of disruptive pigmentary patterns in the leopard frog Rana pipiens II. Wild type and mutant cell specific patterns. J. Morph. 146:35-54.

38. Sokal, R.R., Rohlf, F.J. 1969. Biometry. The principles and practice of statistics in biological research. San Francisco: W.H. Freeman, 776pp.

39. Southwood, T.R.E. 1978. Ecological methods, with particular reference to the study of insect populations. 2nd ed. London: Chapman & Hall, 524 pp.

40. Sterling, W.L. 1976. Sequential decision plans for the management of cotton arthropods in Southeast Queensland. Aust. J. Ecol. 1:265-274.

41. Stinner, R.E., Gutierrez, A.P., Butler, G.D. Jr. 1974. An algorithm for temperature-dependent growth rate simulation. Can. Entomol. 106:519-524.

42. Taylor, L.R. 1961. Aggregation, variance and the mean. Nature (London). 189:732-735.

43. Taylor, L.R., Woiwod, I.P., Perry, J.N. 1978. The density dependence of spatial behavior and the rarity of randomness. J. Anim. Ecol. 47:383-406.

44. Taylor, L.R., Woiwod, I.P., Perry, J.N. 1979. The negative binomial as a dynamic ecological model for aggregation, and the density dependence of k. J. Anim. Ecol. 48:289-304.

45. Taylor, L.R. 1984. Assessing and interpreting the spatial distributions of insect populations. Ann. Rev. Entomol. 29:321-357.

46. Wald, A. 1945. Sequential tests of statistical hypotheses. Ann. Math. Stat. 16:117-186.

47. Wald, A. 1947. Sequential Analysis. New York: J. Wiley & Sons Inc. 212 pp.

48. Waters, W.E. 1955. Sequential sampling in forest insect surveys. For. Sci. 1:68-79.

49. Wilson, L.T., Gutierrez, A.P., Leigh, T.F. 1980. Within-plant distribution of immatures of Heliothis zea (Boddie) on cotton. Hilgardia 48:12-23.

50. Wilson, L.T. 1982. Development of an optimal monitoring program in cotton: emphasis on spidermites and Heliothis spp. Entomophaga 27: (Special Issue) 45-50.

51. Wilson, L.T., Gutierrez, A.P., Hogg, D.B. 1982. Within-plant distribution of cabbage looper, Trichoplusia ni (Hubner) on cotton: development of a sampling plan for eggs. Environ. Entomol. 11:251-254.

52. Wilson, L.T., Barnett, W.W. 1983. Degree-days: an aid in crop and pest management. Calif. Agric. 37:4-7.

53. Wilson, L.T., Gonzalez, D., Leigh, T.F., Maggi, V., Foristiere, C., Goodell, P. 1983. Within-plant distribution of spidermites (Acari: Tetranychidae) on cotton: a developing implementable monitoring program. Environ. Entomol. 12:128-134.

54. Wilson, L.T., Pickel, C., Mount, R.C., Zalom, F.G. 1983. Presence-absence sequential sampling for cabbage and green peach aphid (Homoptera: Aphididae) on Brussels sprouts. J. Econ. Entomol. 76:476-479.

55. Wilson, L.T., Room., P.M. 1983. Clumping patterns of fruit and arthropods in cotton, with implications for binomial sampling. Environ. Entomol. 12:50-54.

56. Wilson, L.T., Zalom, F.G., Smith, R., Hoffmann, M.P. 1983. Monitoring for fruit damage in processing tomatoes: use of a dynamic sequential sampling plan. Environ. Entomol. 12:835-839.

57. Zalom, F.G., Wilson, L.T., Kennett, C.E., O'Connell, N.V., Flaherty, D.L., Morse, J.G. 1986. Presence-absence sampling of citrus red mite. Calif. Agric. 40:15-16.

CHAPTER 7

METHODS OF FIELD DATA COLLECTION AND RECORDING IN
EXPERIMENTS AND SURVEYS

K. L. Bowen and P. S. Teng

Department of Plant Pathology
University of Illinois, Urbana, IL 61801
and
Department of Plant Pathology
University of Minnesota, St. Paul, MN 55108, U.S.A.

Crop loss assessment requires the collection of a large amount of data in order to understand the interactions of factors that may be affecting yield. These data need to be quantitative, and most often are collected through surveys and field experiments (1, 6, 17). Surveys involve the collection of data over a large area or region, commonly by trained investigators called "scouts", to provide information about the crops and varieties being grown, the major pests and stresses, and the relative importance of various factors on the crops (7, 11). Field experiments tend to be more limited in scope, providing data on a few specific factors affecting a particular crop, and their relationship to that crop's yield.

The lack of rapid methods to obtain data on yield-limiting factors has been an obstacle in crop loss work (17). Data collected in the field without direct application to specific objectives or analyses, may be incomplete and never used (5). Standardized forms for the collection of experimental and survey data are needed to provide consistency among workers and scouts. Data that are collected and recorded need to be accurate and in a form ready for processing without undue delay (5). However, entry of large data sets can be extremely time-consuming and frustrating if the format of the data is incompatible with desired analyses. This consideration has led to the development of computer-like devices for data collecting and recording (12, 19). These devices can increase the efficiency of some phases of data collection and lead to the production of the final report in a very short time. The increased ease of data collection and management is dependent on careful planning of the type of data to be collected. In some cases, more than the necessary data may have been collected (5).

This paper describes some of the methods in which experimental and survey data are collected and handled prior to analysis. The advantages and disadvantages of each method are also presented and compared.

PHASES OF DATA COLLECTION

The process of collecting quantitative field data that can be analyzed commonly follows a particular sequence of steps (5, 17). Generally, the first step involves the definition of the format under which data should be collected. In field experiments, this involves deciding on the number of replicates, the sampling procedure, the number of samples, and the variables on which observations are to be made. In surveys, data elements may include field identification, a crop code, stage of crop development, and pest abundance and damage (16). The second step may be coding observations or actual recording. An example of coding done prior to recording is the numbering or labeling of plots and replicates, or the use of numerically coded growth stages (3, 20). Once preliminary coding is complete, field observations are recorded. Some data may need to be further coded, depending on the method of recording used. For example, notes of observations on the incidence of water stress in field experiments could have been recorded as "little" or "severe", etc., and these could be translated to numeric values just prior to statistical analysis.

Field data may be recorded such that analysis is immediately possible, either manually or electronically (especially if an electronic recorder, with programming capabilities or built-in functions is used). It is more common, however, that data recorded in the field require some transcription to another form prior to analysis.

After transcription (if any) is complete, data can be entered into a computer file. This is often the most time-consuming step in the preparation of data for computer analysis (19). Data entry can be done directly with electronic data recorders, or indirectly by keying data in on a computer terminal.

METHODS OF DATA COLLECTION

The Field Book

The field book, or "paper and pencil", method of data recording is the traditional way of recording experiment and survey data (17). Workers using this method can go directly into a field with their equipment (paper and pencil) and write down any information they want to record. One person may do the actual recording while another makes observations, which decreases the time spent in the field and the awkwardness of taking notes and making assessments simultaneously. When recording is complete (and after necessary data coding), data are usually keyed into a computer file for analysis.

The field book is the most common method of recording data on field phenomena by trained personnel. It is low cost, widely available, familiar to workers, and flexible in its use. Editing is done simply by erasing an error and the written records are permanent, providing for future reference to that particular set of data. The major disadvantages of the field book method are that, in inclement weather, paper may become dirty, and notes may become smeared; and there is a fairly high chance of error in transcription of data to the computer for analysis.

In surveys or experimental work, where several people may collect data, preplanning and consistency are important. A format for the data should be established prior to field work, and maintained throughout the season to avoid confusion. All data forms require the appropriate spacing to be left for identification of the current year, the location, crop variety, crop growth stage, pests on which observations are to be made, individual observations desired, and room for comments (16). It is generally desirable to have all information for one survey site or experiment on a single sheet of paper.

The Paper Grid-Digitizer Method

The paper grid/digitizer method, an improvement over the field book method, uses data forms each displaying a "grid" pattern which contains all possible coded values for any variable (17). When taking notes in the field, particular values on each grid are marked off at the appropriate point. The grids, which provide hard copies of the data for storage, are taken back to the laboratory and placed onto an electronic graphics board connected to a microcomputer. A sensor is used to "read" each mark on the grid, and the marks are converted to numbers by the computer (19).

The equipment needed for this method are data forms, a digitizing pad and computer interface (e.g., a Houston Instrument, Hipad Digitizer interfaced to a 64K RAM Apple II Plus microcomputer). Digitizing pads start at $900, and together with a microcomputer, may be prohibitive in cost. Grids, which are taken into the field, are essentially "field books", though they are easier to handle while recording since all data is already printed. Thus, there is familiarity in the handling of these grids and space is delimited for desired data so little is forgotten. The major advantage for using this method is the increased efficiency of data transcription to computer files.

Voice Recordings

Rather than writing observations, assessments and field notes can be recorded on a small portable recorder such as a dictaphone or small cassette recorder. This usually involves playback of the tape and transcription of information to written or printed records before computer entry and analysis, and for record-keeping.

The cost of these machines ranges from the tens to several hundred dollars. Many models are easily carried and operated in one hand, though there is much variation between models (generally, dictaphones are smaller and easier to handle). While these are not considered weather-proof, they stand up to some inclement weather better than paper, and because of their size, can be easily protected from direct rain. Formatting of data prior to entry into a field is unnecessary, since voice recordings of observations are made. Tapes can be rewound and re-recorded if editing is needed, and the size of the data set and number of variables that can be recorded is unlimited, especially if extra tapes are available. Problems may arise if the tape jams during recording and goes unnoticed, but once recordings are made, the tapes serve as a permanent record of the data. Power source varies with make and model -- usually several small batteries are required for field use and AC adaptors are available. The probability of making errors in data sets that are collected through voice recordings is high because of the transcription involved and the inability to easily review data to see what has or has not been recorded. A relatively new development is the availability of devices called "voice synthesizers", which convert voice signals into digital form using a microcomputer. Thus, a tape recorder can be played back in the laboratory and through the voice synthesizer, field readings are converted into numbers on a computer screen.

Electronic Notebooks

Electronic notebooks, electronic data collection systems or data entry terminals are special microprocessor devices which may be used for manual or electronic data collection. These devices generally have a numeric keyboard with alphabetic characters, a liquid crystal display ("LCD"), and a power source which is a rechargeable battery with an alternating current (AC) adaptor. Two such instruments are Electro General's Datamyte and Omnidata's Polycorder, which both have several operating modes for inputting data, editing, and for setting parameters for the instrument's use. Either instrument can be purchased with up to 64K RAM (Random Access Memory). (Note: 1 K RAM is approximately equal to 1000 characters) Information such as data or prompt settings can be entered into these instruments directly from the computer, or data can be output directly to a computer. Both are capable of recording electronic data through analog input modes.

When using electronic notebooks, compatibility must be ensured between experimental design and data collection, so that collected data can be easily recorded. In the field, collected data are entered into the memory via a numeric keypad. When observations are complete, or the memory is full, these instruments can be interfaced to a cassette recorder, for saving data for later transfer to a regular computer file, or data can be transferred directly to a computer for analysis and/or production of hard copies for storage or review.

The cost of an electronic notebooks is over $3000 and back-up services (e.g., the availability of immediate repair services) are limited. These devices tend to have their own fairly simple languages for programming prompts, though some computer expertise may be needed when transcribing data. They weigh about 4 lbs. and their size varies from 8" x 4" x 3" to 13" x 10" x 2", so they seem uncomfortable, at first, while being used in the field. Certain models were designed for agricultural use, and are weather-proof, though the display may be difficult to read in direct sunlight. These devices were designed to be used as notebooks, and allow several "volumes" with different data formats (i.e., different experiments) to be programmed at the same time. Disadvantages, besides the cost and need for a compatible computer interface, are the lack of permanent records in the event of a malfunction, and the inability to add notes or comments beyond the data that fits into the format.

Portable Computers

Portable microcomputers are available in a range of sizes and capabilities from hand-held, calculator-like computers, to full-size "lap computers" with full size keyboards (19). The Panasonic Link, Hewlett-Packard's 41c, and Radio Shack's pocket computers have larger memories and increased programming capabilities over standard calculators, but are very much like calculators in size. Some of these units are available with alpha-numeric characters, an LCD display of up to 24 characters, printer and cassette interfaces. Batteries provide power for these units, though AC adaptors may be available.

"Lap computers" are portable microcomputers that can be programmed (commonly in BASIC language), and are used for data recording, data analysis or for word processing. Several makes and models are on the market, including Radio Shack's TRS80 Model 100 and Model 200, and Epson's HX-20 and PX-8. These portable microcomputers weigh about 4 lbs, are roughly 2" x 12" x 8", have a full-size, typewriter style keyboard with embedded 10-key numberpad, making them fairly easy to work with and very versatile. These are operable by batteries or through an AC adaptor, have LCD displays of 4 or more lines by 20 to 80 characters. Some models have a built-in modem, allowing direct telephone transfer of information; others have a built-in microcassette drive for mass storage or voice recordings, and analog input channels. These units range in price from $399 for the 8K RAM Model 100 to $995 for the 64K RAM PX-8. Peripheral equipment available for these microcomputers can include modems, printers, cassette drives, floppy disk drives, and bar code readers. The technology for lap computers is becoming more sophisticated very fast and at this time, available models resemble full-sized microcomputers in capabilities.

Several groups have developed programs for in-field use of portable microcomputers (9, 10, 13, 14, 15, 18). These programs have allowed estimation of the profitability of fungicide application (18), have made complicated sampling methods more feasible (14), and have allowed data inspection and editing at any point in the data entry (13). Data collected on these devices can be sent directly to a larger computer through direct interfaces or phone lines, thus practically eliminating error in data transcription and allowing the collection of daily observations from survey scouts.

The handling of portable microcomputers in the field may take some getting used to, especially when using versions with a flip-up display. Also, these devices are not weather-proof, though they may withstand inclement conditions if enclosed in a large plastic bag. In some cases, programmed microcomputers have been known to have a very slow response time.

Other Methods

Crop loss data can be recorded with methods other than those presented. Remote sensing, video image analysis and radiometer measures of sunlight reflected off leaves are forms of data recording without human estimation of damage (2, 4, 8, 9, 10). Currently, however, these methods are still being verified through traditional field experiments (1). If there are language barriers between farmers, scouts, and/or researchers, versions of the "field book" or the paper grid/digitizer methods can be used. In these circumstances, forms can be printed with diagrams or photographs of pests and stresses at different levels of severity, and observers can mark the appropriate diagram. Another method described elsewhere in this book (Chapter 22) is the use of "mark-sense" paper, paper specially coated with material and coded so that after being marked in the field, a sheet can be read in the laboratory and the codes translated into digital form by the computer. This technology is still relatively expensive but there are signs that costs of the equipment may decline in the near future.

CHOICE OF METHOD

The traditional method of data collection, the field book, is still widely used in agriculture, and is preferred if access to computer technology is limited. This method is the least expensive and most widely available of the methods presented. The "field book" method of data recording is more familiar to many, and more efficient in the transcription of data than the voice recording method, but less efficient than those methods involving computer technology. However, voice recording of data in the field is less time consuming than field book data entry and may be advantageous for use in bad weather.

The paper grid/digitizer method of data recording, provides an easy, legible and organized manner of recording field data. The completed grids provide documentation of original data and errors are minimal when transcribing data to a computer, since values are recorded electronically. However, programming skills may be needed for modifying existing digitizer software to fit experiments and for designing grids that are compatible for digitizing and for data recording.

Electronic notebooks and portable microcomputers tend to be awkward to handle in the field while recording data, but they do provide an accurate and rapid means of transcribing data to computer files for analysis and for direct transfer of data to a printer for multiple copies. Unlike the digitizer method, these devices are simple to set up, since little programming expertise is necessary. Some microcomputers and electronic notebooks have analog inputs and internal clocks, allowing their use in electronic and automatic data recording from, for example, radiometers (9, 10). Those devices which are programmable may also be useful in small experiments when some results are needed immediately (15, 19).

From the preceding comparisons, it seems that the paper grid/digitizer method is the best system for data recording in terms of ease of recording, ease and accuracy of data transfer, and storage of data. This system is recommended when large amounts of data are to be collected. However, lack of hardware and persons experienced in programming the digitizer or using a digitizing pad may be a real constraint in using this method.

With increasing capability to collect large amounts of data, careful planning prior to data collection is becoming more important. There is a real risk of collecting too much data, such that the database becomes unwieldy and difficult to use (17, 19). This chapter has only addressed the first aspect of a lengthy process of converting data into useful information.

80

LITERATURE CITED

1. Burleigh, J.R. 1980. Experimental design for quantifying disease effect on crop yield. In Crop Loss Assessment, pp. 50-62. Misc. Pub. No. 7, Univ. Minn. Agric. Exp. Stn., St. Paul, Minn. 327 pp.
2. Ellington, J., Phillips, K., Dearholt, D., Kiser, K. 1985. Image Analysis. NCCI Data Acquisition Workshop, Jan. 7-10, Rosemount, Ill.
3. Fehr, W.R., Caviness, C.E. 1977. Stages of soybean development. Special Report 80. Coop. Ext. Ser., Agric. and Home Econ. Iowa State Univ. Exp. Stn., Ames, Iowa. 12 pp.
4. Gerten, D.M., Wiese, M.V. 1984. Video image analysis of lodging and yield loss in winter wheat relative to foot rot. Phytopathology 74:872.
5. Heong, K.L. 1981. The uses and management of pest surveillance data. Malay. Agric. J. 53:65-89.
6. James, W.C., Teng, P.S. 1979. The quantification of production constraints associated with plant diseases. Appl. Biol. 4:201-267.
7. King, J.E. 1980. Cereal survey methodology in England and Wales. In Crop Loss Assessment, pp. 124-133. Misc. Pub. No. 7, Univ. Minn. Agric. Exp. Stn., St. Paul, Minn. 327 pp.
8. Lindow, S.E., Webb, R.R. 1983. Quantification of foliar plant disease symptoms by microcomputer-digitized video image analysis. Phytopathology 73:520-524.
9. Pedersen, V.D. 1984. Multispectral radiometry using a 12-bit analog-to-digital converter interfaced with a portable microcomputer. Phytopathology 74:872.
10. Pedersen, V.D., Fiechtner, G. 1980. A low-cost, compact data acquisition system for recording visible and infrared reflection from barley crop canopies. In Crop Loss Assessment, pp. 71-75. Misc. Pub. No. 7, Univ. Minn. Agric. Exp. Stn., St. Paul, Minn. 327 pp.
11. Ridgway, R.L. 1980. Assessing agricultural crop losses caused by insects. In Crop Loss Assessment, pp. 229-233. Misc. Pub. No. 7, Univ. Minn. Agric. Exp. Stn., St. Paul, Minn. 327 pp.
12. Rouse, D.I., Teng, P.S. 1984. Understanding computers: A modern tool in plant pathology. Plant Dis. 68:365-369.
13. Royer, M.H. 1984. A user-friendly data recording program for the TRS80 Model 100 portable computer. Phytopathology 74:758.
14. Stowell, L.J., Delp, B.R., Grogan, R.G. 1984. Disease distribution and crop loss assessment using a field-portable microcomputer. Phytopathology 74:871.
15. Taylor, S.E. 1985. Data collection methods. Proc. NCCI Data Acquisition Workshop, Jan. 7-10, Rosemount, Ill.
16. Teng, P.S. 1981. Data recording and processing for crop loss models. In Crop Loss Assessment Methods, Supplement 3, ed. L. Chiarappa, pp. 105-109. FAO/Commonw. Agric. Bur., Farnham Royal, UK.
17. Teng, P.S. 1984. Surveillance systems in disease management. FAO Plant Prot. Bull. 32:51-60.
18. Teng, P.S., Montgomery, P.R. 1982. RUSTMAN: A portable microcomputer-based economic decision model for sweet corn rust control. Phytopathology 72:1140.
19. Teng, P.S., Rouse, D.I. 1984. Understanding computers: Applications in plant pathology. Plant Dis. 68:539-543.
20. Zadoks, J.C., Chang, T.C., Konzak, C.F. 1974. A decimal code for the growth stage of cereals. Weed Research 14:415-421.

CHAPTER 8

GENERATING THE DATABASE FOR DISEASE-LOSS MODELING

W. W. Shane and P. S. Teng

Department of Plant Pathology,
Ohio State University, Columbus, OH 43210,
and
Department of Plant Pathology,
University of Minnesota, St. Paul, MN 55108, U.S.A.

To study the effect of disease or other pests on crop yield requires the collection or generation of a database in which there are different pest intensities and corresponding plant/crop yields. Sample survey and field plot techniques have commonly been used to provide the needed database, with the size of the host population ranging from single tillers or plants, to microplots to field plots (34). The literature also shows that most single tiller/plant data have been collected from farmers' fields with natural pest infestations using sample survey techniques, while most microplot and field plot data have been collected from designed experiments (40).

In the context of yield loss assessment, the primary purpose of acquiring pest-yield data is not to compare pest-infested treatments with a control, uninfested treatment as stated by LeClerg (22). Rather, the objective is to enable estimation of a response of yield to different levels of pest infestation (35). There is a fundamental distinction between yield loss experiments and experiments for comparing fungicides or other treatments, and this affects the manner in which the database should be generated. For example, in a yield loss experiment, the desired objective is not to prove that pest intensity A is significantly different from pest intensity B but rather, that there is a significant response relationship between varying pest intensities and crop yield or loss. As noted by Teng and Oshima (39), this distinction points to the need for more treatments (many different pest intensities) and less replication, in order that the full effect of the pest on yield be determined.

Comprehensive reviews of the literature have previously been done (3, 11, 13, 18, 40) and this chapter will emphasize the practical aspects of generating a database for pest-loss modeling.

SINGLE TILLER/PLANT TECHNIQUES

In the single tiller/plant technique, many tillers or plants (50 to 2000) are tagged from one or more fields, assessed for disease intensity and harvested at maturity (27). The tillers or plants are selected for tagging to represent as wide a range of pest intensities as feasible, including a "healthy" reference. Each plant then becomes a single "case" for statistical analyses, resembling a treatment plot or mean value in a plot experiment. Workers have used single tillers or plants in different ways to obtain pest-loss data. Chester (5) documents many examples of "paired-plant" techniques, in which pairs of one healthy and one diseased plant are identified in the same or different fields, disease progression followed during the season and the plants harvested. The members of the pair are selected to be as similar as possible except for the difference in disease. The rationale for pairing is to enable estimation of yield loss for a given pair of plants but minimizing environmental or genetical differences, especially the difference in yield potential. However, Richardson et al. (27) found that pairing did not significantly improve the precision of the disease-loss models they derived from primary tiller data.

Yield potential differences between individual plants or tillers account for much variability in data collected using this method (13). Hau et al. (9) attempted to reduce some of this variability by standardizing all the plants to a reference potential yield when disease was absent. Working with cassava pests, these workers derived a relationship between root yield and healthy green leaf area, therefore enabling them to calculate a potential root yield for each plant from the potential green leaf area (i.e. no symptoms and defoliation). The data reported by Hau et al. (9) for Cassava Mosaic Virus showed that the method enabled a high proportion of the variation in yield loss to be explained. Gaunt et al. (8) used a paired-plant technique for estimating losses in Faba Beans due to Ascochyta, but applied chemicals to produce a wider range of disease severities. Shane (29) modified the technique by grouping wheat tillers of approximately similar bacterial blight severity to derive a mean severity and yield, thereby reducing some of the inter plant variability in potential yield. The single tiller technique has been extensively used by King (20) to develop equations for estimating yield loss due to diseases on wheat and barley in England and Wales.

There are many situations under which single tillers or plants may be used to provide data for characterizing a pest-loss relationship. Richardson et al. (27) argue strongly that their single tiller method has several merits, namely,
1. The method can provide data from a wide range of disease intensities after just one growing season,
2. There is economy of labor and land compared to plot techniques,
3. There is no use of chemicals or other methods to manipulate disease levels, and
4. There is increased precision in estimating the slope of the regression line of yield loss on disease.
In a comparison of single tiller versus plot technique to estimate losses caused by barley leaf scald, Mayfield and Clare (24) found that the single tiller technique did not result in any savings in time or labor. James and Teng (13) have argued that the single tiller method appears suited for late developing epidemics of short duration, where only one yield component is likely to be affected and can account for most of the loss due to disease. This is the case with rust diseases on cereals. However, with early developing epidemics of long duration, development of disease-loss equations requires sequential assessment of severity and not the single disease assessment of late epidemics (19, 27). With the single tiller method, it is often difficult to maintain a "healthy", zero disease reference and workers have resorted to regressing disease intensity with yield to calculate the reference yield. This is done by taking the intercept on the y-axis (yield) when x (disease intensity) equals zero. Gaunt et al. (8) used chemicals in their approach to maintain disease free plants.

There are several problems associated with the single tiller/ plant technique (19, 27). In general, the regression equations (see Chapter 11), have only accounted for a small amount of the variation in yield and loss, i.e. low coefficients of determination. Enhancements to the technique by Hau et al. (9) and Shane (29) have allowed a greater proportion of the variation to be explained. Disease-loss equations derived from single tiller/plants are just what they imply, and it is questionable if such equations can be used for estimating losses in a crop (33). The dynamics of disease progression and yield accumulation measured on single plants are sufficiently different when compared with a population of plants, to raise serious questions about the application of equations derived from this technique. Admittedly, the single plant technique allows for a quick determination of yield loss when resources are not available for experimental work, and may provide a reasonable first estimate of the disease-loss relationship.

Teng (33) pointed out the scalar problems associated with developing models for characterizing disease-loss relationships using data collected from different levels of biological organization, e.g. plant part, individual plant, group of plants, crop. Using individual plants can provide a "quick and simple" estimate while increasing the size of the experimental unit to field plots also introduces increased variability due to environment. As will be discussed later in this chapter, plot experimentation may have inherent representational errors due to interplot interference.

MICROPLOT TECHNIQUES

The term "microplot" is commonly used to describe containerized, miniature plots of the type used to study the effects of nematodes on plant yield (1). This method, used primarily for soil pathogens, offers many advantages of field plot tests and overcomes much of the variability associated with fields. The presence of microorganisms within plots can be controlled so that specific effects can be studied. The basic setup is as follows: 1) <u>establishment of barriers</u>, in which open-ended tile or fiber glass cylinders are embedded in the ground. Barker (1) has devised a tractor "power-take-of" microplot cutter to aid establishment of circular microplots with minimal disturbance of the soil profile. Alternatively, uniform soils can be placed within the barriers. 2) <u>destroy organisms within barrier</u>, in which the soil is treated with methyl bromide or other appropriate additive to kill undesired pests. Sufficient time is needed to allow fumigated plots to aerate and beneficial organisms may have to be reintroduced. 3) <u>introduce plants and pests</u>.

A matter of concern in microplot studies with mobile soilborne pathogens is the issue of numbers versus concentration. Pest intensity is often expressed in terms of concentration, i.e. pest number per unit soil volume. Small microplots may result in fewer numbers of pests reaching the roots compared to larger plots with the same initial concentration.

Microplot experiments usually entail an enormous amount of work but may be the best way to study complex interactions or weak main effects.

FIELD PLOT TECHNIQUES

Of all the techniques reported in the literature, field plot techniques utilizing a standard experimental design are the most common (13, 40). This group includes all those experiments in which suitably-sized (see later) plots, replicated and with two or more treatment levels in the experiment, are used. Data from field plots are commonly collected over three or more seasons and from several locations representing the diversity of farming practices in the area.

Treatment Types and Replications

Paired treatment and multiple treatment experiments have been used to generate the database for disease-loss modeling. In paired treatment experiments, pairs of healthy (protected with chemical) and diseased plots are replicated to provide many levels of disease intensity (4, 21). From each pair of treatments is derived a datum set with known disease intensities and corresponding yield loss. In the multiple treatment experiment, two or more disease intensity levels are found in the same replicated experiment (15, 41). Disease levels are produced by the use of chemicals, differential host resistance or cultural practices (see Chapter 9). The number of replicates in either paired or multiple treatment experiments has been at least three. LeClerg (22) recommended a minimum of four replicates but preferably five or six. Multiple treatment experiments generally allow a wider range of disease levels to be generated than paired treatment experiments (13).

Size and Shape of Plots

The plot size used in disease-loss experiments has to minimize the variation in crop yields encountered in a particular area. The yield of major crops in the U.S.A., measured under experimental conditions, has a coefficient of variation between 5% - 15% (30). Plot size and shape also influence the dynamics of disease epidemics. With disease-loss experiments, it is important that disease progression in treatment plots approximates that in farmers' fields, under equivalent conditions. Recommendations for some crops have been given by LeClerg (22):

Barley and Wheat	8 rows X 6 m long
Cotton	4 rows X 9 m long
Maize	4 rows X 9 m long
Potatoes	4 rows X 9 m long
Rice	10 rows X 6 m long
Sugarcane	6 rows X 17 m long.

In practice, there is a tradeoff between reducing yield variability by increasing plot size, and the potential of increasing the variation due to soil and other factors because of increased plot size. By increasing plot size, it is commonly possible to reduce the number of replicates required to detect a difference between two treatments, when compared with a smaller plot (14).

The actual plot size harvested in an experiment is dictated by design and equipment. For example, Teng and Bissonnette (36) used 30.5 m X 3 row plots to determine the effect of early blight on potato yield and harvested one 15.0 m row from the middle of the plot. Generally, at least 0.3 m from each end are omitted while only the middle rows are harvested.

Experimental Design

Regardless of whether an experiment is paired or multiple treatment, plots have commonly been deployed using one of several standard designs, such as the randomized complete block, Latin Square, complete factorial or split-split plot. Descriptions of these designs are to be found in many statistical textbooks, such as Snedecor and Cochran (30). Experimental designs have various strengths and weaknesses that generally are concerned with the following issues: 1) ease of plot establishment, management, assessment and harvest, 2) control of interplot interference, 3) providing a means to partition off different sources of variation in the data, some of interest, others of nuisance, and 4) ease of statistical analysis. The appropriateness of using balanced designs with complete replication has recently been questioned (39), since many of these designs were originally meant for testing differences between treatment levels and not for estimating responses. In disease-loss experiments, the objective of the database is NOT for testing that one level of disease severity causes significant loss when compared with another. Rather, the objective is, almost without exception, to determine a response relationship between yield loss and increasing disease severities. Response estimation designs have been much researched in agronomic experiments (7) and these emphasize number of treatments rather than number of replications. A disease-loss relationship is a response relationship and we are generally interested in determining the model that best describes this relationship.

Experimental designs included under the category of response surface methods include the incomplete factorial, central composite and rotatable (7). Teng and Montgomery (38) used an incomplete factorial with no replication, in which each treatment plot was a unique epidemic, to generate data for modeling sweet corn yield loss due to common rust. The database gave several regression models with high coefficients of determination, each model had loss as the dependent variable and rust severity and growth stage as the independent variables. The workers felt that the design enabled them to explore a wider range of disease severities at each crop growth stage, given a limited number of plots (38). If a conventional design had been used, the number of treatments would have been less since some plots would have had to be used as replicates. Previously, Teng and Gaunt (37) had noted that many regression models of cereals in the literature were inadequate for explaining a range of disease severities at more than one growth stage. This often lead to meaningless estimations of loss using the models when exceptionally high or low severities were encountered. The use of non-replicated experiments for disease-loss modeling holds great potential, although much research still needs to be done to determine its general applicability.

Intraplot Uniformity

Of equal concern in experimentation is the uniformity of disease occurrence on plant units within each plot. High within-plot heterogeneity of disease severities is due to sparse or non-uniform distribution of initial inocula, differential disease susceptibilities in the plant population, and/or short distance dispersal of secondary inocula (42). An average severity value or other single statistic does not satisfactorily represent a plot of plants with a wide range of severities. This is particularly true if the relationship between severity and yield loss is a non-linear one.

The overestimation of yield loss has been recognized as a problem when severity-loss models derived from relatively uniform plot information are used in fields where the pathogen shows an aggregated distribution (28). As noted by Noe and Barker (25), overestimation is the result of skewness in the frequency distribution of disease or pathogen intensities. Generally, the skewness is due to a greater frequency of pathogen or disease intensities below the mean and a relatively few high severities.

Interplot Interference

The phenomenon of interplot interference occurs in experiments with contiguous treatment plots, in which the treatments represent different levels of disease or insect populations (42). VanderPlank refers to this as the "cryptic error" in field experiments, to distinguish it from the experimental error accountable in the analyses of variance of plot data. Interplot interference exists in almost all disease-loss experiments with foliar pathogens having aerially-dispersed propagules. The concept is described in detail by James (12). When a treatment (plot A) with a certain level of disease is sited next to another (plot B) with higher disease, then plot A often overestimates the treatment effect because of a net gain of inoculum from plot B. Plot B, because of its higher level of disease, has exerted a positive interference on plot A. Conversely, plot B may underestimate the treatment effect because of the presence of plot A, since plot A, with a lower level of disease, exerts a negative interference on plot B.

The degree of interplot interference in field experiments can be quantified using a design proposed by James et al. (16). Interference can be reduced by the use of guard rows of a tall crop, by separating treatment plots, by increasing plot size (2) and by the use of special designs to balance treatment effects (17). The presence of interplot interference means that there is possibility of error when applying models derived from field experiments for estimating loss in farmers' fields (13).

THE SYNOPTIC METHOD

While the majority of research on disease-loss has been with single diseases, some research has been reported on the effect of multiple pests on yield and loss. Stynes (31, 32) developed a "synoptic" method in Australia to determine the effect of multiple factors on wheat yield. In his method, the same portions of selected farmers' fields were intensively sampled during the growing season and weather, soil, crop and pest variables measured at known times. The data were then subject to multivariate statistical analyses to determine models which could account for the contribution of individual factors or groups of factors to yield loss. Earlier, Pinstrup-Andersen et al. (26) had also used a survey procedure in Colombia to partition the contribution of biological, social, economic and environmental factors towards reducing attainable bean yields to actual, observed yields. In this work, farmers were interviewed and entire bean fields sampled for rust, bacterial blight, leafhoppers, angular leafspot, plant population and water availability. The International Rice Research Institute (IRRI) (10) has used partial pest control treatments on portions of farmers' fields to identify constraints on rice production.

Wiese (43, 44) sampled 100 pea fields in Idaho, U.S.A. and on each assessed pests (insects, weeds, diseases, nematodes), weather (hail, frost, precipitation, degree-days), soil characteristics (nutrients, pH, topography, bulk density, penetrability, moisture, temperature), and cultural practices (cultivar, cropping sequence, pesticides, fertilizer, tillage, seeding). The fields were visited at 1-2 week intervals for approximately 4 months and the same 0.1 ha of each field used for measurements. The results enabled development of an overall interactive yield model, with 12 variables, that explained 82% of the variability in seed yields (43). The model, however was year-specific and could not be used to predict losses in another year. A problem common to all the above work (26, 32, 43) is reliance on natural pest infestations, the type and intensity of which varies between years but which may be within a narrow range. Therefore, it is possible for the influence of some pests on crop yield to be underestimated. However, the synoptic approach has much merit in defining the magnitude of actual yield constraining factors under farmer conditions, and in this respect, is a useful tool in pest management (44).

LITERATURE CITED

1. Barker, K.R. 1985. The application of microplot techniques in nematological research. In An Advanced Treatise on Meloidogyne, Vol. II, Methodology, ed K.R. Barker, C.C. Carter, J.N. Sasser, pp. 127-134. Raleigh: North Carolina State University Graphics. 223 pp.
2. Bowen, Kira L., Teng, P.S., Roelfs, A.P. 1984. Negative interplot interference in field experiments with wheat leaf rust of wheat. Phytopathology 74:1157-1161.
3. Burleigh, J.R. 1980. Experimental design for quantifying disease effect on crop yield. In Crop Loss Assessment, ed. P.S. Teng, S.V. Krupa, pp.50-62. Misc. Pub. no. 7, Univ. of Minn. Agric. Exp. Stn., St. Paul, Minn. 327 pp.
4. Calpouzos, L., Roelfs, A.P., Madson, M.E., Martin, F.B., Welsh, J.R., Wilcoxson, R.D. 1976. A new model to measure yield losses caused by stem rust in spring wheat. Tech. Bull. Afric. Exp. Stn. Univ. Minnesota no. 307. 23 pp.
5. Chester, K.S. 1950. Plant disease losses: their appraisal and interpretation. Plant. Dis. Rep. Suppl. 193:189-362.
6. Darwinkel, A. 1979. Ear size in relation to tiller emergence and crop density. In Crop Physiology and Cereal Breeding, ed J.H.J. Spiertz, T. Kramer, pp. 10-15. Pudoc, Wageningen, Netherlands. 185 pp.
7. Dillon, J.L. 1968. The Analysis of Response in Crop & Livestock Production. Oxford: Pergamon Press. 135 pp.

8. Gaunt, R.E, Teng, P.S., Newton, S.D. 1978. The significance of ascochyta leaf and pod spot disease in field bean (_Vicia faba_ L.) crops in Canterbury, 1977-78. Proc. Agron. Soc. New Zealand 8:55-57.

9. Hau, B., Kranz, J., Dengel, H.L., Hamelink, J. 1980. On the development of loss assessment methods in the tropics. In Crop Loss Assessment, ed. P.S. Teng, S.V. Krupa, pp. 254-261. Misc. Pub. no. 7, Univ. of Minn. Agric. Exp. Stn., St. Paul, Minn. 326 pp.

10. IRRI. 1979. Farm-level Constraints to High Rice Yields in Asia : 1974-77. Los Banos, Philippines: IRRI. 411 pp.

11. James, W.C. 1974. Assessment of plant diseases and losses. Annu. Rev. Phytopathol. 12:27-48.

12. James, W.C. 1978. Importance of interplot interference in field experiments involving plant diseases. In Epidemiology and Crop Loss Assessment, ed. R.C. Close et al., pp. 29-1/ 29-15. Lincoln College Press.

13. James, W.C., Teng, P.S. 1979. The quantification of production constraints associated with plant diseases. Appl. Biol. 4:201-267.

14. James, W.C., Shih, C.S. 1973. Size and shape of plots for estimating yield losses from cereal foliage diseases. Exp. Agric. 9:63-71.

15. James, W.C., Shih, C.S., Hodgson, W.A., Callbeck, L.C. 1972. The quantitative relationship between late blight of potato and loss in tuber yield. Phytopathology 62:92-96.

16. James, W.C., Shih, C.S., Callbeck, L.C., Hodgson, W.A. 1973. Interplot interference in field experiments with late blight of potato (_Phytophthora infestans_). Phytopathology 63:1269-1275.

17. Jenkyn, J.F., Bainbridge, A., Dyke, G.V., Todd, A.D. 1979. An investigation into inter-plot interactions, in experiments with mildew on barley, using balanced designs. Ann. Appl. Biol. 92:11-28.

18. Judenko, E. 1973. Analytical method for assessing yield losses caused by pests on cereal crops with and without pesticides. Trop. Pest Bull. 2. London: Centre for Overseas Pest Research. 31 pp.

19. King, J.E. 1976. Relationship between yield loss and severity of yellow rust recorded on a large number of single stems of winter wheat. Plant Pathol. 25:172-177.

20. King, J.E. 1980. Cereal survey methodology in England and Wales. In Crop Loss Assessment, ed. P.S. Teng, S.V. Krupa, pp. 124-133. Misc. Pub. no. 7, Univ. of Minn. Agric. Expt. Stn., St. Paul, Minn. 326 pp.

21. Large, E.C., Doling, D.A. 1962. The measurement of cereal mildew and its effect on yield. Plant Pathol. 11:47-57.

22. LeClerg, E.L. 1971. Field experiments for assessment of crop losses, pp. 2.1/1-11, In FAO Manual on the Evaluation and Prevention of Losses by Pests, Diseases and Weeds, ed. L. Chiarappa. UK: FAO/CAB.

23. Love, H.H. 1943. Experimental methods in agricultural research. Univ. of Puerto Rico Agric. Exp. Stn., Rio Piedras, Puerto Rico. 229 pp.

24. Mayfield, A.H., Clare, B.G. 1978. A comparison between single-tiller and plot methods for estimating losses in yield of barley with leaf scald disease. In Epidemiology and Crop Loss Assessment, ed. R.C. Close, et al, pp. 10-1/ 10-3. Lincoln College Press.

25. Noe, J.P., Barker, K.R. 1985. Overestimation of yield loss of tobacco caused by the aggreagated spatial pattern of _Meloidogyne incognita_. J. Nematol. 17:245-251.

26. Pinstrup-Andersen, P., de Londono, N., Infante, M. 1976. A suggested procedure for estimating yield and production losses in crops. PANS 22:359-365.

27. Richardson, M.J., Jacks, M., Smith, S. 1975. Assessment of loss caused by barley mildew using single tillers. Plant Pathol. 24:21-26.

28. Seinhorst, J.W. 1973. The relation between nematode distribution in a field and loss in yield at different average nematode densities. Nematologica 19:421-427.

29. Shane, W.W. 1985. Population dunamics of _Pseudomonas syringae_ pv. _syringae_ and crop losses associated with _Xanthomonas campestris_ on spring wheat. Ph.D. Thesis, Univ. of Minn., St. Paul, MN. 83 pp.

30. Snedecor, G.W., Cochran, W.G. 1967. Statistical Methods. Ames, Iowa: Iowa State University Press. 6th ed. 593 pp.

31. Stynes, B.A. 1975. A Synoptic Study of Wheat. Ph.D. Thesis, Univ. of Adelaide, Australia. 291 pp.

32. Stynes, B.A. 1980. Synoptic methodologies for crop loss assessment. In Crop Loss Assessment, ed. P.S. Teng, S.V. Krupa, pp. 166-175. Misc. Publ. 7, Univ of Minn. Agric. Exp. Stn., St. Paul, MN. 326 pp.

33. Teng, P.S. 1983. Estimating and interpreting disease intensity and loss in commercial fields. Phytopathology 73:1587-1590.

34. Teng, P.S. 1985a. Crop loss assessment methods: current situation and needs. In An Advanced Treatise on Meloidogyne, Vol. 11, Methodology, ed. K.R. Barker, C.C. Carter, J.N. Sasser, pp. 149-158. Raleigh: North Carolina State University Graphics. 232 pp.

35. Teng, P.S. 1985b. Construction of predictive models : II. Forecasting crop losses. Advances in Plant Pathol. 3:179-206.

36. Teng, P.S., Bissonnette, H.L. 1985. Estimating potato yield responses from chemical control of early blight in Minnesota. American Potato Journal 62:595-606.

37. Teng, P.S., Gaunt, R.E. 1981. Modelling systems of disease and yield loss in cereals. Agricultural Systems 6:131-154.

38. Teng, P.S., Montgomery, P.R. 1981. Response surface models for common rust of corn. Phytopathology 71:895. (Abstr.)

39. Teng, P.S., Oshima, R.J. 1983. Identification and assessment of losses. In Challenging Problems in Plant Health, ed. T. Kommedahl, P.H. Williams, pp.69-81. St. Paul: American Phytopathological Society. 538 pp.

40. Teng, P.S., Shane, W.W. 1984. Crop losses due to plant pathogens. CRC Crit. Rev. in Plant Sci. 2:21-47.

41. Teng, P.S., Blackie, M.J., Close, R.C. 1979. A comparison of models for estimating yield loss caused by leaf rust (Puccinia hordei Otth) on Zephyr barley in New Zealand. Phytopathology 69:1239-1244.

42. VanderPlank, J.E. 1963. 1963. Plant Diseases: Epidemics and Control. New York: Academic Press. 349 pp.

43. Wiese, M.V. 1980. Comprehensive and systematic assessment of crop yield determinants. In Crop Loss Assessment, ed. P.S. Teng, S.V. Krupa, pp. 262-269. Misc. Publ. no. 7, Univ. Minn. Agric. Expt. Stn., St. Paul, MN 326 pp.

44. Wiese, M.V. 1982. Crop management by comprehensive appraisal of yield determining variables. Ann. Rev. Phytopathol. 20:419-432.

CHAPTER 9

METHODS OF GENERATING DIFFERENT LEVELS OF DISEASE EPIDEMICS IN LOSS EXPERIMENTS

D. N. Sah and D. R. MacKenzie

Department of Plant Pathology and Crop Physiology,
Louisiana State University, Baton Rouge, LA 70803, U.S.A.

Much of what is known about yield loss caused by individual plant diseases has been the result of field plot experimentation. There are many reports in the literature of studies that were conducted to demonstrate the nature of the relationship between different levels of disease and the yield of a particular crop (2, 4, 7, 15, 26, 31). To gather such information, many different methods have been developed for field plot experimentation (17).

On rare occasions, events or conditions are such that information can be collected without the need for manipulating various factors. More commonly, however, experimental procedures are needed to generate different levels of disease severity for experimental studies.

Many complex problems are encountered when attempting to design experiments that would alter those factors important to a crop's yield. This chapter will review those factors with the intention of pointing out the limitations and the realities of this type of experimentation.

EXPERIMENTAL DESIGN

Yield loss research almost inevitably requires that regression analysis be used to sort through the significance of the relationship of disease severity to crop yield. This is because replications of the severity-loss relationship are difficult. For example, in a field of four replications, it is unreasonable to expect that all replications will have identical disease severities throughout the growing season. Analysis of variance from a factorial experiment becomes difficult in these cases.

Regression analysis is a powerful tool to evaluate statistically the relationship between disease severity and crop yield (35). Most researchers attempting to characterize the relationship between disease severity and yield loss have used simple or multiple regression (7, 15, 19, 27). Regression does require, however, that the independent variable be measured without error. This creates another dilemma for those experimenting in yield loss relationships. Most commonly, disease is estimated by "eyeball". It is far from being measured without error. Most individuals overlook this difficulty, but the concern remains.

Plot size, the amount of initial inoculum, and other components of the experimental design need careful consideration in the early stages of planning yield loss experiments. Plots too small and in close proximity can have confounded information as inoculum spreads between plots and interferes with the independence of the estimates. Small plots are also difficult for yield estimations because coefficients of variation increase and error terms tend to be quite large.

The use of larger plots also has drawbacks, particularly if the area is heterogeneous or if the disease occurs in foci (see also Chapter 8). Although yield loss estimation may be more precise in large plots, the distribution of disease and its development within the plot can cause significant difficulties. In addition, big plot size will result in large block size, which may increase the error variances because of environmental differences among plots within the block. This problem can be solved, however, by using balanced incomplete block designs (8). Obviously, a compromise of an "intermediate size" plot is often necessary. Romig and Calpouzos (27) used only 25 m^2 wheat plots separated by 35 m of barley in their work to assess yield losses caused by stem rust. However, James et al. (16) used a 60 m^2 plot size for potato and separated the plots by 100 m - 300 m to reduce interplot interference.

The monitoring of environmental factors is critical to a complete understanding of the disease development, the crop's development, and the effects of disease severity on the crop's yield. Unfortunately, key environmental parameters are rarely reported in the literature. We suspect that they are rarely recorded. With the advent of low cost but sophisticated on-site data loggers much more information should be available for interpretation of yield loss experiments. Better methods are needed to statistically digest the weather information in ways meaningful for the interpretation of the yield loss results.

GENERATING DIFFERENT LEVELS OF DISEASE

Commonly, in these types of experiments, all factors are kept constant except disease. Various levels of disease epidemics are generated and one or more factors are manipulated in order to study the effect of disease on yield in one field experiment. Several methods are available to accomplish this goal.

Time of Inoculation

By manipulating the point in time at which various plots are inoculated, it is sometimes possible to exert a measurable effect on the final amount of disease. Agrios et al. (1) found a direct relationship between time of inoculation of pepper plants (Capsicum annum L.) with cucumber mosaic virus (CMV) and yield loss. Groups of "Lady Bell" pepper plants were mechanically inoculated with CMV in a series of eight successive inoculations carried out at weekly intervals. Plants inoculated in early growth stages were significantly shorter in height and produced less fruit yield than plants inoculated later in the season.

Johnson et al. (20) found no difference in percentage of tobacco (Nicotiana sp.) plants infected due to different date of inoculation with tobacco mosaic virus from seven to sixty-three days after transplanting in the field. However, by inoculation of 15%, 30%, 60% and 100% of the plants per plot in seven susceptible cultivars, yield was reduced 7%, 10%, 17% and 30% and value was reduced 7%, 14%, 21% and 36%.

Reddy et al. (26) generated epidemics of bacterial leaf blight of rice with different disease progress curves by manipulating the initial dates and frequencies of subsequent inoculation, and by using a bactericide. Epidemics of bacterial leaf blight of rice were initiated by inoculating plants with Xanthomonas campestris pv. oryzae (Ishiyama) Dye at four different stages of development. Griffiths et al. (12) used greenhouse studies to evaluate the effect of powdery mildew disease (caused by Erysiphe graminis DC. ex Merat f. sp. hordei Marchal) at different growth stages on grain yield of barley.

Often the experiments that vary time of inoculation become quite confusing as the final amount of disease is not that which might be expected. During the course of the growing season environmental factors, random events, and other factors come into play on the course of disease development within the plots. Consider the effects of weather as an example. Delaying the inoculation of plots by a week in early season might be insignificant when compared to weather conditions that could occur later in the growing season. Such conditions could slow down or speed up the epidemic development, allowing one plot to become very similar to another, even though the inoculation was delayed.

Varying the Amount of Initial Inoculum

Other experimental methods use varying amounts of initial inoculum to affect the final outcome of disease. A detailed discussion of the relationship between the amount of inoculum and the amount of disease is provided by VanderPlank (36). Different levels of nematode densities can be obtained by inoculating nematode-free soils with different numbers of nematodes and this is the basic approach used with this type of pest (32, Chapter 8). Yield losses caused by soil-borne fungi are also predicted by establishing the relationship between inoculum density and disease severity (3). However, such experiments are feasible only for plants grown in pots or in microplots of inoculated soil.

There are two basic methods for establishing different nematode density levels. In one method, the soil is fumigated and known numbers of eggs or larvae of nematodes are introduced to establish the range of initial population densities (5, 10). Another approach is to manipulate nematode populations in a field by using host and non-host crops or by using nematicides (5, 11). Similarly, different inoculum densities of soil-borne fungi have been maintained in the field either by crop rotation or by the addition of different amounts of resting spores in the soil (2, 25, 34).

Spreader Rows

Often spreader rows of susceptible crops are used as sources of inoculum to spread disease throughout large areas (15, 31). The complications of gradients of disease make analysis of the results quite difficult. There are, however, analytical procedures that can be used to study the spread of disease in space along with the increase in time. These analytical procedures can be used to determine various types of resistance and perhaps even tolerance to disease. More research on the epidemiology of disease spread is needed to better use this approach to yield loss research.

Manipulation of Environmental Conditions

Some researchers use the known effects of modified environmental conditions to enhance or suppress the development of disease, for example, the use of overhead irrigation to extend periods of leaf wetness and thereby increase disease development (9). This has been very successful in some types of experiments (e.g., downy mildew of corn, potato late blight, bacterial stalk rot of corn) but the technique adds the complication of unknown effects due to moisture in some plots and not in others.

Geographic Areas

Several levels of disease may be obtained by planting the crop variety in different geographic areas that vary in the prevalence of a particular plant pathogen (21). In this approach, however, the effects of disease intensity on yield are complicated by several other variables at each area, such as other diseases, other pests, or different weather and soil properties.

Fungicides

Perhaps the most common research method used to adjust the amount of disease severity between plots is the use of fungicides. The availability of very specific fungicides within the last twenty years has added a new dimension to this method of yield loss research. Several researchers have used combinations of one or more fungicides at different rates and times to generate different levels of disease (4, 15, 16, 27, 31). Unfortunately, complete control of a disease often is not possible with protectant fungicides because new foliage produced between spray applications may become infected. Broad spectrum fungicides may control other diseases besides the particular disease under study, and thereby confound the results. Also, a portion of the measured effect may be attributed to differences in the level of fungicide applied (18, 22). Fungicides applied may produce a phytotoxic effect on plants above their effect on the disease. Some of these problems can be solved by testing the effect of fungicides on the yield of a crop under disease free conditions and using any positive or negative effect as a covariate in the analysis.

Genetic Manipulation

Another fairly common research method in yield loss experimentation is the use of genetic differences to study the effects of disease severity on yield. Sometimes cultivar comparisons can be made, however, the complications of vastly different genetics between cultivars limits the conclusions, as cultivars may have different yield potential due to their different genetic background. In addition, genotype x environment interactions may affect the yield producing process in different cultivars. For that reason, isogenic lines of the same cultivar can sometimes be used in comparisons to study the disease-yield loss relationship (29, 33). Two criticisms are made of this method. First, isogenic lines quite commonly result in an all versus nothing comparison. This tells little about modest amounts of disease which are important to commercial crop production. second, critics argue that isogenic lines are often far from truly isogenic. Gene linkages carried along during back crossing may provide a portion of the measured differences.

Miscellaneous Methods

Other methods that are used to study yield loss plant disease relationships include mutilation of portions of the plant or removal of entire plants to simulate the effects of disease on yield (14, 31). Compensation by the plants, however, may take place, and losses in yield may not be proportionate to the percentage of missing plants or plant parts. Also, removal of plant parts may result in an injurious effect on the metabolism of plants and may not have the same impact on yield as defoliation caused by plant pathogens.

Healthy seeds are sometimes mixed in different proportions with seeds naturally infected with plant pathogens in order to obtain different levels of diseases (6, 28). Shannon & Mulchi (30) used growth chambers to study the effect of ozone on wheat yield. Different types of injury (e.g. by caustic acid, scissors, or flails) are inflicted on plants to measure the plant's response and to draw conclusions about how those plants may have responded to a diseases causing similar injury. Many questions can be raised about the extendability of such information to a plant disease relationship. Very little of this type of research is being done today.

MULTIPLE PEST MODELS

The design of experiments to study multiple pest impact on crop yield is now receiving more attention (13, 23, 24, 37). Recently, it has been realized that crop yield under commercial conditions is often influenced by multiple stresses (38, 39). Plant yield losses may be the consequence of the effects of several agents such as fungi, bacteria, viruses, mycoplasmas, nematodes, insects, weeds, and abiotic factors. These factors interact with each other and may reflect only one aspect of the complexity in the crop ecosystem. Factorial experiments can be used to determine if significant interactions exist. Many levels of each factor are desirable, but as the number of levels and factors increases, the size of the experiment quickly becomes large and may become unmanageable. Additionally, more than three factor interactions are hard to interpret. As a general rule, 2^n factorial experiments are valuable in exploratory approaches to a yield loss problem, where "n" denotes number of treatment factors with two levels of each factor. Once the important factors are determined, expanded experiments can then be conducted with two to three factors at several levels.

Research on multiple pest impact on crops requires close attention to the methods used in order to control one pest in a way that will not interfere with the methods used to modulate another pest. For example, some fungicides are known to have insecticidal activity. Other fungicides are thought to enhance some non-target plant pathogens. In those plots where insect feeding is heavy, weed growth tends to become more luxuriant. Leaf area lost to insect feeding is thus not available for plant disease development. Great care must be given to the details of a multiple pest yield loss experiment.

CONCLUDING REMARKS

Research methodologies in yield loss assessment are, at present, very inadequate for the needs of this important activity. Virtually any method that is used in yield loss assessment is subject to criticism. Researchers must recognize and deal with this reality.

Inasmuch as the research needs to be done even though the methodologies are flawed, there is no choice but to work to develop better methods for yield loss assessment. Until those methods are developed, however, researchers must continue to conduct investigations in the face of criticism. Even with the limitations of existing methods, the knowledge that is obtained is worthwhile and will continue to be a significant contribution to our understanding of crop yield losses.

LITERATURE CITED

1. Agrios, G.N., Walker, M.E., Ferro, D.N. 1985. Effects of cucumber mosaic virus inoculation at successive weekly intervals on growth and yield of pepper (Capsicum annum) plants. Plant Dis. 69:52-55.
2. Ashworth, L.J., Jr., McCutcheon, O.D., George, A.G. 1972. Verticillium albo-atrum: the quantitative relationship between inoculum density and infection of cotton. Phytopathology 62:901-903.
3. Ashworth, L.J., Jr., Huysman, O.C., Weinhold, A.R., Hancock, J.G. 1981. Estimating yield losses caused by soil-borne fungi. In Crop Loss Assessment Methods - Supplement 3, ed. L. Chiarappa, pp. 91-95.Rome: FAO. 123 pp.
4. Backman, P.A., Crawford, M.A. 1984. Relationship between yield loss and severity of early and late leaf spot diseases of peanut. Phytopathology 74:1101-1103.
5. Barker, K.R., Shoemaker, P.B., Nelson, L.A. 1976. Relationship of initial population densities of Meloidogyne incognita and M. hapla to yield of tomato. J. Nematol. 8:232-239.
6. Bockus, W.W., Sim, T., IV 1982. Quantifying Cephalosporium stripe disease severity on winter wheat. Phytopathology 72:493-495.
7. Burleigh, J.R., Eversmeyer, M.G., Roelfs, A.P. 1972. Development of linear equations for predicting wheat leaf rust. Phytopathology 62:947-953.
8. Cochran, W.G., Cox, G.M. 1957. Experimental Design, 2nd ed. New York: John Wiley and Sons, Inc. 617 pp.
9. Cohen, Y., Sherman, Y. 1977. The role of airborne conidia in epiphytotics of Sclerospora sorghi on sweet corn. Phytopathology 67:515-512.
10. Di Vito, M., Greco, N., Carella, A. 1985. Population densities of Meloidogyne incognita and yield of Capsicum annum. J. Nematol. 17:45-49.
11. Ferris, H. 1984. Nematode damage functions: The problems of experimental and sampling error. J. Nematol. 16:1-9.

12. Griffiths, E., Jones, D.G., Valentine, M. 1973. Effects of powdery mildew at different growth stages on grain yield of barley. Ann. Appl. Biol. 80:343-349.

13. Harper, A.M., Atkinson, T.G., Smith, A.D. 1976. Effect of _Rhopalosiphum padi_ and barley yellow dwarf virus on forage yield and quality of barley and oats. J. Econ. Entomol. 69:383-385.

14. Hirst, J.M., Hide, G.A., Stedman, O.J., Griffith, R.L. 1973. Yield compensation in gappy potato crops and methods to measure effect of fungi pathogenic on seed tubers. Ann. Appl. Biol. 73:143-150.

15. James, W.C., Shih, C.S., Hodgson, W.A., Callbeck, L.C. 1972. The quantitative relationship between late blight of potato and loss in tuber yield. Phytopathology 62:92-96.

16. James, W.C., Shih, C.S., Callbeck, L.C., Hodgson, W.A. 1973. Interplot interference in field experiments with late blight of potato (_Phyophthora infestans_). Phytopathology 63:1269-1275.

17. James, W.C. 1974. Assessment of plant diseases and losses. Annu. Rev. Phytopathol. 12:27-48.

18. Jenkins, J.E.E., Melville, S.C., Jemmett, J.L. 1972. The effect of fungicide on leaf diseases and on yield in spring barley in South-West England. Plant Pathol. 21:49-58.

19. Jenkyn, J.F., Bainbridge, A. 1974. Disease gradients and small experiments on barley mildew. Ann. Appl. Biol. 76:269-279.

20. Johnson, C.S., Main, C.E., Gooding, G.V., Jr. 1983. Crop loss assessment for flue-cured tobacco cultivars infected with tobacco mosaic virus. Plant Dis. 67:881-885.

21. Madden, L.V. 1983. Measuring and modeling crop losses at the field level. Phytopathology 73:1591-1596.

22. Melville, S.C., Jemmett, J.L. 1971. The effect of glume blotch on the yield of winter wheat. Plant Pathol. 20:14-17.

23. Noling, J.W., Bird, G.W., Grafius, E.J. 1984. Joint influence of _Pratylenchus penetrans_ (Nematoda) and _Leptinotarsa decemlineata_ (Insecta) on _Solamun tuberosum_ productivity and pest population dynamics. J. Nematol. 16:230-234.

24. Prabhu, A.S., Singh, A. 1975. Appraisal of yield loss in wheat due to foliage diseases caused by _Alternaria triticina_ and _Helminthosporium sativum_. Indian Phytopathol. 27:632-634.

25. Pratt, R.G., Janke, G.D. 1978. Oospores of _Sclerospera sorghi_ in soils of south Texas and their relationship to the incidence of downy mildew in grain sorghum. Phytopathology 68:1600-1605.

26. Reddy, A.P.K., MacKenzie, D.R., Rouse, D.I., Rao, A.V. 1979. Relationship of bacterial leaf blight severity to grain yield of rice. Phytopathology 69:967-969.

27. Romig, R.W., Calpouzous, L. 1970. The relationship between stem rust and loss in yield of spring wheat. Phytopathology 60:1801-1805.

28. Schaad, N.W., Sitterly, W.R., Humayadan, H. 1980. Relationship of incidence of seed borne Xanthomonas campestris to blackrot of crucifers. Plant Dis. 64:91-92.

29. Schaller, C.W. 1951. The effect of mildew and scald infection on yield and quality of barley. Agron. J. 43:183-188.

30. Shannon, J.G., Mulchi, C.L. 1974. Ozone damage to wheat varieties at anthesis. Crop Sci. 14:335-337.

31. Schneider, R.W., Williams, R.J., Sinclair, J.B. 1976. _Cercospora_ leaf spot of cowpea: Models for estimating yield loss. Phytopathology 66:384-388.

32. Seinhorst, J.W. 1981. Methods for generating diffent pest/yield relationship - (Nematodes). In Crop Loss Assessment Methods-Supplement 3, ed. L. Chiarappa, pp. 85-89. Rome: FAO. 123 pp.

33. Slinkard, A.E., Elliot, F.C. 1954. The effect of bunt incidence on the yield of wheat in eastern Washington. Agron. J. 46:439-41.

34. Slope, D.B., Etheridge, J. 1971. Grain yield and incidence of take-all (_Ophiobolus graminis_ sacc.) in wheat grown in different crop sequences. Ann. Appl. Biol. 67:13-22.

35. Teng, P.S. 1981. Use of regression analysis for developing crop loss models. In Crop Loss Assessment Methods - Supplement 3, ed. L. Chiappara, pp. 51-55. Rome: FAO. 123 pp.

36. VanderPlank, J.E. 1975. Principles of plant infection. New York: Academic Press, Inc. 216 pp.

37. Wallace, H.R. 1983. Interaction between nematodes and other factors on plants. J. Nematol. 15:221-227.

38. Weise, M.V. 1980. Comprehensive and systematic assessment of crop yield determinants. In Proceedings, E.C. Stakman Commemorative Symposium on Assessment of Losses Which Constrain Production and Crop Improvement in Agriculture and Forestry, pp.262-269. Misc. Pub. no. 7, Univ. of Minn. Agric. Exp. Stn., St. Paul, Minn. 327 pp.

39. Weise, M.V. 1982. Crop management by comprehensive appraisal of yield determining variables. Annu. Rev. Phytopathol. 20:419-32.

CHAPTER 10

METHODS OF STUDYING THE RELATION BETWEEN DIFFERENT INSECT POPULATION LEVELS, DAMAGE AND YIELD IN EXPERIMENTS AND SURVEYS

P. T. Walker

formerly with Tropical Development and Research Institute, London.
10 Cambridge Road, Salisbury, Wiltshire
SP1 3BW, England

Pest management must have a quantitative base on which to make decisions about chemical, cultural or biological methods of controlling pest attack. Economic decisions require economic data on production and the benefits to be obtained with different amounts of pest attack. The relation between crop yield (y) and pest density or amount of damage (i) can be modeled simply as a function:

$$y = f(i).$$

The assessment of (i) _____ in Chapter 3. Here we look at methods of studying the relati_____ (i) by experiment and survey.

_____LDS

Crop_____ the amount of harvestable economic product_____ fruit, tubers, sugarcane, hay etc., or a_____ugar or protein. The amount can be_____ tree, cane or bunch, but yield is c_____ as area: kilogram/hectare, tons/ac_____r of grains, stems, tillers, plants o_____xpressed per unit of area. Grams per_____tudy of yield should also examine the

Production_____yield per hectare and the area of prod_____quantity, and should be examined and_____roduce may be found by reference to_____nt tests, or such special tests as_____fiber length of cotton. The area o_____ial sampling or survey methods speci_____

Methods of measurin_____sts and experts in the crop itself shoul_____een developed for sowing, harvesting, thresh_____d taking crop yields in censuses, as discussed b_____3), Simaika (35) and in AAB (1). Methods for rice a_____(13) and Khosla (19), and for potatoes by Bastiman et a_____be measured by capacitance, timber by girth calipers, sugar_____, and yields can be estimated by remote sensing from aircraft or sa_____estimates can be very accurate and should not be neglected. The condi_____of the yield, for example, moisture content, whether dehusked, shelled or cleaned is important, as well as the maturity. Variation in these factors can be greater than any variation due to pest attack.

HOW TO ESTABLISH THE RELATION BETWEEN INFESTATION AND YIELD

If yield is measured at a single pest density or damage intensity, no information is available about the effects on yield of lower or higher levels. Two points can show a trend, but even three give no indication whether yields rise or fall between the points. Yields at a wide range of pest or damage levels are required. How are these obtained? There are reviews by Pradhan (27), Chiarappa (9), Judenko (18), Bardner & Fletcher (3), Walker (43, 44) and Singh & Khosla (36) on the topic.

Yields of single plants, paired plants or of several plants, with and without pests or damage, or with different amounts of attack, can be taken at random or in some recognized design such as blocks or latin squares. Plants may be marked in some way, for example with plastic tags. If single plants are used, the proportion of attacked plants in the population must be known to calculate the loss per area. This method has the advantage that plants can be protected and other variables eliminated. The disadvantage is that no account of compensation and increased yield by unattacked plants for those attacked near them is taken. The pattern of attack is unnatural, and yield loss so obtained should be compared with assessments on larger, more natural areas of attack.

Losses in harvested or stored produce may be assessed by comparing the weight of attacked and unattacked grains, seeds, fruit etc. Again, loss per attacked grain, per 1000 grains or per known weight or volume will give the total loss if multiplied up by the percentage of grains attacked, the total weight or volume of produce.

Natural Infestation

This is often used to give a range of pest densities or attack intensities, either in plants, plots, fields or larger areas. The infestation is assessed as in Chapter 3 and yields taken. In some crops, such as cocao, yields may be related to the infestation the year before. Sometimes, infestation rate can be measured at harvest, for example with sugarcane borer or Hessian fly on wheat.

Lim (22) measured loss in rice due to Nilaparvata on naturally infested and uninfested crops, and Rogers (31) assessed loss in sunflower due to natural infestations of bud moth using paired plants, eliminating some variation by stratifying by flower-head diameter. The advantages of this method are that plant responses are natural and without the effects of chemicals, artificial infestation or damage. The disadvantages are the natural variation due to growing conditions, pest complexes and the uncertainty of infestation levels. Partitioning and stratifying the material may help to account for and avoid such variation.

Use of Chemicals

As with diseases, birds, rodents, nematodes and weeds, chemicals (in this case insecticides or acaricides) are often used to obtain a range of pest densities or damage intensities to relate to yields. Chemicals may be combined with artificial infestation or caging to regulate pest attack.

Preliminary response trials are often needed to find which pesticide or technique should be used to obtain different pest attacks. For example, complete control will be needed to find the maximum yield. Different pesticides can be used, at a range of concentrations or number of treatments, perhaps at different times or stages of crop life or of the pest cycle. Crop loss data are sometimes secondary to tests of pesticide action, but no less valuable for that. If there

CHAPTER 10

METHODS OF STUDYING THE RELATION BETWEEN DIFFERENT INSECT POPULATION LEVELS, DAMAGE AND YIELD IN EXPERIMENTS AND SURVEYS

P. T. Walker

formerly with Tropical Development and Research Institute, London.
10 Cambridge Road, Salisbury, Wiltshire
SP1 3BW, England

Pest management must have a quantitative base on which to make decisions about chemical, cultural or biological methods of controlling pest attack. Economic decisions require economic data on production and the benefits to be obtained with different amounts of pest attack. The relation between crop yield (y) and pest density or amount of damage (i) can be modeled simply as a function:

$$y = f(i).$$

The assessment of (i) was examined in Chapter 3. Here we look at methods of studying the relation between (y) and (i) by experiment and survey.

YIELDS

Crop yield may be a general term for the amount of harvestable economic product, either as directly harvested grain, fruit, tubers, sugarcane, hay etc., or as a processed product such as flour, juice, sugar or protein. The amount can be expressed per unit of crop, such as per plant, tree, cane or bunch, but yield is commonly given in terms of a constant base such as area: kilogram/hectare, tons/acre. Experiments must give details of the number of grains, stems, tillers, plants or canes per hectare to enable yields to be expressed per unit of area. Grams per plot are meaningless to the farmer. A full study of yield should also examine the components of yield.

Production is total yield, and can be derived from the yield per hectare and the area of production. Quality is usually as important as quantity, and should be examined and quoted in trials and surveys. The grade of produce may be found by reference to marketing size grades, color, taste or constituent tests, or such special tests as baking quality of flour, oil content of seed or fiber length of cotton. The area of production must be measured, sometimes by special sampling or survey methods specific to the crop (35).

Methods of measuring yield are often specialized, and agronomists and experts in the crop itself should be consulted. Special methods have been developed for sowing, harvesting, threshing and processing field trials and taking crop yields in censuses, as discussed by Dyke, (10), Little & Hills (23), Simaika (35) and in AAB (1). Methods for rice are given by Gomez & Gomez (13) and Khosla (19), and for potatoes by Bastiman et al. (4). Pasture can be measured by capacitance, timber by girth calipers, sugar by optical methods, and yields can be estimated by remote sensing from aircraft or satellite. Eye estimates can be very accurate and should not be neglected. The condition of the yield, for example, moisture content, whether dehusked, shelled or cleaned is important, as well as the maturity.   Variation in these factors can be greater than any variation due to pest attack.

HOW TO ESTABLISH THE RELATION BETWEEN INFESTATION AND YIELD

If yield is measured at a single pest density or damage intensity, no information is available about the effects on yield of lower or higher levels. Two points can show a trend, but even three give no indication whether yields rise or fall between the points.   Yields at a wide range of pest or damage levels are required.   How are these obtained?   There are reviews by Pradhan (27), Chiarappa (9), Judenko (18), Bardner & Fletcher (3), Walker (43, 44) and Singh & Khosla (36) on the topic.

Yields of single plants, paired plants or of several plants, with and without pests or damage, or with different amounts of attack, can be taken at random or in some recognized design such as blocks or latin squares.   Plants may be marked in some way, for example with plastic tags.   If single plants are used, the proportion of attacked plants in the population must be known to calculate the loss per area.   This method has the advantage that plants can be protected and other variables eliminated.   The disadvantage is that no account of compensation and increased yield by unattacked plants for those attacked near them is taken. The pattern of attack is unnatural, and yield loss so obtained should be compared with assessments on larger, more natural areas of attack.

Losses in harvested or stored produce may be assessed by comparing the weight of attacked and unattacked grains, seeds, fruit etc.   Again, loss per attacked grain, per 1000 grains or per known weight or volume will give the total loss if multiplied up by the percentage of grains attacked, the total weight or volume of produce.

Natural Infestation

This is often used to give a range of pest densities or attack intensities, either in plants, plots, fields or larger areas.   The infestation is assessed as in Chapter 3 and yields taken.   In some crops, such as cocao, yields may be related to the infestation the year before.   Sometimes, infestation rate can be measured at harvest, for example with sugarcane borer or Hessian fly on wheat.

Lim (22) measured loss in rice due to <u>Nilaparvata</u> on naturally infested and uninfested crops, and Rogers (31) assessed loss in sunflower due to natural infestations of bud moth using paired plants, eliminating some variation by stratifying by flower-head diameter.   The advantages of this method are that plant responses are natural and without the effects of chemicals, artificial infestation or damage.   The disadvantages are the natural variation due to growing conditions, pest complexes and the uncertainty of infestation levels.   Partitioning and stratifying the material may help to account for and avoid such variation.

Use of Chemicals

As with diseases, birds, rodents, nematodes and weeds, chemicals (in this case insecticides or acaricides) are often used to obtain a range of pest densities or damage intensities to relate to yields.   Chemicals may be combined with artificial infestation or caging to regulate pest attack.

Preliminary response trials are often needed to find which pesticide or technique should be used to obtain different pest attacks.   For example, complete control will be needed to find the maximum yield.   Different pesticides can be used, at a range of concentrations or number of treatments, perhaps at different times or stages of crop life or of the pest cycle.   Crop loss data are sometimes secondary to tests of pesticide action, but no less valuable for that.   If there is a complex of different pests, their effects on yield can be separated by physical methods, caging, timing, etc., but also by using different pesticides, systemic for sucking pests, acaricides for mites, specific lepidoptericides such as Bacillus thuringiensis or virus, or specific soil pests. Wilson et al. (48) used five different insecticides at different times to measure the effects of three different insects on alfalfa yield.

The advantage of the method is that pest populations can be fairly accurately controlled. On the other hand, pesticides such as carbofuran are known to stimulate plant yield, and may also control nematodes or other pests which are reducing yield. The pesticide or its solvent may well be phytotoxic if carelessly applied, or may affect parasitoids indirectly influencing final yield. Pesticides are also often the cause of interplot interference due to drift or other effects.

Artificial Infestation

Pest attack is often increased by putting known numbers of eggs, larvae or adults on the crop. Cages are sometimes used to enclose the population introduced, or to exclude natural populations. Barriers, such as plastic or metal walls may be used to confine the population, for example of cutworm caterpillars, of the experimental plots, or metal containers to hold soil beetle larvae. The natural infestation may need to be removed, as with cabbage butterfly eggs or cereal borer eggs, or by spraying a cage with a knockdown and non-persistent insecticide.

Infestation rates introduced should cover a range of population levels which give the maximum range of responses, either arithmetic: 0, 2, 4, 6, or geometric: 0, 2, 4, 8, or 0, 2, 20, 200. The population should be monitored because the pests may not survive or become established. Special methods of dispensing, for example maize stem borer eggs, Ostrinia have been used. There are many examples: Prasad (28) infested cabbage with Pieris eggs, Todd et al. (41) infested sorghum with Nezara bugs in cages, and Hall & Teetes (15) put different densities of four kinds of bugs on sorghum panicles to assess yield losses.

The spatial distribution pattern of pests is important, both to satisfy biological requirements and to avoid unnatural compensation effects on yield. Fery et al. (11) distributed Heliothis eggs uniformly on tomato, and Lynch et al. (24) distributed Ostrinia eggs on maize in a Poisson distribution.

One disadvantage of the method is the need to collect or breed the pests for infestation, which may be difficult, labor-intensive or costly. Another is the difficulty of timing infestation, often critical, so that it is the same as natural attack. When using cages, the cage may affect the plants and the yield, and these effects should be checked and corrected for. For example, both infested and uninfested plants should be caged, or closed and open cages used (38), or cages removed as soon as possible (46). Light and air movement may be reduced in cages, but not temperature and humidity (45). Catling (8) used floating cages to study losses in deep water rice due to stem borers. The mesh size of cages is important, particularly if birds are to be included or excluded from trials (6). A more natural method is to remove eggs, larvae or adult pests. Judenko (18) removed Pieris eggs from cabbage, and cereal stem borer egg masses can easily be removed. Light, pheromone or other traps can be used to remove populations, although they may have the opposite effect.

Alternatively, pest attack can be increased by spreading infested material in the crop, for example sorghum infested with shoot fly, wheat with Hessian fly, or cereals with lepidopterous stem borers. Such trials may not be popular with farmers or research station managers. Attractant materials such as fish meal can be used to increase sorghum shoot fly infestation.

Simulated Damage

Plants may be damaged artificially to imitate the effects of pest attack on yield. Whole plants can be removed at random or in a regular or grouped pattern, to resemble the attack of the pest. Flowering, fruiting or seed heads can be removed, leaves removed or cut (26), stems damaged or roots cut. There are many examples of simulated damage to beans, cassava, maize and other cereals, sugarbeet and other crops. The advantage of the method is exact control of the amount of damage, which should be measured as leaf area, root length or dry weight, etc. removed or remaining. The time of damage in relation to the crop growth stage or growing conditions is also controlled. It may be difficult, however to damage known amounts of stem or the growing point of roots.

Although the relation between the amount of damage and yield may be easily found, the relation between different pest densities and damage is more difficult because different stages of the pest may be present, they may attack for different lengths of time, and there may be other factors such as temperature and rainfall operating on the pests and the crop. Nevertheless the relation between pests, damage and yield should be established.

The stage of crop growth damaged may be critical, needing a knowledge of the morphology and physiology of yield production, for example, the position of the growing point in cereals in relation to where and when the plant is damaged, the method of storage of sugar in sugarcane in relation to stem borer attack, or the contribution of the flag leaf to grain filling in cereals. These factors will affect the actual shape of the yield-infestation response curve. A study of the effects of pests on the components of yield is thus very important, as for example, was done on wheat (50) and sorghum (47).

Resistant Varieties

Yield loss due to pests can be assessed by comparing the yields of varieties of crop that are susceptible with those resistant to pest attack. Yield differences between varieties in the absence of pests are assumed to be small, or at least known and taken into account.

Some varieties are tolerant, that is produce a yield when attacked by pests, which confuses the situation. Schoonhoven & Pena (32) studied losses in cassava due to thrips by this method, and Harvey & Hackecroft (16) used resistant varieties of sorghum with pesticides to study losses due to aphid.

INTERPLOT INTERFERENCE

However different pest populations are obtained, the treatment of one plot may interfere with the population in other plots nearby. Pesticide may drift in wind and repellency has been mentioned. If the pests are mobile or small and airborne, they will migrate or be carried from one plot to the next, altering the populations intended and making the loss measurements useless or misleading. Similar situations occur in plant disease trials, which are discussed in Chapter 8.

Two common situations are that low populations in one plot act as a "sink" into which pests migrate, resulting in higher populations in the treated plot and lower pest densities in the untreated. If pests cause damage before dying, there may be more damage in the treated than in the untreated plots. Alternatively, if a treatment repels pests, there may be more in the untreated plots then expected.

To prevent interference, larger plots can be used, screened with hessian or burlap, with bushes such as pigeon pea or with other crops, or plots simply placed some distance apart; but the greater the distance between plots, the greater the variation due to soil, climate or other effects. Joyce & Roberts (17) and Reed (29) give examples of planthoppers in the cotton crop, and planthoppers behave similarly in small plots of rice. The problem is also discussed in AAB (1).

LOSS SURVEYS

The type and extent of a survey will depend on its purpose, which may simply be to find the types of loss and their main causes, to find the distribution of losses in different areas or types of farming, or to actually evaluate losses with a view to forecasting crop production or justifying control measures (43, 44). There are several approaches, depending on the variability of pest attack and agriculture, the degree of accuracy required and the resources available.

Direct Loss Surveys

When the distribution of pest attack and types of farming are very variable, and no relation between yield and infestation has been found, actual crop-cutting surveys can be made, yields in areas of different infestation levels measured, averaged, and loss assessed by comparison with the yields expected in the absence of pest attack. Yields in areas of different infestation level should be weighted by the area sampled, that is multiplied by the area before dividing the totals by the total area of each crop. This prevents undue weight being placed on small areas of high infestation or uncommon amounts. Any of the methods above can be used to obtain different amounts of infestation, for example by treating half of each plot with insecticides.

Such surveys have been done in India, on rice by Abraham et al. (2), Seth et al. (33, 34), on maize by Kishen (20) and Singh et al. (37), and on wheat by Kishem et al. (21). Catling et al. (8) surveyed rice losses in Bangladesh, George & Gair (12) loss in wheat by Sitobian aphid in England, and Wood et al. (49) losses in yams and maize caused by termites in Nigeria. Crop yield surveys often done by agricultural census teams can provide a model for loss surveys. In all types of survey, statistical advice on the sampling plan, number and size of samples and confidence limits is essential.

Surveys of Infestation

If there is a reliable model of the yield-infestation relationship, and the infestation and types of farming are fairly uniform, a survey of the pest infestation, as pest density or amount of damage, can be used to assess average yields, and from these, losses. This approach is often quicker and cheaper than sampling yields.

Another form of yield-infestation relationship is a yield loss conversion factor (CF), previously found by experiment, with a different factor for each grade of attack, e.g., heavy = 0.7, moderate = 0.5, and light = 0.2. The average factor, after weighting for the area or number of samples of each grade found, is used to give the actual yield as a proportion of maximum yield 1.0, and hence loss. The factors can be varied according to season, locality or crop variety (5). A similar approach is a "coefficient of harmfulness" (18). Strickland (39) surveyed the loss of Brussels sprouts due to _Brevicoryne_ aphid in the UK, losses due to the corn borer, _Ostrinia_ have been surveyed in the USA using a factor of 3% loss per borer per plant (42), and Grace (14) surveyed sugar losses due to the borer _Diatraea_ in Brazil by this method.

The probability of loss, or risk factor, can be incorporated in loss surveys if the probability of pest attack, the number of attacks over several years, is known. Bullen (7) used a crop vulnerability index to assess the potential damage to crops by the desert locust. The probability of loss can be given as maps of iso-risk and iso-loss (30).

Comparison of Annual Yields

If there is no information on losses from experiments or surveys, losses due to a pest may be estimated by comparing crop production from an area attacked by a pest, or from years when attack is severe with places where there is little attack or years when attack is absent, or before and after attack appeared. Nichols (25) used a "normal year" method. Correction or allowance must be made for other causes of yield variation, such as climate, pesticide use or attack by other pests or diseases. This approach was used to supplement other information on losses of cassava in Africa due to green mite and mealybug (Walker, unpublished).

Use of Economic Indicators

Losses due to pests can also be estimated by examining the economic effects of pest attack. These include a rise in market price of the crop from other areas or countries, increased price and scarcity of seed or planting material, or change to alternative crops or products for food or sale. Again, the effects of factors other than pests must be considered. Losses due to pests are an economic, as well as a biological problem.

CONCLUSION

Other examples of surveys of crop loss and the economic aspects of crop loss and economic thresholds are given in Chapters 12, 18, and 20. Crop loss assessment is often difficult and the results unsatisfying, and it may be asked whether information with a wide margin of error is worth having. Information can always be improved, and the alternative is none at all. Crop loss information remains the essential basis for any economic assessment of the effect of pests on yield.

LITERATURE CITED

1. AAB. 1985. Field trial methods and data handling. Aspects of Applied Biology 10. Associate of Applied Biologists, Wellesbourne, Warwick, U.K. 528 pp.
2. Abraham, T.P., Khosla, R.K. 1967. Assessment of losses due to incidence of pests and diseases in rice. J. Indian Soc. Stasticians. 19:69-82.
3. Bardner, R., Fletcher, K.E. 1974. Insect infestations and their effects on the growth and yield of field crops, a review. Bull. Entomol. Res. 64:141-160.
4. Bastiman, B., Bevis, A.J., Wellings, L.W. 1985. Methods for measuring potato crops. See Ref 1, pp. 199-212.
5. Basu, P.K. 1978. A yield loss conversion factor for peas moderately affected by fusarium root rot. Can. Plant Dis. Survey. 58:5-8.
6. Bruggers, R.L., Ruelle, P. 1982. Efficacy of nets and fibres for protecting crops from grain-eating birds. Crop. Protection. 1:55-65.
7. Bullen, F.T. 1969. The distribution of the damage potential of the desert locust, Schistocerca gregaria. Antilocust Memoir 10. Centre for Overseas Pest Research, Tropical Development & Research Inst., London. 72 pp.
8. Catling, H.D., Shamsul Alam, Miah, S.A. 1978. Assessing losses in rice to insects and diseases in Bangladesh. Exp. Agric. 14:277-287.
9. Chiarappa, L. 1971. Crop Loss Assessment Methods: FAO manual on the evaluation and prevention of losses by pests, diseases and weeds. FAO and Commonw. Agric. Bur., Farnham Royal, UK (loose-leaf + supplements). 123 pp.
10. Dyke, G.V. 1974. Comparitive Experiments with Field Crops. London: Butterworths. 209 pp.
11. Fery, R.L., Cuthbert, F.P, Jr., Perkins, W.D. 1979. Artificial infestation of the tomato with eggs of the romato fruitworm. J. Econ. Ent. 72:392-394.
12. George, K.S., Gair, R. 1979. Crop loss assessment on winter wheat attacked by the grain apgid, Sitobion avenae (F), 1974-7. Plant Pathol. 28:143-149.
13. Gomez, K.A., Gomez, A.A. 1983. Statistical Procedures for Agricultural Research. 2nd ed. London and New York: Wiley. 688 pp.

14. Graca, L.R. 1976. Estimative economica dos prejulzos causados pelo complexo broca-prodridoes na cana-de-acucar no Brasil. Bras. Acucar. 88:12-34.

15. Hall, D.G., Teetes, G.L. 1982. Yield loss-density relationships of four species of panicle-feeding bugs in sorghum. Environ. Entomol. 11:738-741.

16. Harvey, T.L., Hackerott, H.L. 1974. Effects of greenbugs on resistant and susceptible sorghum seedlings in the field. J. Econ. Entomol. 67:377-380.

17. Joyce, R.J.V., Roberts, P. 1959, The determination of size of plot suitable for cotton spraying experiments in the Sudan Gezira. Ann. Appl. Biol. 47:287-305.

18. Judenko, E. 1973. Analytical method for assessing field losses caused by pests on cereals with and without pesticides. Bull. 2. Centre for Overseas Pest Res., Tropical Development & Res. Inst., London. 31 pp.

19. Khosla, R.K. 1977. Techniques for assessment of losses due to pest and diseases of rice. Ind. J. Agric. Sci. 47:171-174.

20. Kishen, K., Sardana, M.G., Khosla, R.K., Dube, R.C. 1970. Field loss caused by pests and diseases in the maize crop. Agric. Situation in Ind. 25:591-593.

21. Kishen, K., Sardana, M.G., Khosla, R.K., Dube, R.C. 1972. Estimates of the incidence of pests and diseases and consequent yield losses in wheat. Ind. J. Agric. Sci. 42:908-912.

22. Lim, G.S. 1980. Brown planthopper outbreaks and associated yield losses in Malaysia. Intnl. Rice Res. Newsletter 5:15-16.

23. Little, T.M., Hills, F.J. 1978. Agricultural Experimentation; Design and Analysis. New York: Wiley. 350 pp.

24. Lynch, R.E., Robinson, J.F., Berry, E.C. 1980. European corn borer: yield losses and damge resulting from simulated narural infestation. J. Econ. Entomol. 73:141-144.

25. Nichols, C.W. 1970. Compiling and reporting crop disease loss data by the "normal year" method. FAO Plant Prot. Bull. 18:25-28.

26. Poston, F.L., Pedigo, L.P. 1976. Simulation of painted lady and green cloverworm damage to soybeans. J. Econ. Ent. 69:423-426.

27. Pradhan, S. 1964. Assessment of losses caused by insect pests of crops and estimation of insect populations. In Entomology in India, pp. 17-58. Entomol. Soc. India Special number, New Delhi. 529 pp.

28. Prasad, S.K. 1961. Quantitaive estimation of damage to cabbage by cabbageworm. Ind. J. Entomol. 23:54-61.

29. Reed, W. 1972. Uses and abuses of unsprayed controls in spraying trials. Cotton Growing Rev. 49:67-72.

30. Rijsdijk, F.H., Zadoks, J.D. 1979. A data bank on crop losses. E.P.P.O. Bull. 9:297-303.

31. Rogers, C.E. 1979. Sunflower bud moth, _Suleima_: behavior and impact of larva on sunflower seed production in the Southern plains. Environ. Entomol. 8:113-116.

32. Schoonhoven, A.V., Pena, J.E. 1976. Estimation of yield losses in cassava following attack from thrips. J. Econ. Entomol. 69:514-516.

33. Seth, G.R., Sardana, M.G., Khosla, R.K., Kalyanaraman, V.M. 1969. Preharvest losses due to pests and diseases in rice. Ind. J. Agric. Sci. 39:1113-1124.

34. Seth, G.R., Sardana, M.G., Khosla, R.K. 1970. Assessment of loss in yield of paddy in Wast Gadavari district. Oryza. 7:1-12.

35. Simaika, J.B. 1982. Estimation of crop areas and yields in agricultural statistics. FAO Economic and Social Development Paper 22, Rome. 186 pp.

36. Singh, D., Khosla, R.K. 1983. Assessment and collection of data on preharvest food grain losses due to pests and diseases. FAO Economic and Social Development Paper 28, Rome. 127 pp.

37. Singh, D., Tyagi, B.N., Khosla, R.K., Avasthy, K.P. 1971, Estimate of the incidence of pests and diseases and consequent losses in yield of maize. Ind. J. Agric. Sci. 41:1094-1097.

38. Sparks, A.N., Chiang, H.C., Burkhardt, C.C., Fairchild, M.L., Weekman, G.T. 1966. Evaluation of the influence of predation on corn borer populations. J. Econ. Entomol. 59:104-107.

39. Strickland, A.H. 1957. Cabbage aphid assessment and damage in England and Wales, 1946-1955. Plant Pathol. 6:1-9.
40. Sylvester-Bradley, R., Grylls, J.P., Roebuck, J.F. 1985. Methods for measuring cereal crops. See Ref. 1, pp. 213-239.
41. Todd, J.W., Jellum, M.D., Leuck, D.B. 1973. Effects of southern green stink bug, Nezara, damage on fatty acid composition of soybean oil. Environ. Entomol. 2:685-689.
42. USDA 1977. Estimation of damage by E.C.B. to grain corn in U.S. in 1976. Coop. Plant Pest. Rep. 2:375-376.
43. Walker, P.T. 1983. The assessment of crop losses in cereals. Insect Sci. and Application 4:97-104.
44. Walker, P.T. 1983. Crop losses: the need to quantify the effects of pests, diseases and weeds on agricultural production, (a review). Agric. Ecosystems Environ. 9:119-158.
45. Way, J.M., Banks, C.J. 1968. Population studies on the active stages of the black beam aphis on its winter host. Ann. Appl. Biol. 62:177-197.
46. Webster, J.A., Smith, D.H. 1983. Cereal leaf beetle, Oulema population desities and winter wheat yields. Crop Prot. 2:431-436.
47. Williams, W.T., et al. 1977. Effect of survey date on growth and yield of three sorghum cultivars. Austral. J. Agric. Res. 28:381-387.
48. Wilson, M.C., Stewart, J.K., Vail, H.D. 1979. Full season impact of the alfalfa weevil, meadow spittlebug and potato leafhopper in an alfalfa field. J.Econ. Entomol. 72:830-834.
49. Wood, T.G., Smith, R.W., Johnson, R.A., Komolafe, P.O. 1980. Termite damage and crop loss studies in Nigeria: pre-harvest losses to yams due to termites and other soil pests. Trop. Pest Management 26:355-370.
50. Wratten, S.D., Redhead, P.C. 1976. Effects of cereal aphids on the growth of wheat. Ann Appl. Biol. 84:437-455.

CHAPTER 11

QUANTIFYING THE RELATIONSHIP BETWEEN DISEASE INTENSITY AND YIELD LOSS

P. S. Teng

Department of Plant Pathology,
University of Minnesota, St. Paul, Minnesota 55108, U.S.A.

The large amount of data generated in disease-loss experiments is difficult to interpret and use, unless it can be synthesized into a quantitative relationship, commonly a model (8). Having a model that captures the essence of how disease affects yield (and loss) will enable further activities such as the formulation of economic thresholds for control. The complexity of the disease-loss relationship is increased with multiple diseases, or when other pest and environmental effects are included as explanatory variables of yield loss. Without using modeling techniques, it is very difficult if not impossible to sort out multiple pest -- multiple environmental variable influences on crop yield.

While many statistical methods are available to quantify the disease-loss relationship, it is important that the underlying biology of the relationship not be sacrificed for statistical reasons. Furthermore, the objective of developing the quantitative relationship is a consideration influencing the type of model to be developed. A model may be considered any representation of a defined system in a form different from the original. Symbolic models, which represent a system or synthesize data about a system, using mathematics, are the most common in plant protection (24). When given known input(s), most models are capable of predicting an outcome with known confidence limits. In loss assessment, the input(s) are commonly disease parameters and the output yield or loss.

There are many approaches for quantifying the disease-loss relationship, some of which have previously been reviewed (7, 8, 18, 26, 31). This chapter will be concerned mainly with the empirical models derived using data generated from field experiments or surveys, although the use of coupled crop-pest models will also be discussed. Empirical models predominate in the disease-loss literature and are most widely used for loss assessment and disease management. Empirical disease-loss models may also be divided into linear and non-linear, according to the statistical technique used. There are many views on the shape of the functional relationship between disease and yield or loss, and this is reflected in the model fitted to the empirical data (26). Chapter 13 gives another viewpoint on disease-loss modeling and should be read in conjunction with this chapter.

LINEAR EMPIRICAL MODELS

The Regression Technique

The technique of least-squares regression fits statistical models based on minimal variance between the model and data (2, 25). Regression models may be univariate (with one independent, disease variable) or multivariate (with two or more independent variables, including non-disease variables). The independent variables are also called predictors. Commonly, the dependent variable, y, is yield or yield loss (%) while the independent disease variable, x, takes several forms, such as a single disease assessment value or an integral value (area under disease progress curve, AUDPC). With a single independent variable, the objective is to develop a model that significantly explains a relationship between y and x.

There are inherent assumptions in using regression analysis for developing disease-loss models (18, 25). In its strictest applications, the technique considers that the independent variables and their variances are "truly" independent; i.e. that the value at one time is not influenced by the value at a previous time. With disease progression data used in modeling, it is common to find serial correlation present, although most workers tend to ignore it (8). Regression also assumes that the data were derived from a normal distribution and have equal variance throughout the range of data values.

Regression disease-loss models are evaluated for significance of the overall relationship using the F-statistic (5). The significance of the regression coefficient (univariate models) or partial regression coefficients (multivariate models) is tested using either the t- or F- statistic. The coefficient of determination, r^2, of value 0 to 1 indicates the proportion of variance in yield or loss explained by the disease variable(s) used. James and Teng (8) considered this the most important statistic for judging field applicability of a model. The r^2 statistic is not tested for significance. Rather, the values of the statistic are deemed satisfactory mainly on subjective grounds. For example, a model which explains 80% of the variance in loss is considered a good model especially if the data used for regression have been derived from multiple locations, several cultivars and over 2-3 seasons. The statistical precision of a model is determined from the S-statistic, the standard error of estimate of the dependent variable, yield or loss. This enables the calculation of a confidence interval, with known probability, for each yield or loss estimate predicted by a regression model.

Single Predictor Models

In single predictor models, there is one disease variable (x) from which yield loss (y) is predicted. The disease variable has been represented in three ways (31, 37): a) as disease intensity at a single, often "critical" time in the crop's life, b) as the number of disease-free days before a predefined, critical injury threshold is reached and c) as an integral value of disease units over a specified period, the area under the disease progress curve (AUDPC).

The first type of single predictor model is exemplified by the equation for estimating loss caused by rice blast,

$$Y = 0.57 \ X,$$

where Y = percent yield loss, X = percent blasted nodes 30 days after heading (11). The second type of single predictor model is exemplified by the equation of Olofsson (17) for estimating loss caused by potato late blight,

$$Y = 234.0 - 1.706 \ X,$$

where Y = potato yield in tons/ha and X = number of blight-free days. Olofsson (17) defined a "critical injury threshold" of late blight severity, in which the period before this threshold was reached was considered blight free. The third type of single predictor model is exemplified by the wheat stem rust AUDPC model first proposed by VanderPlank (33) using the data of Kingsolver et al. (14). VanderPlank (33) showed that yield loss was approximately equal to 1.0 AUDPC for what stem rust on the graphical scales he used. AUDPC is quantified in arbitrary units dependent on the scales used for plotting disease progress curves.

Single point models are best developed for diseases of short duration which affect only one yield component such as grain weight (8). These models assume that loss can be estimated by knowing the amount of disease at just one point in a crop's growth, implying that the course of the disease epidemic before and after the point of disease assessment is relatively stable and predictable between seasons. Some workers have considered that there is a phase in a crop's lifetime when it is particularly sensitive to disease, i.e. a critical point (7). It is more logical, given the use of linear regression in developing most single point models, to simply consider the growth stages identified in the statistical analyses as being points giving the best fit between disease data and yield (loss) data, rather than assigning some physiological significance to them. With AUDPC models, many disease assessments are necessary to derive one AUDPC unit (see Chapter 5), but it is difficult to distinguish between short duration, high infection rate epidemics and long duration, low infection rate epidemics, since both may have comparable AUDPCs.

Multiple Predictor Models

Many diseases increase in severity at variable infection rates (see Chapter 5), making it difficult to estimate their effect on yield by making a single disease assessment. Multiple predictor models use two or more disease assessments (x's) to provide the predictors for estimating yield or loss (y). The disease predictors that have been used are either severity increments during a specified period such as a week (9) or actual severities at specific times (3, 33).

In the first type, James et al. (9) tested nine potato late blight severity increments per week, for weeks between August and October, as potential predictors for yield loss. If the weekly blight increments are denoted X1 to X9, then the generalized equation would have the form,

$$Y = f (X1, X2, X3, X4, X5, X6, X7, X8, X9).$$

The authors further developed two practical equations, one each for early and late infections. This enabled them to account for the highly variable rates of disease progression in Canada, and the effect of late blight during each weekly period on overall yield loss.

The second type of multiple predictor model is exemplified by work on barley leaf rust (33), in which the authors used rust severity data on leaves 1 and 2 at six growth stages for regression. The generalized equations may be represented as,

$$\text{Leaf 1} \quad Y = f (X1, X2, X3, X4, X5, X6),$$
$$\text{and Leaf 2} \quad Y = f (X7, X8, X9, X10, X11, X12),$$

where Y = percent yield loss. The six growth stages used, on the decimal scale (8), were 39/40 (flag leaf ligule/collar just visible - X1, X7), 49/50 (first awns/spikelets visible - X2, X8), 58/59 (emergence of inflorescence completed - X3, X9), 64/65 (anthesis halfway - X4, X10), 73/74 (early milk - X5, X11) and 83/84 (early dough - X6, X12). Various combinations of the X1 to X12 predictors were tested in a stepwise regression procedure to identify multiple point models. The model,

$$Y = - 0.07 - 4.35X3 + 3.26X9 + 0.72X4 - 1.82 X10 + 0.62X5 + 0.42X11 - 0.03X6 + 0.22X12$$

which used severity values from both leaves at four growth stages explained 97.6% of the loss variation, had a standard error of estimate of loss of 2.49% and was considered by the authors to be useful for use in future disease management systems (32).

A refinement of multiple predictor modeling was made by Seck (22), who quantified the influence of leaf position on severity-loss models for estimating loss due to wheat leaf rust. Seck (22) developed the concept of "Effective Severity" (ESV), in which the rust severity was corrected for leaf position by an empirically-determined multiplier,

$$ESV = K_i \quad x \quad DS_i,$$

where K_i = multiplier for leaf position i, DS_i = disease severity on leaf position i. The leaf rust multipliers were, respectively, 0.26, 0.12 and 0.03 for leaf positions 1, 2 and 3 from the top.

A hybrid of the above two types of multiple predictor models was developed for potato early blight by Teng and Bissonnette (27), in which both severity increments and actual severities were used as predictors in the same equation. These authors developed different equations for estimating yield loss in early and late maturing potato cultivars of the form,

Early maturing cultivars
    Y = 0.82 + 0.64 (% blight increment between days 56 and 66)
            + 0.61 (% blight increment between days 66 and 76)
            + 1.35 (% blight increment between days 76 and 86)
Late maturing cultivars
    Y = 2.18 - 4.77 (% blight on day 56) + 0.74 (% blight on day 76)
            + 0.57 (% blight on day 96).

The equations explained 75% and 70% of the variation in loss respectively and were used to estimate statewide losses due to the disease (28).

Linear Multiple Pest Models

Most of the published literature on pest-loss models estimate loss caused by single insects or diseases, even though this is an uncommon situation under field conditions (31). When models for single pests are used to estimate losses in the same field caused by several pests, there is a real danger of either overestimation or underestimation of the total loss. Walker (34) gives examples where the percent loss caused by multiple pests has exceeded 100 %. An early attempt to correct for multiple disease effects was made by Watts Padwick (35),

$$W \% = 100 \ x \ (1 - ((100-W_1) \ (100-W_2) \ (100-W_n))/ \ 100^n),$$

where W = total percent loss, W_1 - W_n = percent loss due to individual diseases.

In Chapter 8, it was noted that some workers have made use of sample survey techniques to generate the database for loss modeling. This recognizes the multiple pest situation as well as the influence of non-pest factors on crop yield. Pinstrup - Andersen et al. (20) reported a pioneering study on bean fields in Colombia, in which they derived an equation utilizing eight factors, which is generalized as follows,

    Yield loss = f (rainfall, % rust, % bacterial blight,
                    Empoasca, certified seed, variable costs,
                    % angular leafspot, plant population).

The authors also represented their equation as a "crop loss profile", showing pictorially how much attainable yield was depressed by each variable until the actual yield was reached (see Chapter 1 for discussion of yield concepts). As noted by Teng (26), one problem with the above equation was that it assumed each factor exerted an independent effect on yield.

An improvement of the Pinstrup - Andersen et al. (20) approach is the "synoptic" approach first proposed by Stynes (23) and subsequently improved by Wiese (36). This approach used principal components analysis to identify groups of variables which may be correlated in their effects on yield. Each group of variables was formed into a principal component and significant principal components were then used as the predictors in a loss equation. The technique of principal components analysis is discussed in Chapter 15. For field peas in Idaho, Wiese (36) developed a model of the form,

Yield = - 2722

+ 30.90	(days from seeding to maturity)
+ 44.13	(ppm soil phosphorus)
- 9.415	(% slope of field site * site location)
- 85.30	(soil moisture factor 1 * soil temperature factor)
- 46.21	(weed incidence factor)
+ 0.00296	(crop plants/sq.acre) X 10^6
- 21.38	(soil moisture factor 1 * soil moisture factor 2)
+ 83.67	(soil moisture factor 2 * soil temperature factor)
+ 38.31	(herbicide phytotoxicity score, 0-9 * weed incidence factor)
+ 11.46	(inches of topsoil)
+ 154.7	(% soil cover by residue of previous crop)
+ 0.1104	(Julian seeding date * days from seeding to maturity).

Although the model explained 82% of the variation in yield in one year, it could not accurately predict yield in another year. Also, no pest factor was significant enough to be included in the overall model. One possible reason is that sample surveys may not provide a wide enough range of values for each predictor variable, leading to an underestimation of the magnitude of the partial regression coefficient.

Field experiments have been used to provide data for modeling multiple pests and in contrast to sample surveys, enable the levels of each factor to be manipulated. King (13), working on potato, used factorial experiments to derive the model for cultivar Abnaki,

$$Y = 23.05 \ L1 + 325.46 \ LB1 - 1.54 \ L3 + 0.06 \ Q - 215.35 \ (L1*LB1),$$

where Y = percent yield loss, L1 = number of Colorado potato beetle (CPB) larvae per stem introduced in week 7, LB1 = late bight severity on day 73, L3 = number of CPB larvae per stem observed on day 94, Q = number of quackgrass plants per plot in week 17 and L1*LB1 is an interaction term. The model explained 84% of the loss variation.

Although the need for multiple pest models is recognized, experiments with multiple pests are very demanding and generally require more land. The use of fractional factorial designs and incomplete replication have been suggested as ways to make use of limited resources for multiple pest experiments (30).

Choice of Models

More than one empirical model can generally be developed from the same database. As such it is important to clearly determine the potential use of a model, and the criteria for judging which model(s) best suit the objectives of model development (26). With linear regression models, increasing the number of disease assessments commonly leads to an improvement in model fit (8). For example, with the barley leaf rust example discussed previously, the proportions of yield loss explained by X3 and X6 were 0.73 and 0.71 respectively when regressed alone, but was 0.82 when both were regressed together (32). The database available for modeling, and implicitly the number of disease assessments made during the growing season, also determines if a single or multiple predictor model can be developed. Much of the work with single tillers (Chapter 8) only allows the development of single predictor models. With multiple predictors, the sign (+/-) and size of the partial regression coefficient need to be considered in relation to the biology of yield loss. James et al. (9) selected multiple predictor models for potato late blight which had only positive coefficients and coefficients of approximately equal value while Burleigh et al. (3) included a negative term in their model for estimating wheat leaf rust losses. Gaunt (6) hypothesized that a negative coefficient is plausible in some crops such as the cereals, i.e. that low severities at certain growth stages may increase yield.

Most of the regression models used for estimating losses from surveys are single predictor models (12), often because resources only allow for one field visit. However, pest management activities in countries like the U.S.A. are increasingly generating sequential data that can be used in a multiple predictor model, e.g. Teng and Bissonnette (28) estimated potato early blight losses from blight severities assessed 3-4 times during a cropping season.

Although empirical models reflect the database from which they are developed, it is necessary to ensure that their use does not exceed the limitations of the original database. In particular, the disease values used as input to estimate loss should not be outside the range of the values used in regression modeling. If they do, then spurious loss estimates may result.

NON-LINEAR EMPIRICAL MODELS

Many biological responses to stress are known to be non-linear, and although the majority of pest-loss models in the literature are linear, this is more a reflection of available statistical techniques than of the actual real nature of the relationship. It has generally been more difficult to fit non-linear models to data; however the continuing improvement in computer software is reducing this limitation. Teng (26) has conceptualized nine possible shapes for the curve relating pest intensity to yield loss, ranging from linear to various S-shapes.

Models with Transformed Predictors

Some workers have transformed the disease variable and then used linear regression to derive a pest-loss model. This implicitly recognizes that the untransformed data show non-linearity and the transformation is used to "straighten" the curve. With wheat stem rust, Romig and Calpouzos (21) derived the equation,

$$Y = -25.53 + 27.17 \, Log_e X,$$

where Y = percent yield loss, X = percent stem rust severity at 3/4 full caryopsis. The model explained 99.3% of the loss variation. The barley mildew model of Large and Doling (15) used a square root transformation of severity as the predictor. Polynomial regression equations can also be derived by transforming the original disease variable, e.g.

Quardratic model $Y = f (X, X^2)$

Cubic model $Y = f (X, X^2, X^3)$

etc. Polynomial equations generally allow for an increase in the proportion of variation explained by the model, if the dataset shows a non-linear relationship when graphed. It is possible to fit a polynomial to almost any dataset, with the real danger of arriving at models with no biological meaning (24).

Weibull Distribution Models

The Weibull distribution function, a flexible model fitting many curve shapes, was modified by Madden et al. (19) for disease-losses. The generalized form of the equation may be written as,

$$L_i = 1 - \exp(-((X_i - d)/b)^S) + u_i ,$$

where L = crop loss proportion, X = disease intensity, d = disease severity below which no loss occurs (minimum threshold), b = slope parameter, s = shape parameter and u = unexplained variance. A single predictor model for potato early blight (19) was found to be,

$$L = 1 - \exp(-((\text{Final severity} - 0.17)/0.57)^{0.51}),$$

in which the minimum disease threshold was 0.17, the slope was 0.57 and shape parameter was 0.51. The concept of a minimum threshold of disease is intrigueing, although difficult to prove in practice because of the high variability in data at low disease severity and low yield loss.

Response Surface Models

The role of the crop in most empirical disease-loss models has been as a growth stage at which disease assessments are done. Calpouzos et al. (4) used two crop-related predictors to develop a model for estimating loss due to wheat stem rust. They linearized disease progress curves by plotting disease severity against crop growth stage and then calculated the slope (X1) of the line. A second predictor, X2, was the growth stage of disease onset. Yield loss was estimated using parameterized values of X1 and X2 and their model is pictorially represented as a three-dimensional response surface (4).

The relationship between disease severity, yield loss and crop growth stage has also been conceptualized as a response surface (29), represented by Yield loss = f (Disease severity, Growth Stage). This is in contrast to the multiple predictor models, which recognize growth stage as part of the disease variable and do not explicitly use it as a predictor, i.e. Yield loss = f (Disease severity at a certain growth stage).

OTHER ASPECTS OF QUANTIFICATION

The above discussion has concentrated on empirical, "best-fit" models derived from field data. In recent years, developments in computer hardware and software have enabled the development of complex crop physiological models capable of relatively precise yield predictions (16). When these crop models are coupled to insect and disease models, they provide a basis for understanding and predicting pest losses (1). As yet, it is difficult to use crop and pest simulation models for estimating field losses because these models require data on many environmental variables to make any prediction. Crop physiological simulation models have remained very much tools to guide research and to further understanding of pest-loss relationships. Simplified crop models with minimal physiological concepts are being developed for use in pest management (10); these require only 2-3 environmental variables to be monitored using simple equipment, and have been shown capable of predicting crop yield in the absence and presence of pest populations. These simpler models are likely to have more practical value for estimating field losses than the complex physiological models. An important issue in quantifying disease-loss relationships is the validity of the final model, both in the context of the biology represented by the model and its statistical soundness. Few empirical loss models undergo testing before they are used, although the process of empirical modeling is often viewed as an evaluative process (26). The criteria used for evaluating a regression model do not give an indication of how the model will predict when used in another year or environment (24). When tested using separate situations, the models of James et al. (9) for potato late blight and Burleigh et al. (3) showed good predictive ability.

LITERATURE CITED

1. Boote, K.J., Jones, J.W., Mishoe, J.W., Berger, R.D. 1983. Coupling pests to crop growth simulators to predict yield reductions. Phytopathology 73:1581-1587.

2. Butt, D.J., Royle, D.J. 1974. Mathematical analysis and modeling. In Epidemics of Plant Diseases, ed. J. Kranz, pp. 78-114. Berlin: Springer-Varlag. 170 pp.

3. Burleigh, J.R., Roelfs, A.P., Eversmeyer, M.G. 1972. Estimating damage to wheat caused by Puccinia recondita tritici. Phytopathology 62:944-946.

4. Calpouzos, L., Roelfs, A.P., Madson, M.E., Martin, F.B., Welsh, J.R., Wilcoxson, R.D. 1976. A new model to measure losses caused by stem rust in spring wheat. Univ. of Minn. Agric. Exp. Stn. Tech. Bull. 307. 23 pp.

5. Draper, N.R., Smith, H. 1966. Applied Regression Analysis. New York: Wiley. 407 pp.

6. Gaunt, R.E. 1980. Physiological basis of yield loss. In Crop Loss Assessment, pp. 98-111. Misc. Publ. no. 7. Univ. of Minn. Agric. Exp. Stn., St. Paul, MN 326 pp.

7. James, W.C. 1974. Assessment of plant diseases and losses. Ann. Rev. Phytopathol. 12:27-48.

8. James, W.C., Teng, P.S 1979. The quantification of production constraints associated with plant diseases. In Applied Biology, Vol. IV, ed. T.H. Coaker, pp. 201-267. London: Academic Press. 285 pp.

9. James, W.C., Shih, C.S., Hodgson, W.A., Callbeck, L.C. 1972. The quantitative relationship between late blight of potato and loss in tuber yield. Phytopathology 62:92-96.

10. Johnson, K.B., Johnson, S.B., Teng, P.S. 1985. A simple potato growth model for crop pest management. Agric. Systems 11:33-45.

11. Katsube, T., Koshimizu, Y. 1970. Influence of blast disease on harvests in rice plant. I. Effect of panicle infection on yield components and quality. Bull. Tohoku Nat. Agr. Exp. Sta. 39:55-96.

12. King, J.E. 1980. Cereal survey methodology in England and Wales. In Crop Loss Assessment, pp. 124-133. Misc. Publ. no. 7. Univ. of Minn. Agric. Exp. Stn., St. Paul, MN 326 pp.

13. King, E.D. 1982. The effects of late blight, Colorado potato beetle and weed populations on yield loss of potato. Ph.D. Thesis, Department of Plant Pathology, Pennsylvania State University. 81 pp.

14. Kingsolver, C.H., Schmitt, C.G., Peet, C.E., Bromfield, K.R. 1959. Epidemiology of stem rust: II (Relation of quantity of inoculum and growth stage of wheat and rye at infection to yield reduction by stem rust.). Plant Dis. Rep. 43:855-862.

15. Large, E.C., Doling, D.A. 1962. The measurement of cereal mildew and its effect on yield. Plant Pathol. 11:45-47.

16. Loomis, R.S., Rabbinge, R., Ng, E. 1979. Explanatory models in crop physiology. Annu. Rev. Plant Physiol. 30:339-367.

17. Oloffson, B. 1968. Determination of the critical injury threshold for potato blight (Phytophthora infestans). Statens Vaextskyddsanst. Medd. 14:81-93.

18. Madden, L.V. 1983. Measuring and modeling crop losses at the field level. Phytopathology 73:1591-1596.

19. Madden, L.V., Pennypacker, S.P., Antle, C.E., Kingsolver, C.H. 1981. A loss model for crops. Phytopathology 71:685-689.

20. Pinstrup-Andersen, P., de Londono, N., Infante, M. 1976. A suggested procedure for estimating yield and production losses in crops. PANS 22: 359-365.

21. Romig, R.W., Calpouzos, L. 1970. The relationship between stem rust and loss in yield of spring wheat. Phytopathology 60:1801-1805.

22. Seck, M. 1985. Modeling the effect of leaf rust (*Puccinia recondita tritici*) on wheat yield, as affected by leaf position and host resistance. M.S. Thesis, Department of Plant Pathology, University of Minnesota. 104 pp.

23. Stynes, B.A. 1975. A synoptic study of wheat. Ph.D. Thesis, University of Adelaide, South Australia. 291 pp.

24. Teng, P.S. 1981a. Validation of computer models of plant disease epidemics: a review of philosophy and methodology. Z. Pflanzenkr. Pflanzenschutz 88:49-63.

25. Teng, P.S. 1981b. Use of regression analysis for developing crop loss models. In Crop Loss Assessment Methods, Supplement 3, ed. L. Chiarappa, pp. 51-55. FAO/CAB.

26. Teng, P.S. 1985. Construction of predictive models: II. Forecasting crop losses. In Advances in Plant Pathology, Vol. 4, ed. D.S. Ingram, pp. 179-206. London: Academic Press. 255 pp.

27. Teng, P.S., Bissonnette, H.L. 1985. Developing equations to estimate potato yield loss caused by early blight in Minnesota. Amer. Potato J. 62:607-618.

28. Teng, P.S., Bissonnette, H.L. 1985. Potato yield losses due to early blight in Minnesota fields, 1981 and 1982. Amer. Potato J. 62:619-628.

29. Teng, P.S., Gaunt, R.E. 1981. Modelling systems of disease and yield loss in cereals. Agric. Systems 6:131-154.

30. Teng, P.S., Oshima, R.J. 1983. Identification and assessment of losses. In Challenging Problems in Plant Health ed. T. Kommedahl, P.H. Williams, pp. 69-81. APS, St. Paul. 538 pp.

31. Teng, P.S. and W.W. Shane. 1984. Crop losses due to plant pathogens. CRC Crit. Rev. in Plant Sci. 2:21-47.

32. Teng, P.S., Close, R.C., Blackie, M.J. 1979. A comparison of models for estimating yield loss caused by leaf rust (*Puccinia hordei* Otth) on Zephyr barley in New Zealand. Phytopathology 69:1239-1244.

33. VanderPlank, J.E. 1963. Plant Diseases: Epidemics and Control. Academic Press: New York. 349 pp.

34. Walker, P.T. 1983. Crop losses: the need to quantify the effects of pests, diseases and weeds on agricultural production. Agriculture, Ecosystems and Environment 9:119-158.

35. Watts Padwick, G. 1956. Losses caused by plant diseases in the colonies. Commonwealth Mycological Institute, Kew. 60 pp.

36. Wiese, M.V. 1980. Comprehensive and systematic assessment of crop yield determinants. In Crop Loss Assessment, pp. 262-269, Misc. Publ. no. 7. Univ. of Minn. Agric. Exp. Stn., St. Paul, MN 327 pp.

37. Zadoks, J.C., Schein, R.D. 1979. Epidemiology and Plant Disease Management. New York: Oxford Univ. Press. 427 pp.

CHAPTER 12

QUANTIFYING THE RELATIONSHIP BETWEEN INSECT POPULATIONS, DAMAGE,
YIELD AND ECONOMIC THRESHOLDS

P. T. Walker

Formerly with the Tropical Development and Research Institute, London.
10 Cambridge Road, Salisbury,
Wiltshire, SP1 3BW England

In a pest management system, decisions are commonly based on the pest density or amount of damage (i) and the crop yield (y), using an economic threshold as a measure. The assessment of both (i) and the relation between (i) and (y) are described elsewhere in this book. Here the types of relationships and the economic importance of crop loss assessment and economic thresholds are examined. The subject has also been reviewed by Walker (67), Reichelderfer et al. (50) and Pedigo et al. (45).

The relation between insects or other causes of loss such as weeds, and their economic effects can be quantified as with plant diseases in the last chapter: crops in crop yield terms, animals in meat or milk yield, humans as loss of earning ability and health.

REASONS FOR QUANTIFYING YIELD LOSS

<u>Allocation of resources</u>   The relative importance of different causes of loss can be compared in order to allocate research and other funds to the most important, i.e. how much to allocate and to what.
<u>Economic decisions in pest management</u>   To compare the value of loss in yield or quality with the cost of controlling or preventing it, and to decide whether to apply controls, where and when.
<u>Food and crop production planning</u>   Projections from the known or forecast chances of pest attack.
<u>Research on how yield is produced</u>   If the cause and mechanism of yield loss are known, its prevention by crop selection, breeding, biotechnology, pest forecasting etc., may be possible.

DEFINITION OF CROP LOSS DUE TO PESTS

Many of the terms have been defined by Chiarappa (10), Judenko (32) and Reichelderfer et al. (50). Crop loss (w) in the presence of pests is best expressed as the percentage reduction in the potential, maximum yield in the absence of pests (m). If the yield in the presence of pests is (y_1):

$$w = \frac{(m - y_1)}{m} \times 100$$

Maximum yield

It may be difficult to decide on a value of the maximum or potential yield (m). Zadoks (in Ref. 13) has discussed the different yields that can be obtained in fields and research stations, with and without various inputs. The decision on which to take depends on the purpose of the assessment: to assess the benefits of a research project, a practical development project, or for general information.

A Crop Loss Profile

A number of different causes of yield loss, (y_1, y_2 etc.) can be expressed graphically as a profile (48), with sections proportional to the fraction of total yield attributed to each cause including diseases, weeds, soil moisture and seed quality. The method fails if there is any interaction between any of the causes of loss, but is useful conceptually.

Pre- and Post-harvest Loss

Yield loss is traditionally divided into loss before and during or after harvest, studied in the field and usually in store, respectively. Different methods of study and control are often used. Many problems however, often start in the field, for example legume beetles, grain weevils, potato tuber moth or carob moth. The Food and Agriculture Organization, United Nations (FAO) has surveyed post-harvest losses (14), and the topic has been reviewed by Adams (1, 2) and Pimentel (46, 47), with bibliographies.

Type and Units of Yield

The yield is usually the quantity of economic product harvested, either the primary product, grain, fruit, stem, etc., or a constituent of it, such as flour, juice, sugar or oil. The quantity, grade or condition of the yield is important. Wheatley (69) divided types of pest damage into those of high or low incidence and high and low severity. High incidence and high severity means a complete loss, while a low incidence of low severity may still be important as cosmetic damage, reducing market value.

Yield and loss can be expressed in terms other than weight, volume or quality. The energy equivalent may be important if fuel, fertilizer or pesticide are the inputs. The monetary value in the local market, as a shadow export price or the value of an alternative crop used for subsistence, may be important. But value and price can change rapidly, depending on the supply, elasticity of demand, exchange rates etc. emphasizing that weight of yield is basic.

Area may be important if efficient production is needed on a limited area. If loss is prevented, the same yield can be grown on less land, releasing the loss-equivalent area, or Ordish area (44) for other use.

HOW YIELD IS LOST

The effects of pests or other causes of yield loss can be better understood as a plastic and dynamic system (Figure 1), which shows where pest attack can influence the final crop yield. Yields of individual plant parts such as tillers (modules) interact to produce the plant or genetic yields, themselves interacting to produce the crop yield (24). Thus reduction in one part of the system may be compensated by increase in another part. With inputs of values and rates, the system becomes a production system that can be simulated.

Yield may be lost in the following ways:
<u>Establishment</u>, germination and early growth can be affected, leading to reduced plant stands or low-yielding plants.

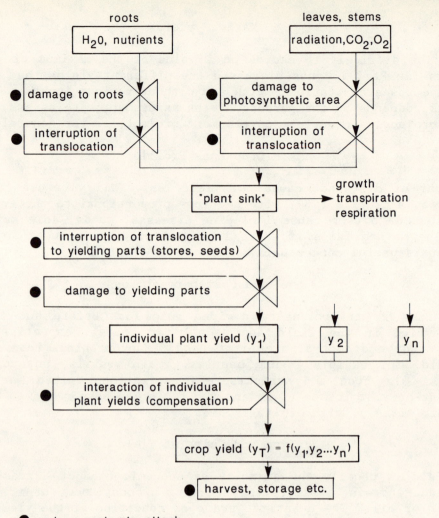

● = where pests etc. attack

Figure 1. How yield is lost: a yield production system showing (●) where pests can cause yield loss.

<u>Photosynthetic area</u> can be reduced or made ineffective, either directly by mining or eating, or indirectly by covering leaves with honeydew, mold etc.

<u>Uptake of water or nutrients</u> by the roots can be reduced by soil pests.

<u>Translocation</u> of water and nutrients or photosynthate to or from roots, stems and leaves to fruit, seeds, or storage organs such as tubers or the stem in sugarcane can be interrupted by stem and petiole borers etc.

<u>Storage organs</u> such as canes, tubers or roots can be damaged in the field or in store.

<u>Reproductive parts</u> such as flowers, seeds, fruit etc, can be damaged, resulting in loss in quality or quantity.

<u>Secondary loss</u> due to pests or diseases attacking the already damaged part, or by disease carried by vectors.

<u>Spoilage</u> and loss of quality by lesions, holes, spots, excreta, parts of insects etc.

<u>Harvesting and processing loss</u> due to difficulties in handling, cleaning or processing, for example fire ants, sticky cotton lint, moth webbing etc.

THE RELATION BETWEEN INFESTATION AND YIELD

One measurement of infestation and yield is of very limited value, two permit the establishment of a trend, but three or more are needed to give a useful indication of how yields relate to a range of infestations, as found in practice. The relation between infestation and yield is commonly expressed as a regression function, usually negative, of yield (y) on infestation or damage (i):

$$y = m - b (i)$$

where (m) is the maximum, potential yield in the absence of pests, or where (i) = 0, and (b) is the rate of loss of yield. Loss (w) is defined as above, and is sometimes studied and quoted, usually positively related to the infestation rate. If yield increases at low infestation, however, loss becomes negative, which is why the expression as yield may be preferred.

The form of the regression function or model, is important when predicting how yield, and hence benefit, changes at different infestation or damage levels found or resulting from control measures. It is important when evaluating economic action thresholds at which to control pest attack. The relation may be simple or very complex, composed of several different relationships due to effects on different plant parts or by different pests or other causes. The relation may vary with the time of attack, the stage of pest and crop, the method of assessment or the growing conditions (7, 56, 67).

Straight Line Relation

Yield decreases in direct proportion to unit increase in infestation or damage. This occurs when one pest or egg batch damages one plant or plant part, for example cereal midge attack. There is no compensation by the plant or its other parts for the attack, and so no pest threshold level below which yield is not reduced.

Sigmoid Relation

If a full range of infestations is examined, there is often a lower threshold below which yields are not reduced, perhaps due to compensation by the plant or other plants for the attack. Some yield is often produced at high levels of attack, producing a sigmoid curve, the rate of loss varying along the curve. Sometimes only the upper, convex upward part is found, often when there is early attack on vegetative parts, leaves etc. In other cases an upwardly concave curve, when attack is late, on reproductive and yielding parts such as fruit or cotton bolls, is found. Compensation cannot occur, and the rate of loss greatest at low infestations. Justesen & Tammes (33) and Tammes (59) discuss the reasons for such yield responses.

Fitting a sigmoid curve is difficult, and often the central section is assumed to be linear for practical purposes, perhaps at infestation rates above a threshold.

Severe Yield Loss at Low Infestations

A very small number of pests can sometimes cause a disproportionate amount of loss, for example the vectors of rice grassy stunt disease, the brown planthopper, Nilaparvata sp. (12). Gradients of disease attack are discussed by Thresh (61). Loss of yield due to cosmetic damage, for example scale insects on fruit, is similarly related to pest density. Control must be very effective and is often very costly.

Relation to the Logarithm of Pest Density

If the effects of pest density are multiplicative, rather than additive, yield may be related to the log or power of pest density, for example that between cotton seed yield and the number of whitefly nymphs per leaf (i) as log (i + 1). Transformation of the data may be necessary before analysis, and detransformation when presenting the results.

Increasing Yield with Increasing Infestation

In some cases, low infestations cause an increase in yield, which then falls with greater infestation. This may occur if pests destroy the apical growing point, producing more fruiting points as in cotton, or more tillers as in sorghum. An increase in yield will depend on good growing conditions afterwards. If leaf area is in excess, loss of foliage to pests will allow more light to reach the leaves and increase yield. Pest attack may stimulate growth and metabolism of the plant, or increase drying or maturity, raising the concentration of a yield product such as sugar per unit weight of sugarcane. The anomalous situation of pests selectively attacking better and higher yielding plants may give a positive relation between yield and infestation (25).

No Relation between Infestation and Yield

Often no relationship can be found. This may be due to averaging very variable data, thus hiding any real effects. The variation may be reduced or accounted for by changing the size or shape of field plots, using more replicates, stratifying to produce more uniform conditions of crop, farming or pest attack, or improving pest infestation, control or assessment techniques or yield measurement.

QUANTIFYING THE RELATION BETWEEN YIELD AND INFESTATION

Linear models of the response of yield (y) to infestation (i) of the form $y = m - b (i)$ have been mentioned. Sigmoid relationships are difficult to fit, but logarithmic or power models are often derived, more for practical than for conceptual reasons. If several factors, such as other pests or diseases, are operating, a multifactorial regression (20) may separate the contributions of each factor, as used by Khosla (34) to quantify losses in rice:

$$y = m_{i_1 i_2} - b_{i_1 i_2}(i_1 - \bar{i}_1) - b_{i_2 i_1}(i_2 - \bar{i}_2)$$

where $(b_{i_1 i_2})$ is the partial regression coefficient of (y) on (i_1) at constant (i_2) etc.

Most relationships can be fitted by a polynomial:

$$y = m - a_i - b_i^2 - c_i^3 \ldots$$

the variance being partitioned into linear, quadratic, cubic etc. components until an accurate prediction of (y) is obtained from (i) (36, 37).

The usefulness of a model to predict yield from infestation can be extended if the factors governing the pest population development can be included, such as temperature, as day degrees, or rainfall. For example loss of forage has been forecast from the grasshopper population (30), which can be modeled from temperature summation (18). Such models can take into account the distribution and probability of pest attack and a non-linear response of pest development to temperature (17).

Duration of Pest Attack

The relation between yield and pest attack not only depends on the numbers of pests but on the duration of attack. Models have quantified the relationship between yield and the number of 'bug days' for the brown planthopper, _Nilaparvata_, mite days or aphid days, together with cumulative temperature.

Mixed Crops

The presence of more than one crop in space or time -- multiple cropping-- may complicate the yield-infestation model. One approach is the agronomic one of expressing yield of mixtures in terms of the pure stand of one of the crops on the same area, the land equivalent ratio (70). The duration and proportion of each crop in the system can be incorporated in an area time equivalent ratio (31). The difference in yield at different pest densities will provide the relation between yield and pest attack on different area or time crop mixtures.

Missing plants, Compensation and Interaction

The distribution pattern of attack in a field affects the relation between yield and infestation. If attack is spaced out, unattacked plants yield more due to removal of competition for light, water and nutrients, and there may be no loss of yield over an area until high rates of infestation occur. A large pest attack may be economically tolerated. If attack is in large groups, however, compensation is impossible and loss in yield may occur at low infestations.

Compensation effects can be quantified from the yield of an undamaged plant surrounded by groups of damaged plants, perhaps in groups of five potato plants (35) or 'cylinders of influence' round tobacco plants (53) attacked by cutworm.

The effect of missing plants can be quantified from the hyperbolic relation often found between plant weight and planting density, the 2/3 thinning rule (55), a simple model of which is given by Hardwick & Andrews (23). The difference between actual and expected yield of potatoes after attack has been used to show difference in compensatory ability (3).

The interaction between different causes of loss, of which pests, diseases and weeds are only a few, can really only be quantified by multivariate methods because of the great variability of the yield response, which is in fact a response surface. Population ecology and weed science are providing some useful leads (8).

Statistical Distribution of Loss

Related to the distribution of pests is the frequency distribution of loss in space and time. This may be non-random because of the cyclical nature of pest attack and non-random climatic events which often control it. The distribution is important in assessing the risk of loss and the demand for pesticides. Tanner (60) found the summed frequency of loss, as a percentage of total loss, followed the same curve in many cases, when expressed as multiples of the average loss. The distribution can be linearized by taking logarithms. Differences between the actual and the expected distribution can be used to explain why such differences exist (63).

CROP LOSS AND ECONOMIC THRESHOLDS

The economics of pest management are discussed in chapters 18 and 20, but it is useful to see how data on yield loss can be used in decision-making. To clarify pest thresholds, examine the rise in infestation (i) with time (t) in Figure 2A, at the same time following the fall in yields (y) with rise in infestation (i) in Figure 2B. First the warning threshold is reached (i_1), then the action threshold (i_2), sometimes called the economic threshold, at which control action must be taken to stop the infestation rising to the damage threshold (i_3), also called the economic injury level, where economic loss occurs. This must be defined, and is usually taken as the point where the ratio of benefit (B), as yield increase due to pest control, to the cost (C) of control is less

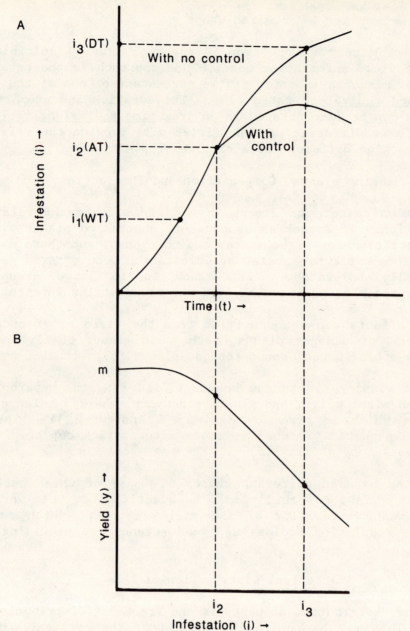

Figure 2. (A) Change in infestation (i) with time (t)
(B) Simultaneously, the relation between infestation (i) and yield (y)
showing pest threshold levels (i_1, i_2, i_3)

than one, although other B/C ratios may be demanded.

On the yield curve (2B), there may be little loss of yield at the action threshold (i_2), but an economic loss at the damage threshold (i_3).

The action threshold (AT) is a dynamic value of the infestation rate, and depends on many factors. These can be expressed in a conceptual model (9) which includes many of the factors, their exact position depending on their positive or negative value and how they are defined. A similar model has been suggested by

Pedigo et al. (45):

$$AT = \frac{t \times T \times P \times SC \times CF}{y \times V \times y/i \times C \times F \times CE \times R \times S}$$

120

The factors which support a high threshold and less control are:

$\quad$ t = the time between the infestation, decision, control and harvest
$\quad$ T = cumulative temperature, which affects infestation, control and yield
$\quad$ P = crop resistance or tolerance to pest attack
SC = survival coefficient: pest infestation development and survival
CF = critical factor: the attitude of society to pest damage, need for food, use
$\qquad$ of chemicals, the environment etc.

Factors which favor a low threshold and more control are:

$\quad$ y = quantity of yield
$\quad$ V = value of yield, depending on (y) and the elasticity of demand
y/i = yield-infestation relationship, the production or loss function
$\quad$ C = cost of control (V and C relationship reviewed by Walker 65)
$\quad$ F = profit or margin required
CE = control efficiency
$\quad$ R = farmers perception of pest attack and yield expectation: attitude to risk
$\quad$ S = crop stress, water status, growing conditions etc.

Other factors may become important, for example Gonzalez & Wilson (21) said that the amount of biocontrol should be included. Hearn & Room (28) included yield development factors, the number of cotton flowers and bolls, in the action threshold. Increased model complexity may give better yield prediction at the expense of more limited application of the model to a range of situations, quite apart from the cost of measuring all the inputs. Cost and applicability remain critical when assessing the usefulness of any model.

The Determination of Economic Action Thresholds

A value developed elsewhere may be taken and modified after field testing; some appropriate value taken and again tested in the field, or the value may be calculated empirically from its constituent factors. However it is derived, it will need to be tested, adjusted, and its acceptance by farmers examined over several years (29).

Threshold Theory and its Application

The development of economic thresholds has been reviewed by Headley (26, 27), Stern (57), Andow & Kiritani (5), Norton (42) and Pedigo et al. (45). The perception of pest attack and their expectations of a good harvest by farmers requires socio-economic study (58), and may differ from that of research workers (15, 16). Actions by farmers depend on their background, experience of local conditions and their attitude to risk. Without this information, much research effort may be wasted.

Decisions on Pest Management

Information on crop loss, the probability of pest attack, and other factors can be used to make decisions in models which may be deterministic or stochastic, depending on probabilities (6, 38, 39, 41, 43). Linear, dynamic programming (68) may be used, or a decision tree, in which the costs and probabilities of different outcomes can be compared (62). Another approach is to arrange the outcomes of different decisions in a pay-off or decision matrix. Inputs can be varied in a simulation model (54). Examples of some of these techniques are given by Reichelderfer et al. (50). The systems approach is reviewed by Ruesink (51) and Getz & Gutierrez (19). Other management techniques such as optimization and cost-benefit analysis are often used in pest management decision-making, after discounting costs and benefits to net present values. Needless to say, the

121

unquantifiable 'externalities' associated with food shortages, health or pesticide use and the environment might raise problems.

CROP LOSS INFORMATION

General works on crop loss due to pests by Ordish (44), FAO (13) and Chiarappa (10) have been supplemented by reviews by Bardner & Fletcher (7) and Walker (66, 67). World crop loss was summarized by Cramer (11), Walker (64), Ahrens et al. (4) and Reed (49), who also give references for loss in individual crops. Information on the USA will be found in Pimentel et al. (47), Schwartz & Classen (52) and Pimentel (46).

CONCLUSION

The loss of crop yield to pests, and indeed loss of other human and animal resources, must be quantified in order to make decisions on their prevention or control. Single observations are too limiting, and a comprehensive model over the whole response surface may be the only way to use all the information in quick decisions in different situations. Automatic data processing can make this possible, but input of pest and yield data remains a basic requirement.

LITERATURE CITED

1. Adams, J.M. 1964. A review of the literature concerning losses in cereals and pulses since 1964. Tropical Sci. 19:1-28.
2. Adams, J.M. 1977. A bibliography on post-harvest losses in cereals and pulses, with particular reference to tropical and sub-tropical countries. Report G110, Tropical Products Institute (TDRI), London. 23 pp.
3. Adams, M.J., Lapwood, D.H. 1983. The effect of _Erwinia_ (blackleg) on potato plants. 2. Compensatory growth. Ann. Appl. Biol. 103:79-85.
4. Ahrens, C., Cramer, H.H., Mock, M., Peschel, H. 1983. Economic impact of crop losses. Proc. 10th Int. Cong Plant Protect. Brighton. 1:65.
5. Andow, D.A., Kiritani, K. 1983. The economic injury level and the control threshold (in rice). Jap. Pesticide Info. No. 43:3-9.
6. Austin, R.B., ed. 1982. Decision-making in the practice of crop protection. Monog. 25 Brit. Crop Protect. Council, Croydon. 238 pp.
7. Bardner, R., Fletcher, K.E. 1974. Insect infestations and their effects on growth and yield of field crops. Bull. Entomol. Res. 64:141-160.
8. Begon, M., Mortimer, M. 1986. Population ecology: a unified study of plants and animals. 2nd ed. Oxford: Blackwell. 204 pp.
9. Chiang, H.C. 1983. Factors to be considered in refining a general model of economic threshold. Entomophaga 27:99-103.
10. Chiarappa, L. 1971. Crop loss assessment methods: FAO manual on the evaluation and prevention of losses by pests, diseases, weeds. FAO and Commonw. Agric. Bur., Farnham Royal, UK. Looseleaf + 2 supplements + supplement 3, 1981. 123 pp.
11. Cramer, H. 1967. Plant protection and world crop production. Pflanzenschutz. Nachr. Bayer. 524 pp.
12. Dyck, V.A. 1974. Pest damage to plots and economic thresholds. Int. Rice Res. Inst. Los Banos, Philippines.
13. FAO. 1967. 1. Papers presented at the FAO symposium on crop losses. 330 pp. 2. Background papers. 97 pp. Rome: FAO.
14. FAO. 1977. Analysis of an FAO survey of post-harvest crop losses in developing countries. Rep. AGPP: MISC/27. Rome: FAO. 147 pp.
15. Farrington, J. 1977. Research-based recommendations versus farmer practice: some lessons from cotton spraying in Malawi. Exp. Agric. 13:9-15.
16. Farrington, J. 1977. Economic thresholds of insect pest infestation in peasant agriculture: a question of applicability. PANS 23:143-148.

17. Feldman, R.M., Curry, G.L. 1982. Operations research for agricultural pest management. Operations Res. 30:601-618.
18. Gage, S.H., Mukerji, M.K. 1971. A perspective of grasshopper population distribution in Saskatchewan and interrelations with control. Environ. Entomol. 6:469-479.
19. Getz, W.M., Gutierrez, A.P. 1982. A perspective on systems analysis in crop production and insect pest management. Annu. Rev. Entomol. 27:447-490.
20. Gomez, K.A., Gomez, A.A. 1983. Statistical procedures for agricultural research. 2nd ed. London and New York: Wiley. 688 pp.
21. Gonzalez, D., Wilson, L.T. 1982. A food web approach to economic thresholds/1 a sequence of pests/predaceous arthropods on California cotton. Entomophaga. 27:31-43.
22. Hall, D.C., Norgaard, R.B. 1973. On the timing and application of pesticides. Amer. J. Agric. Econ. 55:198-201.
23. Hardwick, R.C., Andrews, D.J. 1983. A simple model of the relationship between plant density, plant biomass and time. J. Appl. Ecol. 20:905-914.
24. Harper, J.L. 1977. Population biology of plants. London and New York: Academic Press. 892 pp.
25. Harris, P. 1974. Possible explanations of plant yield increases following insect damage. Agro-ecosystems 1:219-225.
26. Headley, J.C. 1972. Defining the economic threshold. In Pest Control Strategies for the Future, pp, 100-108. Washington, DC: Nat. Acad. Sci. 376 pp.
27. Headley, J.C. 1972. The economics of agricultural pest control. Annu. Rev. Entomol. 17:273-278.
28. Hearn, A.B., Room, P.M. 1978-9. Analysis of crop development for cotton pest management. Protect. Ecol. 1:265-277.
29. Herdt, R.W., Castillo, L.L., Jayasuriya, S.K. 1984. Economics of insect control in rice in the Philippines. In Judicious and Efficient Use of Insecticides on Rice, pp. 41-56. Proc. FAO/IRRI Workshop 1983. Los Banos, Philippines: IRRI. 180 pp.
30. Hewitt, G.B., Onsager, J.A. 1982. A method of forecasting potential losses from grasshopper feeding on northern mixed prairie forages. J. Grassland Management 35:53-57.
31. Hiebsch, C. 1978. Comparing intercrops with monoculture. In Annu. Rep. 1976-1977, pp. 187-200. Raleigh, N. Carolina: Soil Sci. Dep. State Univ.
32. Judenko, E. 1983. Glossary of terms used in the application of the analytical method for assessing losses caused by pests. Rocz. Nauk Roln. E. 10:9
33. Justesen, S.H., Tammes, P.M.L. 1960. Studies of yield losses. 1. Self-limiting effect of injuries or competitive organisms on yield. Tijdschr. Plziekt. 66:281-287.
34. Khosla, R.K. 1977. Techniques for assessment of losses due to pests and diseases of rice. Indian J. Agric. Sci. 47:171-174.
35. Killick, R.J. 1979. The effect of infection with potato leaf roll virus on yield and some of its components. Ann. Appl. Biol. 91:67-74.
36. Lynch, R.E. 1980. European corn borer: yield losses in relation to hybrid and stage of corn development. J. Econ. Entomol. 73:159-164.
37. Mead, R., Curnow, R.N. 1983. Statistical Methods in Agricultural and Experimental Biology. London: Chapman and Hall. 335 pp.
38. Mumford, J.D., Norton, G.A. 1984. Economics of decision-making in pest management. Annu. Rev. Entomol. 29:157-174.
39. Norton, G.A. 1976. Pest control decision-making, and overview. Ann. Appl. Biol. 84:444-447.
40. Norton, G.A. 1976. Analysis of decision-making in crop protection. Agroecosystems 3:27-44.
41. Norton, G.A. 1982. A decision-analysis approach to integrated pest control. Crop Protect. 1:147-164.
42. Norton, G.A. 1984. Economic considerations in I.P.M. programmes. Proc. Malay. Plant Protect. Symp. (in press).

43. Norton, G.A., Mumford, J.D. 1983. Decision-making in pest control. Appl. Biol. 8:87-117.

44. Ordish, G. 1952. Untaken Harvest: Man's Loss of Crops from Pest, Weed and Disease. Constable, London. 171 pp.

45. Pedigo, L., Hurchins, S.H., Higley, L.G. 1986. Economic injury levels in theory and practice. Annu. Rev. Entomol. 31:341-368.

46. Pimentel, D., ed. 1981. Pest Management in Agriculture. 1. Boca Raton, Florida: CRC Press. 597 pp.

47. Pimentel, D., et al. 1980. Environmental and social costs of pesticides: a preliminary assessment. Oikos. 34:126-140,

48. Pinstrup-Andersen, P., Londono, N. de, Infante, M. 1976. A suggested procedure for estimating yield and production losses in crops. PANS 22:359-365.

49. Reed, W. 1983. Crop losses caused by insect pests in the developing world. Proc. 10th Int. Cong. Plant Protect., Brighton 1:74-79

50. Reichelderfer, K.H., Carlson, G.A., Norton, G.A. 1984. Economic guidelines for crop pest control. Plant Prod. Protect. Paper 58, Rome: FAO. 93 pp.

51. Ruesink, W. 1976. Status of systems approach to pest management. Annu. Rev. Entomol. 21:27-37.

52. Schwartz, P.H., Classen, W. 1981. Estimates of losses caused by insects and mites to agricultural crops. See Ref 46, pp. 15-77.

53. Shaw, M.J.P. 1980. Growth compensation for insect pest damage in field tobacco. Zimbabwe Sci. News. 14:97-100.

54. Shoemaker, C.A., Huffaker, C.D., Kennet, C.E. 1979. A systems approach to integrated pest management of a complex of olive pests. Environ. Entomol. 8:182-189.

55. Solbrig, O.T. 1980. Demography and evolution in plant populations. In Bot. Monog. 15, pp. 35-42. Oxford: Blackwell. 222 pp.

56. Southwood, T.R.E., Norton, G.A. 1973. Economic aspects of pest management strategies and decisions. In Insects: Studies in Population Management, ed P.W. Geier, L.R. Clark, D.J. Anderson, H.A. Nix. Ecol. Soc. Aust. Mem. 1. Canberra. 295 pp.

57. Stern, V.M. 1973. Economic thresholds. Annu. Rev. Entomol. 18:258-280.

58. Tait, E.J. 1978. Factors affecting the usage of insecticides and fungicides onfruit and vegetable crops in Great Britain. 1. Crop specific factors, pp. 127-142. 2. Farmer specific factors, pp. 143-151. J. Environ. Management 6:127-151.

59. Tammes, P.M.L. 1961. Studies of yield losses. 2. Injury as a limiting factor of yield. Tijdschr. Plziekt. 67:257-263.

60. Tanner, C.C. 1962. The distribution of crop losses from pests around the average loss. PANS B. 8:110-117.

61. Thresh, J.M. 1976. Gradients of plant virus diseases. Ann. Appl. Biol. 82:381-406.

62. Valentive, H.T., Newton, C.M., Talerico, R.L. 1976. Compatible systems and decision models for pest management. Environ. Entomol. 5:891-900.

63. Walker, P.T. 1965. The distribution of loss of yield in maize and of infestations of the maize stem borer in East Africa. Meded. Landbouwhogesch. Gent. 30:1577-1587.

64. Walker, P.T. 1975. Pest control problems, pre-harvest, causing major losses in world food supplies. FAO Plant Protect. Bull. 23:70-82.

65. Walker, P.T. 1977. Crop Losses: some relations between infestation, cost of control and yield in pest management. Meded. Landbouwwet. Gent. 42/1 919-926.

66. Walker, P.T. 1983. The assessment of crop losses in cereals. J. Ins. Sci. Appl. 4:97-104.

67. Walker, P.T. 1983. Crop losses: the need to quantify the effects of pests, diseases and weeds on agricultural production. Agric. Ecosystems Environ. 9:119-158.

68. Watt, K.E.F. 1963. Dynamic programming, look-ahead programming and the strategies of insect pest control. Can. Entomol. 95:525-536.

69. Wheatley, G.A. 1974. Insect problems in crops for processing. Pestic. Sci. 5:101-112.
70. Zandstra, H.G., Price, E.C., Litsinger, J.A., Morris, R.A. 1981. A Methodolgy for On-Farm Cropping System Research. Los Banos, Philippines: IRRI. 149 pp.

CHAPTER 13

EMPIRICAL MODELS FOR PREDICTING YIELD LOSS CAUSED BY A SINGLE DISEASE

Choong-Hoe Kim and D. R. MacKenzie

Department of Plant Pathology and Crop Physiology,
Louisiana State University,
Baton Rouge, Louisiana 70803, U.S.A.

Until fairly recently in the history of science, the Reductionist Method has been the standard scientific method for experimental design. By this approach all variables are held constant while one factor is varied and tested for significance. For example, in an experiment to test the effects of increased nitrogen fertilizer on corn yield, all of the weeds are controlled, while all of the diseases are excluded. A statistical analysis is then performed on treatments over nitrogen level. The nitrogen level affects weed growth and pathogen development, which, in turn affects the crop's yield.

As more complex questions are asked with more variables, the reductionist method and its experimental design become more and more inappropriate. The reductionist scientists (who typically use the analysis of variance) often become entangled in factorial experiments that yield incomprehensible interactions (e.g., blocks by replications, by varieties, and by fertilizer treatment). It is very difficult to interpret high order interactions in single factor analysis.

A new order of thinking recently evolved called systems science (9, 27, 28, 29). By this form of reasoning, it is held that the whole is more than the sum of its parts. The systems science approach deals with a combination of factors (i.e., interactions) to study a complex process, rather than with individual factors.

Systems science in biology is relatively new. Because of the discoveries of systems science, some old rules of biological science are being rewritten. This is upsetting to some scientists and exciting to others. What do those changes mean to epidemiology? Yield loss and crop loss forecasting have become encouraged by this approach. Systems science will permit the development of a far better understanding of the nature of the complex relationships that exist between factors, as they exist in the real world. The most useful tools of systems science will undoubtedly be modeling and simulation. Such methodologies allow for testing the consequences of changes on a representation of the real world with minimum efforts and costs.

A model is a stylized representation of the real world. Models are said to have great worth if they can be verified, as they simplify complex biological systems; however, verification is often difficult. Simple description cannot accommodate all the variability of a system that occurs in nature. Consequently, winning acceptance of a "stylized representation" by the scientific community is arduous (11, 20).

A significant number of models have been proposed that attempt to predict yield loss and these have been reviewed by James (6) and Teng and Gaunt (21). Most loss modeling has used empirical models that originate in or are based on observations, previous experience, and experimental data. Empirical models differ significantly from theoretical models that are, as the name implies, based solely on theory. Empirical models derive their appeal from their method of construction (e.g. regression), and because verification activities are more easily accomplished than in theoretical models.

CROP LOSS ASSESSMENT

For many years it has been assumed that as sufficient knowledge was generated on the relationship of disease to yield loss in crops, all of that information could be aggregated over areas for regional crop loss assessments. It is now apparent that this assumption is false.

For reasons not well understood, individual yield assessments for specific fields, when gathered over an entire region, give a very inaccurate estimate of a region's crop productivity. As systems science asserts, the whole is more than the some of the parts.

There is a need for accurate assessments of crop losses for several applications. One often cited use would be for research planners to better use program resources (research efforts and funds). Although significant, this application palls in contrast to the use of crop loss information in assessing and managing the world's food resources.

There is a dearth of information on regional and national production of the major crop commodities in most countries. Most production estimates are simply calculated by applying a coefficient to last year's estimate. Economists call such calculations macromodeling. Little, if any, biology is included in the "adjustment". This methodology must be changed if accurate estimates of crop production are to become available. The complexity of this research will require new scientific methods and the development of extensive data bases that are appropriate to specific model development. As knowledge accumulates, more interdisciplinary efforts will be needed to assess the multiple stresses that affect a crop. The complexities of the research will also require the sophisticated tools of statistics and computer science to deal with these significant questions in meaningful ways.

SELECTION OF THE COMPONENTS FOR A MODEL

The construction of any model to predict the consequences of events yet to occur in nature is an extremely difficult task (for example, weather forecasting). Even more complicated to predict are biological events, especially those subject to environmental variables, as in the case of plant growth modeling. At a third level of complexity is the task of trying to predict yield loss caused by a single disease because: (1) the pathogen is influenced by the environment, (2) the status of the host in turn is influenced by the pathogen, and (3) the host and the pathogen interact. It is essential for modeling to identify clearly the variables to be considered and to understand their relationship.

The selected variables may be screened based on the relative importance of their contribution to yield loss. Selection of the items that must be used to develop models for predicting yield loss is difficult. Individual variables may be easily evaluated. The major problems lie with interpreting the interactions between those variables. For example, the amount of nitrogen in the soil will directly affect the yield of a crop, the plant's predisposition to the pathogen, and it may even directly affect the pathogen itself. These are undoubtedly poorly understood interactions and are difficult to describe mathematically. Similar illustrations can be given for other factors such as temperature, relative humidity, and soil moisture.

Ideally, a "good" model for the prediction of yield loss in a crop includes the many abiotic stresses (e.g. drought, heat, cold, and injury). It includes several diseases, many insects, some weeds, and perhaps some nematodes. Critics of research on yield loss models are quick to point out that contemporary yield loss models are not "real world situations" inasmuch as most models focus on a single disease. Such single factor models do not consider all the stress to which the host is actually subjected. We accept that criticism as valid, but we hasten to point out that dealing with such complex associations greatly compounds the task of model construction to a point beyond current modelling abilities.

Selection of factors

Some environmental factors are obviously more important in disease development than are others. The selection of individual variables for use in the construction of a model is usually judged in terms of the magnitude of their effect on the resulting variation. This can be conventionally determined with the coefficients of determination (R^2) of the variables in the regression model; however, the selection of variables is often arbitrary. Temperature and relative humidity are the most commonly selected variables. Sometimes variable selection is based only on a scientist's ability to measure specific factors easily. The common use of relative humidity in disease forecasting is an unstated admission that the measurement of leaf wetness was, until recently, too difficult. There are many other examples.

The selection of biological factors for inclusion is also a necessary step in model construction. The stage of host growth, plant density, and the amount of disease present are some of the common factors. The reasons for some choices are sometimes obscure. Cultivar differences (e.g. inherent resistance of host plant) are often ignored. Similarly, pathogen population differences (e.g. frequency of virulent isolates) are glossed over, not because they are insignificant, but simply because it is too difficult to include them in the model as variables.

All of the above considerations point out the need for more research, but these facts should not be used as an argument against research efforts. It is hoped that as the biology of more complex situations of crop stress is understood, more representative yield loss prediction models will evolve.

THE CLASSIFICATION OF MODELS

Single disease yield loss models can be divided into four types. These divisions are arbitrary, and are intended only to simplify the task of their description. These four categories are: 1. Forms of regression, 2. Forecasting plus regression, 3. Factor reduction through regression, and 4. Simulation plus regression.

Forms of regression

There is a long history of research efforts that have attempted to look at the relationship between varying amounts of disease and a crop's yield. Much of that work has been with small grain crops for foliar disease (4, 7, 18).

These earlier attempts most commonly used a form of simple linear regression having one independent variable (level of disease). Yield loss was the dependent variable. If a simple linear regression model was not appropriate for the data, transformation of one or more of the variables was often attempted to produce a straight line for the linear regression analysis (5, 13, 17). Various transformations such as square roots, logarithmic, logistic, reciprocal, or other transformations were said to improve the "goodness of fit" for a linear relationship. But little attention was paid to the effects of the transformations on error variances, on normality of residuals, and to other very appropriate statistical questions (21). A transformation may lead to a relatively simple method of estimating yield losses. But transformation also may often obscure the fundamental relationships between variables (16).

Many of these early research efforts focused on a single critical level of disease and its effect on crop yield (1, 7, 25). The "critical point models" (6) have some interpretive worth when a level of disease at a "critical point" is related to yield loss over several years. Most models, however, are considered to be of little applicable worth inasmuch as the "critical point" is commonly nearly coincident with time of harvest. This precludes the direct application of critical point models to yield loss forecasting.

Another limitation of critical point models is the intuitive feeling by some scientists that early season foliar destruction may not be equal to the physiological impacts of late season foliar destruction on yield. Thus was born the "multiple point model". In this model the individual contributions of disease severity to yield loss is looked at over selected time intervals (6). An increase in the amount of potato late blight in the fourth week would not have the same contribution to yield loss if it had occurred during the eighth week of the growing season. Multiple point models view the change of disease from week to week and incorporate that relationship into total yield loss by employing partial regression coefficients in a multiple regression equation (8). The multiple point models attempt to adjust yield loss for the dynamic relationships between crop/epidemic development.

The multiple point model also has its limitations. Most striking are the difficulties in its application to forecasting yield losses in the early phase of an epidemic. It would seem that an epidemic's development over a full season is required to evaluate the disease relationship to yield.

Forecasting plus regression

If the major criticism of the critical point and multiple point models for yield loss assessment is their inability to predict crop loss, then one alternative is to predict the amount of disease during the course of an epidemic and use that estimation to mathematically evaluate the yield loss model.

Foliar diseases of small grains can be "projected" to final disease severity with variable accuracy using the presently employed disease forecasting system (2, 12). Some methods currently being researched use repeated disease scouting assessments. These can be used to epidemiologically "project" the course of an epidemic -- given the assumption of no drastic change in the progression of disease increase. The final amount of disease severity would then be interpreted for its yield loss equivalent through a disease/critical point yield loss model.

Other researchers are working on the application of multiple point models to forecast yield loss. For example, Wallin's Severity Values have been used to forecast potato late blight disease (26). Severity Values are convenient "integrators" of important environmental factors that contribute to late blight epidemics. The contribution of extended periods of high relative humidity and temperature is interpreted daily on an arbitrary scale of 1 to 4 as an index of weather effect on disease development. When viewed over a period of time, potato late blight can be "forecast". In addition to using that information to schedule fungicide applications prior to disease occurrence, Severity Values could be used to "probe" the future with hypothetical weather patterns. From that weather information one could calculate the expected change in disease for extended periods, and then interpret those disease changes through a multiple point model to obtain an estimation of yield loss. Admittedly, these methods tend to make assumptions from assumptions and to extend substantially into the future.

The application of this research is to disease management and disease-control decision theory. At present much of the application of disease management lacks well-defined action thresholds (points beyond which disease management strategies should be initiated) because of the lack of estimates of yield loss for certain crop diseases. Yield loss prediction models will contribute to this area of need.

This type of information may also be of value in the area of crop loss assessment. "As used herein, yield loss is not synonymous with crop loss" (14).

Factor reduction through regression

Another very interesting research approach to yield loss forecasting is the use of multiple regression. The multiple-point yield loss models previously mentioned use multiple regression of different levels of disease as the independent variables. Actually, any biological and environmental factors and their combinations can be used as variables to estimate yield loss. By this method one can assess the significant factors to be used to predict yield loss earlier in the season (22). Some of the unimportant variables can be screened out by stepwise procedures. Some variables may be highly correlated and may also be reduced. The statistical questions to be asked are: of the given factors that can be observed or measured during the growing season, which factors are significantly associated with yield loss at harvest?

Several statistical models have been developed by this approach for some foliar diseases of small grain; most notably, wheat (15, 19). Some of the statistically identified variables obtained from that research are not intuitively apparent (3) and the statistical theory of regression implies no biological cause and effect relationship between such variables. Some of the relationships simply defy explanation. Be that as it may, the method works; therefore it has practical application.

Simulation plus regression

As knowledge has increased in epidemiology, the scientific understanding of the effects of various environmental factors on disease development has been greatly improved. The complex relationships among various factors in a biological system has inevitably led to simulation as a tool to describe how various factors affect one another in the infection process. The relationship between environmental factors (e.g., temperature, light, humidity, and moisture) and the infection processes (e.g., conidial germination, germ tube growth, appressorial formation, and penetration,) can now be described as mathematical equations for several plant/pathogen systems (10, 23, 24). When the infection process is taken over small units of time, and is cycled through a step by step computer simulation, very complex relationships between factors can be interpreted for an entire epidemic process.

The epidemiological interpretation of a biological process through simulation is now being linked to yield loss regression models to provide prediction capabilities. As environmental events change, computer simulations can probe the consequences of that change and interpret those outcomes for their effects on crop yield.

It would be unfair to leave the reader with the impression that the state of this research is of immediate practical application. There remains a tremendous amount of work that needs to be done to unravel the relationships between the many individual factors of the environment, the host, and the pathogen. As these relationships are better understood, it will be possible to probe the future for expected outcomes and to make decisions on how best to deal with existing and emerging problems.

LITERATURE CITED

1. Basu, P.K., Brown, N.J., Crete R., Gourley C.O., Johnston, H.W., Pepin, H.S., Seaman, W.L. 1976. Yield loss conversion factors for fusarium root rot of pea. Can. Plant Dis. Surv. 56:25-32.
2. Bourke, P.M.A. 1970. Use of weather information in the prediction of plant disease epiphytotics. Ann. Rev. Phytopathol. 12:345-370.
3. Butt, D.J., Royle, D.J. 1974 Multiple regression analysis in the epidemioliogy of plant disease. In Epidemics of Plant Disease, ed. J. Kranz, pp. 78-114. Berlin: Springer-Verlag. 170 pp.
4. Burleigh, J.R., Roelfs, A.P., Eversmeyer, M.G. 1972. Estimating damage to wheat caused by Puccinia recondita tritici Phytopathology 62:944-946.
5. Gregory, L.V., Ayers, J.E., Nelson, R.R. 1978. Predicting yield losses in corn from southern corn leaf blight. Phytopatholohy 68:517-521.
6. James, W.C. 1974. Assessment of plant diseases and losses. Ann. Rev. Phytopathol. 12:27-48.
7. James, W.C., Jenkins, J.E.E., Jemmett, J.L. 1968. The relationship between leaf blotch caused by Rhynchosporium secalis and losses in grain yield of spring barley. Ann. Appl. Biol. 62:273-288.
8. James, W.C., Shih, C.S., Hodgson, W.A., Callbeck, L.C. 1972. The quantitative relationship between late blight of potato and loss in tuber yield. Phytopathology 62:92-96.
9. Kranz, J., Hau, B. 1980. Systems analysis in epidemiology. Ann. Rev. Phytopathol. 18:67-83.
10. Kranz, J., Mogk, M., Stumpf, A. 1973. EPIVEN-ein Simulator fur Apfelschorf. Z. Pflanzenkr. Pflanzenschutz 80:181-187.
11. Kranz, J., Royle, D.J. 1978. Perspectives in mathematical modelling of plant disease epidemics. In Plant Disease Epidemiology, ed. P.R. Scott, A. Bainbridge, pp. 111-120. Oxford: Blackwell Scientific Publications. 329 pp.
12. Krause, R.A., Massie, L.B. 1975. Predictive systems modern approaches to disease control. Ann. Rev. Phytopathol. 15:31-47.
13. Large, E.C., Doling, D.A. 1962. The measurement of cereal mildew and its effect on yield. Plant Pathol. 11:47-57.
14. MacKenzie, D.R., King, C. 1981. Developing realistic crop loss models for plant disease. In Proceedings of the International Crop Loss Symposium, August, 1980, pp. 85-89. St. Paul, Minnesota.
15. Nelson, L.R., Holmes, M.R., Cunfer, B.M. 1976. Multiple regression accounting for wheat yield reduction by Septoria nodorum and other pathogens. Phytopathology 66:1375-1379.
16. Neter, J., Wasserman, W. 1974. Applied Linear Statistical Models; Regression Analyses of Variance, and Environmental Designs. Homewood, Ill: Richard D. Irwin Inc. 842 pp.
17. Romig, R.W., Calpouzos, L. 1970. The relationship between stem rust and loss in yield of spring wheat. Phytopathology 60:1801-1805.
18. Richardson, M.J., Whittle, A.M., Jacks, M. 1976. Yield-loss relationship in cereals. Plant Pathol. 25:21-30.
19. Sallans, B.J. 1948. Interrelations of common root rot and other factors with wheat yields in saskatchewan. Scientific Agriculture 28:6-20.
20. Teng, P.S. 1981. Validation of computer models of plant disease epidemics: a review of philosophy and methodology. Z. Pflanzenkr. Pflanzenschutz 88:49-63.
21. Teng, P.S., Gaunt, R.E. 1980. Modelling systems of disease and yield loss in cereals. Agric. System 6:131-154.
22. Verma, P.R., Morrall, R.A.A., Tinline, R.D. 1976. The effect of common root rot on components of grain yield in Manitou wheat. Can. J. Bot. 54:2888-2892.
23. Waggoner, P.E., Horsfall, J.G. 1969. EPIDEM, a simulator of plant disease written for a computer. Conn. Agric. Exp. Stn. Bull. 698. 80 pp.
24. Waggoner, P.E., Horsfall, J.G., Lukens, R.J. 1972. EPIMAY, a simulator of southern corn leaf blight. Conn. Agric. Exp. Stn. Bull. 729. 84 pp.

25. Wallen, V.R., Jackson, H.R. 1975. Model for yield loss determination of bacterial blight of field beans utilizing aerial infrared photography combined with field plot studies. Phytopathology 65:942-948.
26. Wallin, J.R. 1962. Summary of recent progress in predicting late blight epidemics in United States and Canada. Am. Potato J. 39:306-312.
27. Watt, K.E.F. 1970. The systems point of view in pest management. In Concepts of Pest Management, eds. R.L. Rabb, F.E. Guthrie, pp. 71-79. Raleigh: North Carolina State Univ. Press. 242 pp.
28. Witz, J.A. 1973. Integration of systems science methodology and scientific research. Agric. Sci. Rev. 11:37-48.
29. Zadoks, J.C. 1971. Systems analysis and the dynamics of epidemics. Phytopathology 61:600-610.

CHAPTER 14

EMPIRICAL MODELS FOR PREDICTING YIELD LOSS CAUSED BY
ONE TYPE OF INSECT: THE STEM BORERS

P. T. Walker

Formerly with the Tropical Development and Research Institute, London.
10 Cambridge Road, Salisbury, Wiltshire
SP1 3BW, England

Stem borers occupy a distinct ecological niche. They often cause yield loss in similar ways, and the methods of population and loss assessment are also similar. Many are lepidoptera, but they may also be coleoptera, such as the coffee, tea and fruit borers, diptera such as sorghum shoot fly, Atherigona, or other cereal borers, Oscinella or Diopsis, or hymenoptera, such as the sawflies.

Crops commonly attacked are cereals and sugarcane, but often woody tree or bush crops such as tea, coffee and fruit are also attacked. They are usually less mobile than other pests, and their population densities and damage more easily assessed, so that models of crop yield loss and infestation are more easily developed.

INFESTATION AND DAMAGE ASSESSMENT

Direct Methods

The population of egg batches, larvae or pupae per unit of stem, stem internode, tiller, branch, plant or tree, or hill or clump of plants, is usually counted or expressed as percentage incidence. Area of ground is preferred as a base unit, and the pest population can be derived from the number of stems, tillers or plants per area of ground. Ratings of degrees of attack can be used but if so, information is probably lost.

Indirect Methods

The number of adults caught in light, pheromone, color or other type of attraction trap is frequently used to assess the population, but because of the difficulty of relating such counts to the absolute pest population per ground unit, they are of little use in modeling until such relationships can be found. A relation between pheromone trap catches and cereal borer populations, and hence yield, seems nearest development.

Indirect Methods of Damage Assessment

The number of attacked plants, tillers, ears or heads is often counted or percentaged under the terms deadhearts, whiteheads, etc. This is usually quicker and cheaper than pest assessment. The number or percent incidence or borer tunnels or exit holes can be counted, or the length of tunnels in stems measured. Again, the relation between such indirect measures of attack must be related to absolute pest density per ground area in order to compare infestations in place or time.

When assessing plant stand losses caused by borers, the early destruction of tillers or plants may lead to incorrect assessment unless later counts are corrected for missing tillers or plants (7). Similar formulae for correcting the percentage of tillers of rice infested with rice stem borer have been given by Gomez (14) and in FAO (13), and in Chapter 3.

THE RELATION BETWEEN DIFFERENT PEST ASSESSMENT METHODS

If a simpler, cheaper or quicker method of assessment is used, its relation to the absolute population per area of ground should be found. Examples have been given by Walker (35), and include that between ratings of maize leaf attack and percent plant attack by Chilo partellus, found to be linear by Kalode & Pant (18), or the curved relationships between the number of Ostrinia nubilalis larvae per 100 plants and the percent maize plants bearing eggs (11), or between percent sugarcane stems bored and percent nodes bored (27), or percent rice hills infested with borer and percent tillers infested. There are many more. A curved relationship would be expected if a finite unit such as a percentage is related to an infinite one, such as insect population. The relationships need to be worked out in the field in order to develop more efficient sampling methods when establishing a yield-infestation model.

YIELD MEASUREMENT

Model output may be the weight of unprocessed product such as maize cobs, grain, rough rice, sugarcane etc., but is usually in terms of dried or cleaned grain, paddy, sugar etc. Clear definition of the yield is necessary.

METHODS OF LOSS ASSESSMENT

Natural Infestations

Models are frequently based on surveys of natural borer infestations and the yields obtained from a range of borer densities or amounts of damage. In India, Kishen et al. (19) used natural infestations of Chilo partellus on maize, Graca (15) natural infestation of Diatraea on sugarcane in Brazil, and Ho (16) Maliarpha borer attack on rice in Kenya.

Chemical Methods

Insecticides are commonly used to obtain a range of infestation rates, or often simply attacked and unattacked plots, for example in insecticide trials on the borer, Busseola fusca in Kenya and Tanzania (34), studies on the rice borer, Scirpophaga on rice in Bangladesh (8), or on the sugarcane borer, Diatraea, by Mathes et al. (21) in the USA.

Artificial Infestation

Borer eggs can be removed and known infestation levels replaced, for example of Ostrinia, the corn borer by Lynch (20) in the USA, or of the African maize borer, Busseola fusca in Ethiopia by Tchekmenev (30). Breniere (6) artificially infested rice with Maliarpha eggs in Madagascar, Appert (3) used the larvae of the same insect to derive yield-infestation models. They emphasize the importance of monitoring the population of borers which finally establish in trials.

Simulated Damage

It is difficult to exactly imitate the position, amount or timing of borer attack artificially. Loss of leaf area is seldom important. Complete loss of plants can be corrected for, as can loss of tillers, as mentioned above (14). Than et al. (31) simulated deadhearts and whiteheads in rice, as caused by stem borers, and Chiang (10) made artificial tunnels in maize stems.

Comparison of Resistant and Susceptible Cultivars

Patch et al. (24) compared the regression of yields of different maize varieties on the infestation rates of <u>Ostrinia nubilalis</u> in the USA, and Van Halteren (33) the yields of rice varieties attacked by <u>Scirpophaga</u> in Indonesia. This method was also used by Avasthy and Krishnamurthy (5) to work out relationships between yield and infestation rates for the sugarcane internode borer in India.

PREDICTIVE MODELS

How precise, comprehensive and generally useful a model is depends on the data and resources available for sampling and processing, and what its purpose is, whether to quantify losses, to forecast crop production on a large scale, or to identify the need for pesticides. It may be required to partition losses and their interaction into individual causes, to give an overall linear regression over a wide area, or to produce a number of yield reduction factors, or correction factors for estimating yield loss in certain areas, seasons, crops or cultivars or pest generations.

For example, Tchekmenev (30) fitted linear regressions of yield per plant (y) on the number of 1st instar <u>Busseola</u> larvae per plant (x):

$$y = 43.6 - 6.62x \quad (r = 0.96).$$

This can be interpreted as 15% loss per larva per plant, which can be compared with other rates of loss per larva elsewhere, or used to calculate loss from surveys of the number of larvae in an infestation.

The actual loss in a linear model depends on the maximum potential yield at x = 0, so there will be different rates of loss in crops of different potential yields. Walker (34) found the following regressions in Tanzania in maize attacked by <u>Busseola fusca</u> borer, (x) being the percentage plants attacked, transformed to angles:

$$y = 45.1 - 0.55x \quad \text{(high yielding crop)}$$

$$y = 14.5 - 0.23x \quad \text{(low yielding crop)}.$$

Different regressions may be found in different aged crops. Sarup et al. (28) obtained the following regressions for yield (y) on the number of eggs of <u>Chilo partellus</u> per maize plant (x) in India:

$$y = 0.21 - 0.009x \quad \text{(plants attacked at 10 days old)}$$

$$y = 0.19 - 0.005x \quad \text{(plants attacked at 17 days old)}$$

$$y = 0.50 - 0.01x \quad \text{(plants attacked at 25 days old)}.$$

Indications on the susceptibility of plants of different ages to attack can be obtained, here 17 day old plants being more tolerant to attack, and permitting a higher economic action threshold then.

The model may not be linear, for the reasons given earlier, and sigmoid relationships are often found, the rate of yield loss varying along the curve. There was a sigmoid response of yield to attack by the African maize stem borer, <u>Busseola fusca</u> in Nigeria, found in trials by Usua (32). Chatterji (9) found the upper part of a sigmoid curve, concave downwards, for maize and <u>Chilo partellus</u> in

India, while Atwal et al. (4) found the lower part, concave upwards, for yield and tunnel length. The relation between percent deadhearts caused by <u>Scirpophaga</u> on rice in Bangladesh (x) and yield (y) was fitted by a logarithmic function by Catling et al. (8):

$$y = 100 - 11.6 \log x.$$

More than one cause of loss can be incorporated in multifactorial models. The yield of sugarcane in USA was related to percent "joints bored" (x_1) by <u>Diatraea</u> <u>saccharalis</u> borer and to the position of the attack up the cane (x_2) by McGuire et al. (22):

$$y = 2.7 - 0.01x_1 + 0.06x_2.$$

In India, Abraham and Khosla (1) related rice yield (y) to many variables, those contributing significantly to the regression being percent whiteheads (x_1), percent infested earheads (x_4) due to borer attack, and attack by a disease, <u>Helminthosporium</u> (x_5):

$$y = 3655 - 40.3x_1 - 32.2x_4 - 303.8x_5.$$

From this, an avoidable loss of 204 kg/ha $\pm$ 32 (SE) was calculated, which, after deducting costs, gave a net return of 76 rupees/ha $\pm$ 22 (SE).

Attack by borer on different parts of the plant can be examined, for example Rose (26), quoting Wall, who measured the percent yield reduction in maize due to <u>Busseola</u> <u>fusca</u> borer in Zimbabwe. Loss of 43% was found after stalk attack only, 13% after leaf attack and 49% with both leaf and stalk attack. Second generation attack on cob tips caused 10% loss. In the USA, loss in maize due to the European corn borer, <u>Ostrinia</u>, has been divided into loss in ear weight (A) plus waste at harvest (B), quantified by Chiang et al. in Chiarappa (12) as:

$$A = an_1 + bn_2 + cn_3 \qquad\qquad B = N (z + t)$$

where (n_1 etc.) are the numbers of 2nd generation larvae in the first three nodes of the plant respectively, (N) is the percent stem breakage, (z) is the percent which is machine-harvestable, and (t) is the effect of weathering.

Most relationships can be fitted by a polynomial function, determined by partitioning the variance into linear, quadratic or cubic components. Thus, rice yield (y) was related to the percent borer-infested stems per hill (x) in Japan (17) by:

$$y = 100 - ax - bx^2$$

In Iowa, USA, Lynch (20) related maize yield to the numbers of <u>Ostrinia</u> borer egg masses/plant, and found significant linear, quadratic and cubic components in different years of study, the model varying with maize variety, time of infestation, crop stage, place and year. The rate of loss was certainly not proportional to rise in infestation.

The importance of knowing the spatial distribution of attack in a field has been mentioned, as this will alter the yield/infestation model. In his study, Lynch (20) infested the plots with <u>Ostrinia</u> in a Poisson distribution in order to standardize this aspect of the model.

CONCLUDING REMARKS

Yield/infestation models can be used to predict yields if the amount of infestation or damage is known. For example, Pimentel and Shoemaker (25) used a linear model to predict the effect of not using pesticides on maize and cotton on yields, prices and land area used for growing the crops, a study which included the expected loss of maize due to corn borer, _Ostrinia_. If a temperature accumulation model for borer development (2) can be incorporated, or if borer populations can be predicted from rainfall, as they can in Africa, or from the numbers in an earlier generation, as in Canada, a complete model to predict yield, financial return or other requirement is possible, as has been done, for example, with soybean and other crops.

LITERATURE CITED

1. Abraham, T.P., Khosla, R.K. 1967. Assessment of losses due to incidence of pests and diseases on rice. J. Indian Soc. Agr. Stat. 19:69-82.
2. Anderson, T.E., Kennedy, G.G., Stinner, R.E. 1982. Temperature dependent models of European Corn Borer developed in North Carolina. J. Econ. Entomol. 11:1145-1150.
3. Appert, J. 1970. _Maliarpha separatella_: observations nouvelles et rappel des problemes entomologiques du riz a Madagascar. Agron. Trop. 25:329-367.
4. Atwal, A.S., Chahal, B.S., Ramzan, M. 1970. Insecticidal control of the maize borer, _Chilo zonellus_. Indian J. Agr. Sci. 40:110-116.
5. Avasthy, P.N., Krishnamurthy, T.N. 1968. Distribution and sampling of sugarcane internode borer damage. Proc. 13th ISSCT, Taiwan: 1285.
6. Breniere, J. 1982. Estimation des pertes dues aux ravageurs du riz en Afrique de l'Ouest. Entomophaga. 27:71-80.
7. Calora, F.B., Ferino, M.P., Glass, E.H., Abalos, R.S. 1968. Systemic granular insecticides against rice stem borers with consideration of the pattern and characteristics of the infestation. Philipp. Entomol. 1:54-66.
8. Catling, H.D., Alam, Shamsul, Miah, S.A. 1978. Assessing losses in rice to insects and diseases in Bangladesh. Exp. Agr. 14:277-287.
9. Chatterji, S.M. 1968. Research on insect pests of maize, with special reference to stalk borers. Ann. Rep. Indian Agr. Res. Inst. 1967-8.
10. Chiang, H.C. 1964. The effects of feeding site on the interaction of the European Corn Borer, _Ostrinia nubilalis_ and its host, the field corn. Entomol. Exp. Appl. 7:144-148.
11. Chiang, H.C., Hodson, A.C. 1959. Distribution of 1st generation egg masses of European Corn Borer in corn fields. J. Econ. Entomol. 52:295-299.
12. Chiarappa, L. ed. 1971. Crop loss assessment methods: FAO manual. FAO, Rome and Commonwealth Agric. Bureaux, Farnham Royal. (looseleaf and three supplements). 123 pp.
13. FAO. 1979. Guidelines for integrated control of rice insect pests. Plant Prod. and Protect. Paper 14, FAO, Rome. 115 pp.
14. Gomez, K.A. 1972. Techniques for field experiments with rice. Int. Rice Res. Inst., Los Banos, Philippines. 46 pp.
15. Graca, L.R. 1976. Estimativa economica dos prejulzos causados pelo complexo broca-prodridoes na cana-de-acucar no Brasil. Bras. Acucar. 88:12-34.
16. Ho, Dang Thanh. 1982. Incidence and damage due to stem borer in rice ecosystems in Kenya, East Africa. Paper to Upland Rice Workshop, Bouake, Ivory Coast (mimeo).
17. Ishikura, H. 1967. Major insect pests of the rice plant. John Hopkins. 729 pp.
18. Kalode, M.B., Pant, N.C. 1966. Studies on the susceptibility of different varieties of sorghum, maize, and bajra to _Chilo zonellus_ under field and cage conditions. Indian J. Entomol. 28:448-464.
19. Kishen, K., Sardana, M.G., Khosla, R.K., Dhube, R.C. 1970. Field losses caused by pests and diseases in the maize crop. Agr. Situation in India. 25:591-593.
20. Lynch, R.E. 1980. European corn borer: yield losses in relation to hybrid and stage of corn development. J. Econ. Entomol. 73:159-164.
21. Mathes, R., Baum, R.J., Charpentier, L.J. 1965. A method of relating yields of sugar and sugarcane borer damage. Proc. 12th ISSCT. 1388-1396.
22. NcGuire, J.U., Mathes, R., Charpentier, L.J. 1965. Sugarcane yields affected by borer infestation and position of injury on stalk. Proc. 12th ISSCT. 1368-1396.

23. Panwar, V.P.S., Sarup, P. 1979. Relationship between successive dates and sowing maize and damage caused by <u>Chilo</u> <u>partellus</u> affecting grain yield. J. Entomol. Res. 3:9-24.

24. Patch, L.H., Still, G.W., Schlosberg, M., Bottger, G.T. 1942. Factors determining the reduction in yield of field corn by the European Corn Borer. J. Agr. Res. 65:473-482.

25. Pimentel, D., Shoemaker, C. 1974. An economic and land-use model for reducing insecticides on cotton and corn. Environ. Entomol. 3:3-20.

26. Rose, D.J.W. 1976. The relation between maize pests and crop loss. Rhodesia Agr. J. 73:139.

27. Ruinard, J. 1971. Nature and assessment of losses caused by sugarcane borers. Entomophaga 16:175-183.

28. Sarup, P., Sharma, V.K., Panwar, V.P.S., Siddiqi, K.H., Marwaha, K.K., Agarwal, K.N. 1977. Economic threshold of <u>Chilo</u> <u>partellus</u> infesting maize crop. J. Entomol. Res. 1:92-99.

29. Singh, D., Khosla, R.K. 1983. Assessment and collection of data on pre-harvest food grain losses due to pests and diseases. Econ. Social Development Paper 28, FAO, Rome. 127 pp.

30. Tchekmenev, S.Y. 1981. The effect of the maize stalk borer, <u>Busseola</u> <u>fusca</u> on the growth, percentage of broken panicles, and yield of maize. Beitr. Trop. Landwirt. Vet. 19:91-95.

31. Than, Htun, Arida, G., Dyck, V.A. 1976. Population dynamics of the yellow rice borer, <u>Tryporyza</u> <u>incertulas</u> and its damage to the rice plant. Int. Rice Res. Inst. Los Banos, Philippines (mimeo).

32. Usua, E.J. 1968. Effect of varying populations of <u>Busseola</u> <u>fusca</u> larvae on growth and yield of maize. J. Econ. Entomol. 61:375-376.

33. Van Halteren, P. 1979. The insect pest complex in South Sulawesi, Indonesia. Meded. 305, Lab. Landbouwhogesch. Wageningen. 79:33-47.

34. Walker, P.T. 1960. The relation between infestation by the maize stalk borer, <u>Busseola</u> <u>fusca</u> and yield in East Africa. Ann. Appl. Biol. 48:780-786.

35. Walker, P.T. 1981. The relation between infestation by lepidopterous stem borers and yield in maize: methods and results. Europ. Plant Protect. Bull. 11:101-106.

36. Walker, P.T. 1983. Crop Losses: the need to quantify the effects of pests, diseases and weeds on agricultural production. Agr. Ecosystems and Environ. 9:119-158.

CHAPTER 15

THE USE OF PRINCIPAL COMPONENTS ANALYSIS AND CLUSTER ANALYSIS IN CROP LOSS ASSESSMENT

W. W. Shane

Department of Plant Pathology,
Ohio State University, Columbus, OH 43210, U.S.A.

A major activity of scientists is the search for patterns in systems. This is followed by the attempt to predict and explain the occurrence of the patterns by inductive or deductive reasoning. The causal connections between biological entities in systems such as an orchard or potato field are often obscure due to the multiplicity of complex and subtle interactions. The researcher is often woefully aware of the inadequacies of standard statistical techniques such as correlation, simple, and multiple regression techniques to elucidate the underlying structures.

This chapter is concerned with the use of the multivariate statistical methods called principal components analysis (PCA) and cluster analysis to study relationships between disease intensities and crop yield/quality. These two techniques are among the most popular of the multivariate approaches in recent times. PCA and cluster analyses serve as exploratory techniques for dealing with masses of unwieldy data. PCA is concerned with combining variables while cluster analysis usually deals with combining observations for a more compact representation of data sets. (Cluster analysis of variables is sometimes done with a result that bears a close resemblance to PCA.) The aim of this chapter is to give an intuitive feel for the techniques and how they can be used to advantage in crop loss research. The reader is referred to several excellent texts on the topic (6, 7, 14). Most of the larger statistical software packages for mainframe and more recently, microcomputers, contain routines for calculation and graphical presentation of PCA and cluster analysis (e.g., BMDP (5), SPSS (12), and SAS (15)). These statistical approaches are practical only with the aid of computer programs due to the large number of computations involved.

PRINCIPAL COMPONENTS ANALYSIS

One major difficulty with multiple regression analysis of complex biological systems is that the predictors (independent variables) are sometimes redundant. Trends in the data can be obscured by such redundancy. Multiple regression is hazardous in cases when the independent variables are strongly correlated among themselves. PCA is a statistical procedure that can be used to address these problems. The strategy of PCA is to construct new variables, called factors, which are composites of the original variables. Often, such factors present a more concise description of the trends in the data than did the original variables.

An example may be useful here to show how new factors are derived. Suppose we have data that are the measured heights and arm lengths of a group. The variables, height and arm length, can be combined in an equation to form a single new variable for each person that can perhaps be called 'size'. The new variable

Table 1. Sample data for Principal Components Analysis -- Yield and quality data for sugar beet plots.

| Plot no. | Sugar content (%) | Yield (Tons/ acres) | Impurities in raw beets (ppm) | | |
			Sodium	Potassium	Amino nitrogen
1	11.9	15.3	720	1780	910
2	13.3	17.2	550	1970	890
3	14.7	20.1	410	1960	810
4	.	.	.	.	.
			etc.		

Table 2. Standardized sample data for beet plots.

Plot	Sugar %	Yield	Sodium	Potassium	Amino nitrogen
1	-1.564	-1.346	1.488	-0.987	0.843
2	-0.435	-0.676	0.507	-0.378	0.476
3	0.693	0.345	-0.300	-0.410	-0.989
4	0.774	0.522	-0.819	0.134	-1.172
5	.	.	.	.	.
			etc.		

or 'factor' as it is called in PCA, can be thought of as a 'blend' of the original two variables. A major trend expressed by this factor is the range of body sizes in the group of people.

A further example is as follows: consider the data in Table 1 which are yield and quality data for raw roots harvested from sugar beet plots subjected to various treatments. Amino nitrogen, sodium, and potassium are impurities within the beet which result in loss of sugar in the refining process. The most valuable beets have high tonnage and sugar content (%) and low impurities. If only two variables were measured, e.g., yield and sugar content (%), the data from each plot could be plotted on a 2 dimensional graph (Figure 1) with yield on one axis and sugar % on the other.

In the same fashion we can imagine additional variables such as sodium concentration in beet samples as other axes such that each observation or case (data from one plot) is a point plotted in multi-dimensional space (Figure 2). Next, as in regression, we fit a line (factor axis) through the greatest trend displayed by the points. However, in PCA we can then seek a second line, perpendicular to the first, which indicates the next greatest trend. Subsequent trend lines (factor lines) are defined similarly until all the variance in the data is exhausted. In PCA the hope is that the major trends in the data can be explained with only a few factors, so-called principal components.

In practice, PCA is usually applied to data that have been standardized within variables to avoid biasing the analysis towards one variable (Table 2). Various statistical packages accomplish this in different ways, usually by transforming each variable to a standard mean and variance or by using a correla-

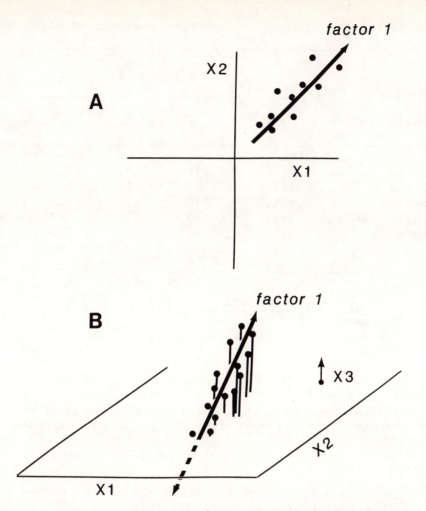

Figure 1. Factor line fitted through: A) two dimensional scatter plot of variables X1 and X2;   B) three dimensional scatter plot of variables X1, X2, and X3.

tion or covariance matrix for the analysis. Many statistical software packages will standardize the variables by default. All variables used in the analysis should be quantitative rather than qualitative. A quantitative variable can take on intermediate values such as a disease severity of 1.5, that is meaningless with a qualitative variable, also known as a class variable, such as cultivar or fungicide type.

Bartlett's Sphericity Test

An initial question to be answered is whether the data should be analyzed by factor analysis at all. Factor analysis is not justified if little correlation exists between variables in the data set. A simple generalized test for correlation among variables is that by Bartlett (2). However, at the present time few software packages include this statistic.

Output for PCA

The initial result from PCA of a data set is the factor matrix, shown in Table 3 for the beet data. Five new factors were obtained (the number of factors calculated is always equal to the number of original variables); however, only the first two 'explained' appreciable trends in the data.

The eigenvalue under each factor column indicates the variability attributed to the factor. The eigenvalues sum to a value equal to the number of variables in

Table 3. Factor matrix for principal components analysis of sugar beet data.

Variables	1	2	3	4	5
			Factors		
Loadings					
Yield	.73818	.01592	-	-	-
Sugar %	.18565	-.85198	-	-	-
Sodium	-.84847	.28670	-	-	-
Potassium	.79457	.47130	-	-	-
Amino N.	-.02106	.72114	-	-	-
Eigenvalues					
	1.93	1.55	0.84	0.51	0.17
Variation explained (%)					
	38.6	31.0	16.7	10.3	0.03

the analysis, which for the beet data example is 5. If there were no trends in the data, each of the five factors would have an eigenvalue of approximately one. Thus, in most analyses, factors with eigenvalues of less than one are usually not considered any further. For the beet data, factor 1 corresponded to 38.6% of the variability (trends).

Factor Patterns

The derived factors can be used in various fashions: 1) use factor values as variables in multiple regression, 2) interpret factors as new concepts, 3) compare factors among data sets, and 4) classify objects using factors as new scales. Most often, factors are examined to reveal independent trends in the data. Individual factors are interpreted by noting those variables that have extreme positive or negative values (so-called high or low 'loadings'). For example, factor 1 of the beet analysis indicates that yield and potassium tend to be high in a plot when sodium is low and visa versa. In factor 2, sugar % contrasts against amino nitrogen. In other words, sugar % is high when the impurity amino nitrogen is low. The trends indicated by factors 1 and 2 are independent of one another due to the method of computation used in PCA. This points out one of the advantages of PCA compared to multiple regression. In PCA, weaker independent trends can be revealed by 'lifting' off a stronger trend first.

A common occurrence in PCA is that the first factor for a data set will have positive loadings (coefficients) of approximately equal magnitude for all the variables. This is so-called 'size' factor, which means that the predominant trend in the data is that certain individuals tend to be large, medium, or small in all the attributes (variables) measured -- e.g., short plants with small leaves, tall plants with large leaves, but generally no short plants with big leaves or visa versa. For example, Madden and Pennypacker (11) used PCA to examine data from 18 tomato early blight epidemics (Alternaria solani). Each case was an epidemic for a plot; the variables for each epidemic were the disease severities at eight assessment dates. Factor 1 in their analysis accounted for 85% of the variation, and was a size factor indicating that there were severe, moderate, and less severe epidemics.

The next factor fitted after a size factor tends to be bipolar -- that is, the factor loadings for the variables in a factor tend to be half negative and half positive. Factor 2 in the analysis of Madden and Pennypacker was bipolar and contrasted the beginning of the season with the end of the season (a rough inter-

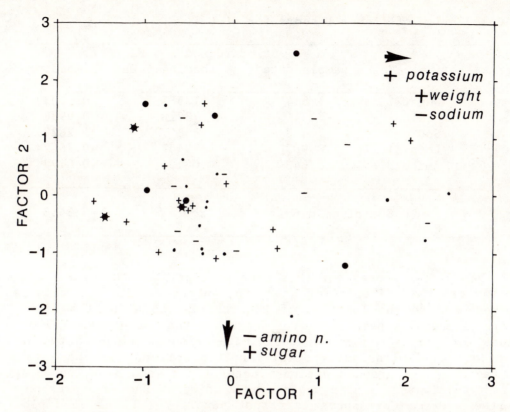

Figure 2. Factor scores for plot harvest data of beet cultivar ACH 14. Symbols give Cercospora leaf spot severities (%) 1 month before harvest as follows:

• = 0 - 1; — = 1 - 2.5; + = 2.5 - 5; ● = 5 - 10; ★ = >10

pretation of this factor is "slope"). Although the trends indicated by factors 1 and 2 are obvious upon re-examination of the data, one of the values of PCA is to induce the researcher to take such an overview from a multi-variable standpoint.

Rotation

A 'size' factor may not be of interest and bipolar factors are often difficult to interpret. For this reason the researcher may elect to use a technique in PCA called 'rotation' to enhance/reveal other trends in the data set. The rotation approach puts a premium on fitting trend lines (factor lines) that maximize contrasts among variables. The rotated factor analysis often presents a simpler picture (so-called simple structure) of trends in the data set; however these may be a different set of trends than displayed by the unrotated analysis. The most popular rotation techniques are the Varimax and Promax methods. As a consequence of rotation some of the variance (magnitude of eigenvalues) is generally shifted away from the first principal components (factors) to those fitted later in the analysis.

Factor Scores

A very useful step in PCA is to calculate factor scores, also called composite indices, for each of the original observations. Each observation will have a new score for factor 1, factor 2, etc., which can be considered as new variables. The factor scores are computed by multiplying the factor loadings by the corresponding standardized variables and then taking the sum of the products. For example, the factor 1 score for plot 1 of the beet data is calculated as shown in Table 4.

Table 4. Calculation of factor 1 score for plot 1.

Variable	Standardized sample data	Factor loading	Data X loading
Sugar %	-1.564	.18565	-0.2903
Yield	-1.346	.73818	-0.9936
Sodium	1.488	-.84847	-1.2625
Potassium	-0.987	.79457	-0.7842
Amino N.	0.843	-.02106	-0.0177
Sum of column is factor score =			-3.3483

The other factor scores for this beet plot or for other plots are calculated in a similar fashion. The factor score for an observation (in this case, a beet plot) corresponds to the location of the observation along the factor axis (see Figure 1B). A useful aspect of factor scores is that observations can be graphed with values for one factor on one coordinate axis and a second factor on another axis in order to simultaneously display multiple trends in the data set. For example, factor 2 was graphed against factor 1 for each beet plot (Figures 2, 3) for two beet cultivars. The symbols used to display factor scores on the figures indicate the average Cercospora leaf spot severity (%) for each plot one month before harvest. As can be seen the trends expressed by factors 1 and 2 do have a biological basis, namely, disease. The analysis has pinpointed an interesting trend which could be the subject of further experiments -- the apparent positive relationship between beet weight and potassium impurities and their inverse relationship with sodium content, all as influenced by disease severity.

PCA is used most frequently to derive new variables which provide overviews about the relationships among the original variables. In one of the most extensive uses of PCA to date in the plant sciences, Stynes [20] used the technique to summarize soil and disease information for a survey of 42 wheat farms in South Australia. As one example, Stynes was able to express 55% of the information in a data set of 11 soil parameters in terms of two factors. The first factor accounted for 41% of the variation in the site data and described a range of coarse to fine textured soils which accompanied an increase in bulk density and water-holding capacities. The second factor accounted for 15% of the variation and summarized a trend of salinity and soil pH. In those sites where salinity decreased slowly within the season the soil pH tended to be high. The principal components analysis provided a means to summarize general trends involving numerous variables.

In other studies the interest may be focused on the objects themselves. For example, Adams [1] used PCA to express genetic diversities of dry bean cultivars. The data set consisted of 36 seed, agronomic, or chemical characteristics (variables) measured for 22 bean cultivars (objects). Factor scores for each cultivar were calculated for the six most significant factors. Similarities among cultivars were indicated by similarities of their factor score profiles. To compare two cultivars, the difference between each corresponding factor score was squared, the squared values for each score were summed, and the square root of this sum was taken to derive a single number. This number represented the Euclidean distance between two cultivars in the six factor coordinate system. This adaptation of PCA is more appropriate than the use of the coefficient of linear correlation (r) because through PCA the redundancy of the original variables is reduced.

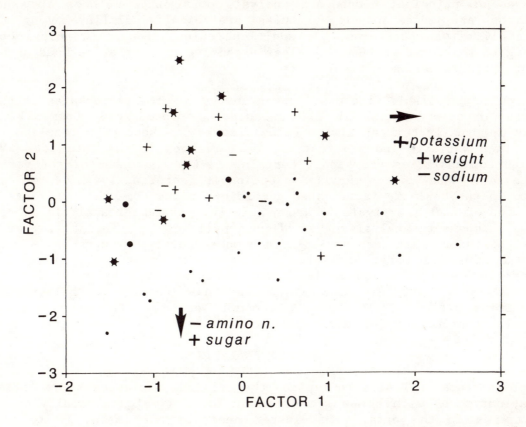

Figure 3. Factor scores for plot harvest data of beet cultivar Betaseed 1230. Symbols give Cercospora leaf spot severity (%) 1 month before harvest as follows:

• = 0 - 5; − = 5 - 10; + = 10 - 20; ● = 20 - 40; ✹ = > 40.

PCA versus ANOVA

The PCA approach as implemented in most statistical packages provides no way to conveniently handle known but unwanted sources of variation in the data set (See reference 14 for further discussion on this point). For instance, in PCA there is no convenient way to allow for the effects of spatial autocorrelation as in randomized complete block designs in the analysis of variance (ANOVA). Blocks are qualitative variables, i.e., numbers used to label blocks are arbitrary. If the block influence is appreciable, the factor line(s) will be "pulled" to this trend in the data during the process of fitting in the PCA analysis. However, since "block" is not included as a formal variable in the analysis the result is a weaker trend as indicated by the factor loadings for the variables actually in the analysis. It is this "free floating" characteristic of PCA, i.e., fitting of factors to trends without consideration of experimental setup, that often makes PCA analyses unique for each data set. This can present a problem for the researcher wishing to find the same general trends among several data sets.

An assumption implicit in PCA is that a linear coordinate system is appropriate for each of the measured variables, although this may not be true in all instances. Conversely, some variables such as soil hydrogen ion concentration are traditionally measured on a non-linear scale (e.g., pH). For example, no evidence is available for or against the hypothesis that the association between beet sugar content and the other variables (sodium, potassium, amino nitrogen) is linear across the range of observed sugar concentrations. The appropriateness of linear or non-linear scaling of data destined for PCA is an issue deserving closer attention.

PCA is only one of many types of factor analysis procedures available. The advantages of principal components analysis over other methods are that fewer assumptions are made and the results are usually similar among different statistical software packages. In addition, calculation of principal components is straightforward and is generally less expensive than other methods such as Rao, alpha or image factoring.

An option available with most factor analysis software packages is the use of estimates of communalities in the calculation procedure. Communality for a variable is usually but not always calculated as the squared multiple correlation between that variable and the rest of the variables in the data set. Variables with high communalities are highly correlated with the other of the variables in the data set. When the communalities option in factor analysis is employed, each variable in the analysis is weighted according to its communality (range 0 to 1). This method of factor analysis is known as principal factor analysis as opposed to principal components analysis. Use of communalities in principal factor analysis follows the theory that each variable is composed of two components -- the portion that is common with other variables and that portion which is unique to itself (non-correlative). The principal factor analysis will often produce a result similar to PCA. Unless the researcher has reason to believe that the communalities differ widely across the variables, the PCA approach is probably preferable to principal factor analysis.

CLUSTER ANALYSIS

The approach in PCA is to replace the original variables for a collection of observations/cases with a new set of fewer factors which hopefully captures the major trends in the data. In cluster analysis the intent is to reduce the complexity of a data set by grouping the members into clusters. Members of a cluster share more characteristics among themselves than with those in other clusters. The uses of cluster analysis are many and include data simplification, prediction, and classification.

Cluster analysis is a relatively objective means to classify objects. In contrast to traditional hierarchical classification schemes such as dichotomous keys, cluster analysis is a numerical method that does not give special consideration to specific variables in the process of determining group membership. A botanical dichotomous key is a hypothesis which designates crucial characters believed to be useful for taxonomy. Cluster analysis is often used in lieu of, or in search of a workable hypothesis for classification. Cluster analysis provides a concise and straightforward technique to summarize similarities among great numbers of objects which have a great number of measured variables.

Although there are many types of cluster analysis, the same three steps are common to all. These are selection of variables, calculation of inter-object similarity, and grouping of objects by a clustering technique. These three steps will be considered separately in the following discussion.

Selection of Variables

The subjectivity of cluster analysis lies in the selection of the variables and the design of the measurement scale. For instance, numerical taxonomy of bacteria is based strongly on the ability of strains to utilize various substrates for growth, in particular, carbohydrate compounds (19). A broader range of tests emphasizing additional types of compounds and their metabolic pathways would likely result in a different clustering of the same strains (16). Redundancy or interdependence of variables in cluster analysis can be countered to some extent by increasing the types and number of variables measured (17). In any case, the researcher should choose variables that are directly relevant to the problem at hand.

Similarity Measurements

The definition of "similarity" is central to the theory and use of cluster analysis. Similarity among objects is measured as the sum total of one to many elements (variables) measured for each of the objects. The two major classes of similarity measurements can be termed "matching" and "distance".

MATCHING MEASUREMENTS refer to binary variables that assume only one of two possible values, e.g., on/off, positive/negative, growth/no growth. Similarity between two objects can be quantified as the number of "simple" matches, positive or negative for the variables; this number is called the matching coefficient of Sokal and Michener (18). In some instances a "positive" match is considered a stronger indication of inter-object similarity and should be weighed more heavily than a "negative" match. For instance, the capacities of two organisms to parasitize a plant (positive match), indicating the presence of the requisite metabolic systems, may perhaps be considered a sign of greater similarity than the inabilities of both to obtain nutrients (negative match).

DISTANCE MEASUREMENTS refer to those variables that have magnitude, and usually more than two possible states. The term "distance" is used because each variable can be represented as a dimension in Euclidean space. As in PCA, variables are usually transformed to standard means and variances before analysis to avoid giving undue weight to any variables.

Clustering Techniques

The number of clustering techniques has increased so rapidly in recent times that one is tempted to suggest that a cluster analysis of the available clustering techniques is needed to help orientate the researcher. Numerous references are available for further details beyond those given here (4, 6, 8). Two major clustering approaches are agglomerative and hierarchical. Agglomerative methods start with each object as a separate cluster and proceeds by grouping clusters together according to some decision rule until the desired number of clusters is obtained. Hierarchical methods work in reverse--all objects start in one cluster which is subdivided, followed by the division of the subclusters. Agglomerative techniques usually have an advantage over hierarchical methods in that assignment of an object to a cluster is not necessarily permanent and may change later in the analysis. In both agglomerative and hierarchical methods similarity among objects is judged on the basis of the tenacity of the objects in the course of the analysis.

Cluster Analysis in Crop Loss Work

Initial applications of cluster analysis in plant pathology were for taxonomic classification of organisms. A more recent application of the technique has been for the organization of soil, water, and pathogen data. In a study of potato growing regions in the North Central United States, Johnson et al. (9) found cluster analysis to be a useful means to define six production areas based on soil permeability, pH, and water-holding capacity. This overview provided a basis for recommendations concerning the allocation of funding for cooperative multi-state research.

Cluster analysis has also been used to categorize the favorableness of environmental conditions for pathogen activity. Sutton et al. (21) used cluster analysis to assign infection prediction indices to 64 combinations of temperature and wetting duration for Botrytis squamosa on onion. Onion leaves were inoculated with spores of B. squamosa and incubated at one of eight temperatures and eight durations. Percentage of inoculation sites with lesions was determined for each of the treatments following a standard incubation treatment. Cluster analysis was used to organize the temperature and wetness duration treatments into 3 clusters based on the degree of infection, i.e, low, moderate, and high infection. These three catagories provided a simple summary that was deemed more suitable for pest management decisions than would be provided by a response surface model.

Analysis of Disease Epidemics

Cluster analysis has been used in plant pathology to consolidate disease epidemic information. Disease intensities and distributions can be measured in many ways. Poon et al. (13) used cluster analysis based on downy mildew disease frequency, intensity, and extensity data to classify 20 sugarcane factory districts in Taiwan. The analysis provided a concise summary of relative disease activity for the districts from a very complex data set collected over 15 years.

Kranz (10) was among the first researchers to recognize the usefulness of cluster analysis to categorize disease epidemics. He used cluster analysis to group 40 plant and pathogen combinations into groups of relatively similar epidemics. Kranz envisioned that examination of epidemics that clustered together would reveal common underlying biological causes. In a similar approach Campbell et al. (3) used cluster analysis to identify similarities among bean hypocotyl rot epidemics measured with six variables including Weibull curve parameters and area under the disease progress curve. One of the often difficult tasks with cluster analysis is to understand what makes each cluster unique. Kranz (10) suggested that cluster analyses for classification of disease epidemics curves may be improved by including information from artificially generated curves of known types into the data set. With this approach, a disease epidemic curve would be classified according to the known curve it clustered with in the analysis.

CONCLUDING REMARKS

Although many factor and cluster techniques have been developed, in reality only a few are currently available in most general mainframe or microcomputer statistical packages. However, standard cluster and PCA techniques are adequate for most applications. A shortcoming in some cluster analysis packages on mainframe and microcomputer packages is the restricted number of cases that can be handled at one time.

The researcher should resist the temptation to assign meaning to weak trends or clusters. Users of these techniques should be aware that most data sets, no matter what the quality, will exhibit trends in PCA or cluster analyses. The techniques will not salvage a data set of dubious quality. A data set represents a "slice" of a larger causal framework; however the scope of the data set may be so small that none of the causal links are "captured".

Principal components and cluster analyses are statistical methods that assume very little in terms of hypotheses of cause and effect or experimental design. This is either a blessing or a curse, depending on the structure and quality of the data. Both techniques are very useful as means to investigate trends among variables or objects. In most cases the analyses will lead the investigator to look for possible underlying biological bases of the observed patterns or clusters.

LITERATURE CITED

1. Adams, M.W. 1977. An estimation of homogeneity in crop plants, with special reference to genetic vulnerability in the dry bean, _Phaseolus vulgaris_. Euphytica 26:665-679.
2. Bartlett, M.S. 1950. Tests of significance of factor analysis. Br. J. of Psychol. 3:77-85.

3. Campbell, C.L., Pennypacker, S.P., Madden, L.V. 1980. Progression dynamics of hypocotyl rot of snapbean. Phytopathology 70:487-494.

4. Dillon, W.R., Goldstein, M. 1984. Multivariate Analysis: Methods and Applications. New York: John Wiley & Sons, Inc. 584 pp.

5. Dixon, W.J., Brown, M.B. 1979. BMDP Biomedical Computer Programs. 4th Edition. Los Angeles, California: University of California Press. 880 pp.

6. Green, P.E. 1978. Analyzing Multivariate Data. Hinsdale, Illinois: Dryden Press. 519 pp.

7. Harmon, H.H. 1976. Modern Factor Analysis. 3rd edition. Chicago, Illinois: University of Chicago Press. 487 pp.

8. Hartigan, J.A. 1975. Cluster Algorithms. New York: Wiley. 351 pp.

9. Johnson, S.B., Teng, P.S., Bird, G.W., Grafius, E., Nelson, D., Rouse, D.I. 1985. Analysis of potato production systems and identification of IPM research needs. Final Report of the Potato Task Force, NC-166 Technical Committee on Integrated Pest Management. 145 pp.

10. Kranz, J. 1974. Comparison of epidemics. Annu. Rev. Phytopathol. 12:355-374.

11. Madden, L., Pennypacker, S.P. 1979. Principal component analysis of tomato early blight epidemics. Phytopathol. Z. 95:364-369.

12. Nie, N.H., Hull, C.H., Jenkins, J,G., Steinbrenner, K., Bent, D.H. 1975. SPSS: Statistical Package for the Social Sciences. New York: McGraw-Hill, Inc. 675 pp.

13. Poon, E.S., Leu, L.S., Liu, C., Cheng, W.T. 1982. Pathogenographic studies of sugarcane downy mildew in Taiwan I. Historical analysis, regional pathogeopraphic analysis, regional pathogeopraphic classification and some considerations of disease attributues of the epidemics. Ann. Phytopathol. Soc. Jpn. 48:153-161.

14. Rummel, R.J. 1970. Applied factor analysis. Northwestern University Press. 617 pp.

15. SAS Institute Inc. SAS User's Guide: Statistics. 1982 edition. Cary, North Carolina: SAS Institute Inc. 584 pp.

16. Sands, D.C., Schroth, M.N., Hildebrand, D.C. 1970. Taxonomy of phytopathogenic pseudomonads. J. Bacteriol. 101:9-23.

17. Skerman, V.B.D. 1967. A Guide to the Identification of the General of Bacteria. 2nd edition. Baltimore, Maryland: Williams & Wilkins Co. 303 pp.

18. Sneath, P.H.A., Sokal, R.R. 1973. Numerical Taxonomy. San Francisco, California: W.H. Freeman and Co. 573 pp.

19. Stanier. R.Y., Palleroni, N.J., Doudoroff, M. 1966. The aerobic pseudomonads: a taxonomic study. J. Gen. Microbiol. 43:159-271.

20. Stynes, B.A. 1975. A synoptic study of wheat. Ph.D. Thesis. University of Adelaide, South Australia. 291 pp.

21. Sutton, J.C., Rowell, P.M., James, T.D.W. 1984. Effects of leaf wax, wetness duration and temperature on infection of onion leaves by _Botrytis squamosa_. Phytoprotection 65:65-68.

CHAPTER 16

A MECHANISTIC APPROACH TO YIELD LOSS ASSESSMENT BASED ON CROP PHYSIOLOGY

R. E. Gaunt

Microbiology Department, Lincoln University College
of Agriculture, Christchurch, New Zealand

Yield may be defined in many ways, depending on the perspective required (28). Theoretical yield is the calculated maximum yield based on physiological criteria and processes, and is used as a reference point by plant physiologists interested in the improvement of biomass production. Attainable yield is that yield obtained when a crop is grown under optimal conditions using available technology, and is often measured in small experimental plots (5). It is a useful reference yield for plant breeders and physiologists interested in optimizing production in a given environment, and for entomologists and plant pathologists interested in minimizing production losses. Yield loss and crop loss are the differences between attainable and actual yield on local and regional scales respectively. They are measures of the effect of any constraint to attainable yield, though the terms are used mostly by people interested in plant protection and not by agronomists, plant breeders and physiologists.

The theme of this chapter is the physiological basis of production and yield, and therefore the physiological basis of yield loss. I will first examine the current knowledge on limitations to production and the balance of factors which control the several aspects of production. This is followed by a discussion of the effect of pests on production systems at the sub-plant, plant and crop level. Finally, I will examine the information that may be derived by a physiological and mechanistic analysis of yield losses induced by pests.

BIOMASS AND YIELD PRODUCTION

The production of biomass is an integration of genotype, environment and genotype/environment interactions with time. Potential biomass production in most major agricultural crops has been changed little by breeding in the last 80 years (1, 23), though there have been improvements in some recently developed horticultural crops. The major change genetically has been in the proportion of biomass which is incorporated into the harvested portion of the crop, or the harvest index. This change has reduced the stability of yield of modern cultivars, and actual yields are more directly influenced by constraints (21), including pests, than with the land races where yields, though lower, were more stable.

The environment, which includes inputs of radiation, temperature, water, nutrients and pests, is often a major limiting factor to both biomass production and yield. Thus crop yields in the semi-arid areas of the world are constrained relative to the areas where water is not limited. Many field trials are designed to observe the effect of increased inputs such as water, nutrients and pesticides. The results normally range from positive to nil responses and, in a few cases, to negative responses (6). This range of responses is testimony to the complexity of the interactions which occur between production and the environment. It is clear that continued description of these responses in different seasons or localities does not increase our understanding of the processes involved. For example, the production of high yields in high input systems, such as the Laloux (12) and Schleswig-Holstein (7) cereal systems, compared to lower input systems has been described in many investigations. Some general understanding of the relationships between yield and inputs is derived from such studies, but the fundamental physiological explanation of the observed effects is usually not addressed.

A further complexity is the existence of genotype/environment interactions, which have a major influence on production and yield. It is well established that genotypes are better suited to some environments than others, hence breeders and agronomists expend enormous effort in cultivar evaluation trials.   The same interaction has been the subject of fewer investigations by physiologists who have sought to explain the observed interactions. Pest constraints are often not examined comparatively in cultivar evaluation trials, despite the fact that a large proportion of the breeding effort is directed at pest, especially disease, resistance in many crops. For a cropping system, the genotype and some aspects of the physical environment often are determined by breeders and agronomists, with due reference to the local conditions and economic climate.   One of the objectives of yield loss assessment may be to determine the pest constraint aspects of the environment.   The interactions between physical inputs and production should be understood before any analysis of the mechanism of yield losses is undertaken.

THE BALANCE BETWEEN GROWTH AND DEVELOPMENT

Ultimately, biomass production and yield are limited by the capacity of the plant production system to convert light energy into carbohydrates, which are then used for growth and development. The conversion of light energy and the utilization of that energy involves the processes of photosynthesis, respiration, mineral and water uptake, transport, structural development and storage. Some of the energy incorporated into carbohydrates in the green tissues of the plant is translocated through symplast and apoplast pathways to the phloem, where it is actively loaded into the transport system. After short or long distance movement by mass flow, the energy is unloaded actively and transported, again through both apoplast and symplast systems, to the site of utilization.   At the site of utilization, the energy is converted into structural or storage materials.   At many positions in this system, some energy is used in maintenance respiration to allow the various processes to occur.   The net result is the growth and development of the plant, which are regulated in complex and interacting ways.   At some growth stages, especially during early crop growth, source production is limited directly by the genotype or environment.   More often the limitation is internally regulated by feedback control mechanisms associated with growth and development.   Yield loss is generally more concerned with the interactions of growth and development with energy assimilation than with assimilation per se.

At early growth stages, growth is often limited by the capacity of the crop to intercept light and convert this into carbohydrates. Once the canopy has closed (LAI > 3), leaf area is no longer the limiting factor and it is the efficiency of energy capture per unit land area and the duration of biomass production which determine the final biomass and yield. As growth proceeds during the season, tissue differentiation occurs and it is the separation of tissues into assimilation (source), structural, transport and storage units which determines the potential yield. In most plants, there is an early transition from vegetative to reproductive or perennative growth, though some crop plants (e.g., lettuce) do not reach this stage before harvest. Those crops which do make the transition may be classified into sexually reproductive or asexual perennative types.   This classification has important implications for the understanding of their growth and development.   In the former, there is a differentiation and development of specialized tissues which will eventually set a maximum potential yield for the crop.

Cereals and Legumes

The potential yield of cereals is determined by the number of ears per unit land area, the number of grains per ear and the potential grain size. After grain set, the only flexibility in yield is in the realization of the yield potential by grain filling (9). In legumes, the development of yield potential is similar in principle to the cereals, but differs in detail and in the fact that they are mostly indeterminate in development (10, 20). After a vegetative growth phase, during which stem branching may occur, flowering is initiated in the axils of recently formed leaves and this continues often for a considerable period of time. Overlapping with continued flower production, flowers are pollinated and pods are set from a proportion of the pollinated flowers, sequentially from the first formed flower axils. As pod set continues, seed number per pod is determined in the first formed reproductive axils and the process of seed filling is initiated. Thus, in some species at least, there may be simultaneous flowering, pod set and seed set and seed filling. The number of flowers produced per axil and the rate of flower production is initially controlled by the ability of the plant to provide assimilates for these processes, whereas pod set, pod abortion and seed set are determined by assimilate production in the region of development. Seed filling is determined mostly by assimilate production by the subtending leaf and the pod itself. Evidence suggests that continued flowering and pod set is regulated by some type of feedback mechanism related to the number of pods already set. There appears to be great flexibility in the pod setting process in terms of the position of pod set, but the final number appears to be rather inflexible and the regulation is not understood fully. During the final stages of seed filling, the legume crop often has an excess of assimilate production capability and senescence, as in many plants, is probably controlled by the demand for assimilates by the seed. A similar production system operates in cotton (20) and several other major crops.

Potatoes and Beets

These crops have different growth strategies to the cereals and legumes, related to the fact that the harvested portion of the crop is a vegetative perennation organ. After a period of leaf area expansion and root development, storage tissue is differentiated as a swollen tap root or, in the case of potatoes, as a number of swollen tubers. There is some evidence to suggest that only the earlier formed potato tubers eventually accumulate sufficient starch to reach harvestable size, but generally the growth and development is not rigidly structured as in the cereals and the legumes. The accumulation of starch in the harvested portion is directly related to the current capacity to produce and transport carbohydrates and thus any constraint to these processes will have a direct effect on yield. Plant senescence is thought to be regulated by daylength and temperature and thus the yield potential is not predetermined by prior development as described above for the grain crops.

Development Strategies

Final yield is determined by the interaction of growth and development. If growth is reduced by a constraint, development continues in most plant species, since development is controlled by daylength, temperature or other factors not directly related to growth. This often causes a reduction of yield potential and/or yield, though the indeterminate and asexual perennative types are less sensitive than the determinate types. We have seen above that plant species have different development patterns, but within a single species, genotype and physical inputs also influence development and the interaction with growth. For example, two cereal cultivars with a similar yield in a particular environment may achieve that yield in very different ways. The winter wheat cultivars Norman (989/10) and

Table 1. Yield and yield components of two winter wheat cultivars grown at Cambridge, U.K., 1977-8.

Cultivar	Yield t DM/ha	No. of ears/m2	% Survival of shoots	No. of grains/ear	No. of grains /spikelet in central spikelets	Mean grain dry weight (mg)
Norman (989/10)	7.54	366	45	47.5	3.52	43.6
Talent (Benoist 10483)	7.23	715	61	36.4	2.72	27.8

Data derived from Austin et al. (1).

Talent (Benoist 10483), with yields of 7.57 and 7.23 t/ha respectively, produced their yields with different contributions from the yield components, as shown in Table 1 (1).

The development of cv Norman implies that there was a greater contribution to yield from apical growth than in the cv Talent, in which there was a greater contribution from tiller production and survival. The different development strategies have important consequences for an understanding of pest constraints to yield, as reviewed by Gaunt (9) and discussed briefly in the next section. There are also differences in the development of a single cultivar grown under different input systems with low and high yield targets. For example, it was shown that sowing date had a marked effect on the yield of both spring and winter barley genotypes (8). Autumn sowing increased yield relative to spring sowing, mostly because of an effect on the number of ears per unit land area. The growth and development of the crops were very different, and the relative contribution of the individual yield components was changed. Again, there are implications for pest sensitivity in the crops with different yield potentials which must be understood before any prediction or analysis of yield loss can be undertaken.

THE EFFECT OF PESTS ON PRODUCTION

During the infection process, plants with reactive defense mechanisms divert energy to those mechanisms at the expense of the normal processes of growth and development described above. There is little evidence available on the cost of these reactions to the plant, but the indications are that it is low relative to the total production and to the normal maintenance respiratory losses. However, there may be some yield loss in resistant plants continually challenged by pathogen inoculum, even though disease never develops (24). There is also some evidence that leaf surface saprophytes may induce defense reactions on a sufficiently large scale to be a significant energy cost to plants in the field (26). Again, there is little information available on the cost of pathogen growth in successful infections, though there has been some investigation of the cost of symbiotic associations, which may be similar at least to biotrophic diseases. Silsbury (22) provided evidence that the cost of symbiotic bacteria in legume root nodules in terms of carbohydrate consumption was low, both in absolute terms and in relation to the beneficial effects of the association. The cost of rust diseases may be comparable to the symbionts. Bowen (4) and Kosuge (11) reached similar conclusions from the available evidence, but Bethenfalvey et al. (3) suggested that the cost might be significant. Clearly more research would be useful in this area.

There has been much more attention given to the cost of successful infections by pathogens in relation to the expression of disease symptoms. Diseases were classified by McNew (19) into six categories on the basis of their effect on plant processes. These were 1) destruction of food reserves; 2) prevention of seedling metabolism; 3) interference with food (energy) procurement; 4) interference with upward translocation; 5) destruction of food manufacture; and 6) diversion of food stuffs. Though he classified diseases with reference to mechanisms of parasitism, the concepts are equally valid for other pests and in the context of yield losses and are probably more useful than the detailed studies of metabolism of infected tissues. Though the latter have provided much useful information on the mechanism of disruption of plant metabolism at the tissue level, the relation of these processes to plant growth, development and yield production are mostly not sufficiently understood to provide insights into yield reduction. Yield loss attributable to the pest constraints described is an integrator of the physiological effect of the pests with genotype and environment. This integration of multiple inputs to a production system is governed, in a complex manner, by the most limiting factors, and thus any disruption will have a direct effect on yield potential or yield only if the process disrupted is limiting at that time. Thus, for example, the disruption of food procurement will only affect yield directly if growth and development at that time is proportional to procurement. If some other factor is limiting, say the rate of transfer of carbohydrate into the endosperm of a cereal, then disruption to the supply of sucrose to the site of transfer will not directly reduce yield until the supply is reduced below the level of the rate of transfer. However, it should be remembered that the interactions are dynamic and not confined to a single time. This means that in the long term, disruption to a process which has no direct effect at the time of disruption may eventually cause losses by reducing a process later in the growth cycle which is limiting. For example, Lim & Gaunt (13) showed that foliar diseases in barley early in the season reduced the size of leaves which emerged later, and that this reduction in leaf size may reduce yield (14, 15).

THE ANALYSIS OF DISEASE EFFECTS.

The relationship between pests and yield loss has been analyzed mostly by statistical methods. Many models (Chapters 11-14) rely on the identification of a pest, quantification of incidence and/or severity (Chapters 2, 3) and the use of statistically derived descriptions of the relationship between pest and yield. These models are extremely useful when used for survey and prediction purposes on a regional scale within the environment and genotype for which the model was developed (Chapter 11). However, they are less well suited for use with specific fields, with different genotypes/environments or as a basis for pest control strategies. They are unsuited because the model is based on statistical relationships which are not necessarily causal and cannot therefore cope with the variations which occur between individual fields nor with the changed relationship in different environments. Similarly, the models will not necessarily identify the period when the pest is actually causing the yield loss (25). The aim of scientific investigation is ultimately to explain relationships rather than describe them, as with the empirical models. At present, there are no comprehensive mechanistic yield loss models which are usable for loss prediction nor for disease management, but a knowledge of yield production can be used to improve the interpretation of empirical models. Some examples illustrate this point further.

Cereals have received the most attention from crop physiologists and people interested in plant protection and pest management. The physiological basis of yield loss in cereals was reviewed recently (9, 25), so only general points are made here. The final yield, and therefore the effect of pests on yield losses, is determined by the balance of growth within the individual crop considered. A post factum analysis of yield components gives considerable information on the effect of disease on yield production. For example, one can conclude that an epidemic which reduces yield solely because of a reduction in mean grain weight has influenced the processes which determine maximum grain size and/or has reduced the ability of the crop to provide carbohydrate materials for the grain filling process. This identifies the period of constraint as being after the boot stage (GS 55). If the weight loss is associated with grain shrivelling, an indicator of insufficient supply (2), the constraint period can be narrowed down further as having occurred after about GS 73. One must caution that this interpretation ignores the role of pre-boot stage accumulation of assimilated carbohydrates stored in the stem. Recent evidence suggests that these reserves may play a significant role in grain filling only under some conditions. On the other hand, if the yield reduction is attributed solely to an effect on the number of grains per ear, it is implied that the constraint occurred during an earlier period of growth, when grain number is determined. Both effects on yield could occur as a result of an epidemic present during specific periods of growth, or present during the whole of the reproductive phase of growth. If the former, little confusion is likely to arise from empirical modeling, but in the latter case an empirical model may correlate disease severity to yield loss at a growth stage which is not related to the cause of the loss.

Once one has an understanding of the growth strategy of a plant species, this physiological knowledge may be used in the interpretation or modeling of disease/yield relationships. In the series of disease curves depicted in Figure 1a, critical point empirical models (Chapters 11, 13) at times x, y and z may relate severity at several growth stages to yield or yield loss, each with high significance of fit and other statistical parameters. The investigator may decide to use the best, statistically, of these models, develop a multiple point model or select a model based on a knowledge of the yield components most affected by the pest. The last option may result in the choice of a model which is not the best from a statistical viewpoint, but which is more related to the causal mechanism involved. The final choice is determined by the objectives of the model. The situation is illustrated further in the series of epidemics in Figure 1b, which is typical of investigations (14) where the epidemic is manipulated to provide some information on the growth stage at which the pest is most damaging. Epidemics A and D represent the extremes of nil and full control, and epidemics B and C represent intermediate epidemics where the duration is controlled rather than the rate of infection as in Figure 1a. If one assumes that treatments B and C are associated with intermediate yields compared to A and D, the best critical point empirical model will be fitted at the time x. However, from a physiological perspective this model will not describe the relationship, taking into account the effect of epidemics B and C on plant processes during the early (y) and late (z) growth stages respectively. An examination of the components, on the other hand, will reveal these effects and the information can be used to select other types of models, possibly multiple point, for disease management purposes. The situation is complicated by the interactions which may occur between early and late epidemic effects on yield (14, 15).

YIELD LOSS PREDICTION AND DISEASE CONTROL

The same approach may be used in a predictive manner rather than post factum, and it is in this area that there is greatest potential for the development of mechanistic models. An analysis of the pattern of plant growth and the development of yield potential can provide information on the major constraints to yield production during the growth cycle for a single crop in the specific genotype/environment in which it is produced. For example, it may be predicted that the crop of Norman wheat grown in the environment referred to earlier (1) may be limited by photosynthate production during tiller development. Thus a disease at that time which reduces photosynthesis or diverts assimilates to the site of infection, will have an effect on yield even at low severities. On the other hand, grain set and filling may be limited by the potential number of grains determined earlier and not by the availability of assimilates. Thus disease during

155

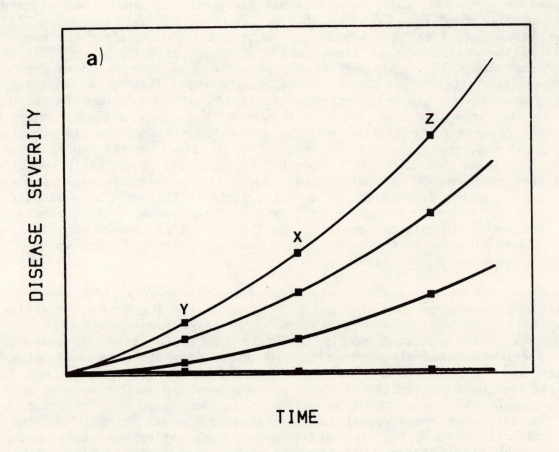

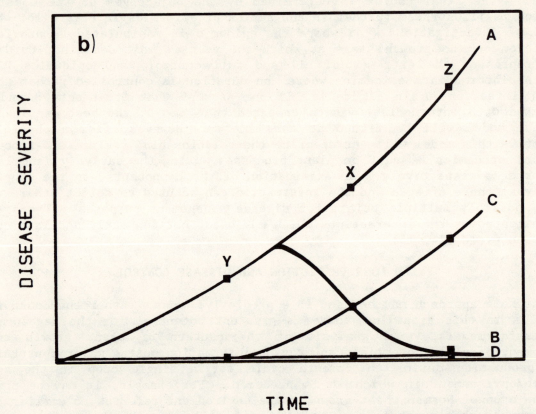

Figure 1. Development of epidemics expressed as disease severity with manipulation of a) rate of infection and b) duration. See text for explanation of x, y, z.

the grain filling period may not reduce yield except at high severities. The same argument will not apply to the cv Talent, where the yield was similar but the growth strategy was quite different. This emphasizes the specificity of mechanistic pest/yield loss relations and also illustrates the limitations of empirical models. Unfortunately, it also illustrates the limitations of current physiological models, as there are few examples of adequate models being available for predictive purposes. However, there is often sufficient information available for an intelligent interpretation of the likely periods of constraint as an aid to disease management. Predictive plant growth models are being developed (27) which will identify the specific constraints to production from genotype/environment data, and these may be used in plant protection as described by Loomis and Adams (16, 17).

At present, pest control or management with variable inputs, such as chemical sprays, is usually based either on an insurance approach or by prediction. Most prediction systems are based on mechanistic models of pest development plus an empirical model of pest/yield loss relationships. As described above, this approach has the inherent hazard of not recognizing the sensitivity of specific crops to pest constraint. This may lead either to the application of chemicals when they are not necessary or to sprays not being applied when they are required. These models could be improved at present by including physiological interpretations of empirical modeling, as in the EPIPRE system, which includes information on crop sensitivity, and in the near future plant growth models will become available which could be linked to, or could include, the mechanistic pest development models.

CONCLUSIONS

The physiological analysis of yield losses is not yet a common part of pest: yield loss studies, partly because there are few plant growth models which have been designed to include pest constraints. As discussed above, physiological knowledge can be used to interpret empirical models, but one of the obstacles to widespread use of this information is that it is very specific to genotype and environment. There are no simple rules which can be used to describe specific situations, though there are general rules for establishing the nature of pest/yield loss relations. The information discussed above, and contained in more detail in the references cited, can be used to give a good estimation of the likely constraints to yield throughout the growth cycle of the crop, and these can be used as a basis for defining thresholds for disease constraints. This approach, coupled with the measurement of pest intensity based on a host parameter (Chapter 2), provides an adequate database for yield loss prediction in specific situations and for disease management. The mechanistic approach will become increasingly useful and accurate as plant growth models are designed by multidisciplinary teams to include pest occurrence as a normal constraint to production along with other factors such as water, temperature and radiation.

LITERATURE CITED

1. Austin, R.B., Bingham, J., Blackwell, R.D., Evans, L.T., Ford, M.A., Morgan, C.L., Taylor, M. 1980. Genetic improvements in winter wheat yields since 1900 and associated physiological changes. J. Agric. Sci., Camb. 94:675-689.
2. Bayles, R.A. 1977. Poorly filled grain in the cereal crop. I. The assessment of poor grain filling. J. Natn. Inst. Agric. Bot. 14:232-240.
3. Bethlenfalvay, G.J., Pacovsky, R.S., Brown, M.S., Faller, G. 1982. Mycotrophic growth and mutualistic development of host plant and fungal endophyte in an endomycorrhizal symbiosis. Plant and Soil 68:43-54.
4. Bowen, G.D. 1978. Dysfunction and shortfalls in symbiotic responses. In Plant Disease: An Advanced Treatise, Vol. III, ed. J.G. Horsfall, E.B. Cowling, pp. 231-256. New York: Academic Press. 675 pp.

5. Chiarappa, L. 1971. Crop Loss Assessment Methods. FAO manual on the evaluation and prevention of losses by pests, diseases and weeds. Farnham, England: Commonw. Agric. Bur. 255 pp.

6. Dougherty, C.T., Langer, R.H.M. 1974. An analysis of a nitrogen-induced depression of yield in irrigated 'Kopara' wheat. N.Z. J. Agric. Res. 17:325-331.

7. Effland, H. 1981. Un systeme intensif en Schleswig-Holstein. Perspect. Agric. 45:14.

8. Ellis, R.P., Russell, G. 1984. Plant development and grain yield in spring and winter barley. J. Agric. Sci.,Camb. 102:85-95.

9. Gaunt, R.E. 1980. Physiological basis of yield loss. In Crop Loss Assessment, ed. P.S. Teng, S.V. Krupa, pp. 98-111. Proceedings of Stakman Commemorative Symposium. Misc. Pub. no. 7. Univ. of Minn. Agric. Exp. Stn., St. Paul, Minn. 327 pp.

10. Hicks, D.R. 1978. Growth and development. In Soybean Physiology, Agronomy and Utilization, ed. A.G. Norman, pp. 17-44. New York: Academic Press. 249 pp.

11. Kosuga, T. 1978. The capture and use of energy by diseased plants. In Plant Disease: An Advanced Treatise, Vol. III, ed. J.G. Horsfall, E.B. Cowling, pp. 86-116. New York: Academic Press. 675 pp.

12. Laloux, R., Poelaert, J., Falisse, A. 1975. Fumure azotee des cereales. Revue de l' Agric. 5:1173-1202.

13. Lim, L.G., Gaunt, R.E. 1981. Leaf area as a factor in disease assessment. J. Agric. Sci., Camb. 97:481-483.

14. Lim, L.G., Gaunt, R.E. 1985. The effect of powdery mildew (_Erysiphe graminis_ DC. f.sp. hordei) and leaf rust (_Puccinia hordei_ Otth.) on spring barley in New Zealand. I. Epidemic development, green leaf area and yield. Plant Pathol. 34:44-53.

15. Lim, L.G., R.E., Gaunt. 1985. The effect of powdery mildew (_Erysiphe graminis_ DC. f.sp. hordei) and leaf rust (_Puccinia hordei_ Otth.) on spring barley in New Zealand. II. Apical development and yield potential. Plant Pathol. 34:54-60.

16. Loomis, R.S., Adams, S.S. 1980. The potential of dynamic physiological models for crop loss assessment. In Crop Loss Assessment, ed. P.S. Teng, S.V. Krupa, pp. 112-117. Proceedings of Stakman Commemorative Symposium. Misc. Pub. no. 7. Univ. of Minn. Agric. Exp. Stn., St. Paul, Minn. 327 pp.

17. Loomis, R.S., Adams, S.S. 1980. Integrative analysis of host-pathogen relations. Ann. Rev. Phytopathology. 21:341-362.

18. McArthur, J.A., Hesketh, J.D., Baker, D.N. 1975. Cotton. In Crop Physiology: Some Case Histories, ed. L.T. Evans, pp. 297. Cambridge University Press. 374 pp.

19. McNew, G.L. 1960. The nature, origin and evolution of parasitism. In Plant Pathology: An Advanced Treatise, Vol. II, ed. J.G. Horsfall, A.E. Dimond. pp. 19-69. Madison: Univ. of Wisc. Press. 715 pp.

20. Pate, J.S. 1977. The pea as a crop plant. In The Physiology of the Garden Pea, ed. J.F. Sutcliffe, J.S. Pate. pp. 469-484. New York: Academic Press. 500 pp.

21. Patterson, H.D. 1980. Yield sensitivity and straw shortness in varieties of winter wheat. J. Natn. Inst. Agric. Bot. 15:198-204.

22. Silsbury, J.H., Smith, S.E., Oliver, A.J. 1983. A comparison of growth efficiency and specific rate of dark respiration of uninfected and vesicular-arbuscular mycorrhizal plants of _Trifolium subterraneum_ L. New Phytol. 93:555-566.

23. Silvey, V. 1981. The contribution of new wheat, barley and oats to increasing yield in England and Wales 1947-78. J. Natn. Inst. Agric. Bot. 15:399-412.

24. Smedegaard-Peterson, V., Tolstrup, K. 1985. The limiting effect of disease resistance on yield. Ann. Rev. Phytopathol. 23:475-490.

25. Teng, P.S., Gaunt, R.E. 1980. Modelling systems of disease and yield loss in cereals. Agricultural Systems 6:131-154.

26. Tolstrup, K. 1984. Saprophytic fungi in the phylosphere of barley and their effects on the plants growth and grain yield. PhD. Thesis. R. Vet. Agric. Univ., Copenhagen. 122 pp.
27. Weir, A.H., Bragg, P.L., Porter, J.R., Rayner, J.H. 1984. A winter wheat crop simulation model without water or nutrient limitations. J. Agricultural Sci., Camb. 102:371-382.
28. Zadoks, J.C., Schein, R.D. 1979. Epidemiology and Plant Disease Management. Oxford, England: Oxford University Press. 427 pp.

CHAPTER 17

THE SYSTEMS APPROACH TO PEST MANAGEMENT

P. S. Teng

Department of Plant Pathology,
University of Minnesota, St. Paul, MN 55108, U.S.A.

Pest management implies, _inter alia_, the maintenance of pest populations at levels of acceptable loss without total dependence on any single control method, and with due consideration of the economic, social and environmental consequences of control inputs. In practice, pest management programs require a knowledge of a) the effect of man and environment on crop and pest population dynamics, b) the effect of pests on yield at different crop development stages, c) the economics of applying control inputs, and d) systems for acquiring, synthesizing and delivering information in an efficient and timely manner. Our modern concept of pest management relies heavily on quick responses to manipulate pest populations before any economic loss is caused. To facilitate this quick response in turn demands that knowledge is available for (a) - (c) above and that a physical mechanism exists for (d). The "systems approach" is a problem-solving philosophy and methodology that is particularly useful for guiding the knowledge generation to support pest management, and for synthesizing the information into useful forms for delivery.

BASIC CONCEPTS

The systems approach represents a holistic view of life, and proposes that a biological system cannot be properly understood or managed by ad hoc knowledge on its components alone. It subscribes to the view that the components of a system interact with each other and are influenced in that interaction by external factors, that a change in one part of the system produces changes in other parts, and that the "whole is more than the sum of its parts". Nearly all biological systems are "open" systems, in the sense that material flows occur into and out of the system; pest management systems have a biological subsystem that is influenced by the external environment. Further, pest (insect, disease, weed) populations exhibit many of the complex, dynamic interactions typical of biological systems and often the only way to adequately understand how these systems function is to build a model of the system.

The approach recognizes a hierarchical organization in natural systems. For example, with a disease epidemic, one level of organization is the population, a second lower level has the pathogen and the host as subsystems, while a third lower level has subsystems for pathogen (germination, sporulation, etc.) and host (leaf, stem, etc.) (27). When modeling a system, it is necessary at the outset to be clear which levels in the hierarchical organization are being addressed. A conceptual boundary is therefore used to distinguish between the system proper and the system environment. Within the conceptual boundary lie all the state variables that constitute the structure of the system. The system proper may be described and quantified by using state variables, so that at any point in time, the value of a state variable is known (19). External to the boundary lie the driving variables that influence the rate at which the system proper functions.

The system environment and the system proper are linked through state and rate equations. A system model may therefore be viewed as a series of equations which collectively describe how the system (and its components) respond to the environment.

A model may be considered any representation of a system in some form other than the original. Thus many types of models exist and it is difficult to have a single system for classifying all models (18). The models most often found in pest management literature are symbolic, mathematical models. Models represent systems, where a system is a collection of objects united by regular interaction to perform an identifiable function. The world is divided into "systems" and "non-systems".

Although the systems approach, as a problem-solving methodology, often leads to the construction of a model, the model itself is not necessarily the product of practical value in pest management (14). Rather, the process of constructing the model, which requires a rigorous examination of the knowledge base on any system, has been shown valuable in generating guidelines for resource allocation. There is therefore a distinction between the systems approach as a philosophy and its use as a methodology (18). Although no consensus exists on terminology, systems research is an encompassing term for activities which have been variously called systems analysis, simulation modeling and system modeling. The systems approach has also been actioned as systems analysis (analysis of system structure and behavior), system control (manipulation of input), system design (restructuring of existing system or structuring of non-existent system) and system synthesis (major rebuilding of system through modeling).

SOFT AND HARD SYSTEMS APPROACHES

Two types of systems have been recognized by some workers: -- 1) those with goals not clearly recognizable and outcomes ambiguous and uncertain, i.e. purposeful or "soft" systems, and 2) those with clear goals and/or predictable outcomes, i.e. purposive or "hard" systems (1, 3). A soft system is exemplified by the activities that collectively represent a pest management decision system while a hard system is exemplified by a pest-host population system. The two types of systems are reflected in the approaches taken to study and manage them (1). The important difference between the two approaches and systems is that a hard system lends itself to building a quantitative, simulation model while a soft system has more unstructured aspects and may be difficult to model. Soft systems are more typical of the unstructured problems associated with human activities, and are more amenable to the building of "expert systems" than quantitative models. Teng (20) has argued that a soft systems approach and its products has more use in the short term for improving pest management than a hard approach and models, while Checkland (3) believes that the hard approach is an extension of the soft.

Both approaches share some common steps and a discussion of these steps will reinforce appreciation of the differences between them. In applying the systems approach, some of the following steps are evident: 1) Specifying and bounding the system in relation to identified problems and objectives, 2) Evaluating the historical and current knowledge about the system, 3) Developing an initial (conceptual) system model, 4) Collecting data and deriving state/rate equations to describe the system, 5) Structuring a detailed system model for computer modeling, 6) Translating the model into a selected language for computer simulation, 7) Sensitivity analyses, verification and validation of model performance, and 8) Model experimentation (19, 23). Steps 1-3 are always present in both the soft and hard systems approaches, and have been called systems analysis (19). Systems analysis is particularly suited for helping scientists improve their understanding of any system, for showing the relationship between research on different system components, for revealing weaknesses and strengths in current knowledge of the system, for promoting interdisciplinary problem-solving, and for guiding the allocation of scarce resources.

PEST MANAGEMENT

Pest management philosophy is in continuous transition, and as such any definition reflects ideas prevalent at that time. Rabb (12) considered pest management to be "the reduction of pest problems by actions selected after the life systems of the pests are understood and the ecological as well as economic consequences of these actions have been predicted, as accurately as possible, to be in the best interest of mankind". This suggests the need to conduct a systems analysis of any pest system before it can be managed, including an evaluation of the non-biological consequences of pest management. Further, a pest management system is a subsystem of a bigger production system (6) and is in turn divided into biological and decision subsystems (11). To practise pest management requires that all available knowledge on the pest, crop and environment be synthesized into a useful form, where control options and their consequences may be systematically evaluated according to economic, social and environmental criteria. A system model is often the only means of synthesizing all the knowledge available for pest management, and together with computer technology, provides a capability for timely evaluation of management options (25).

Inherent in the practice of pest management is the rational use of multiple control methods, and minimal dependence on any single control method, such as a chemical. The need to use multiple control methods in any pest system led to the development of the concept of Integrated Pest Management (IPM), also known as Integrated Pest Control (IPC) (17). The Food and Agriculture Organization (FAO)/ United Nations Environmental Program (UNEP) Panel of Experts on Integrated Pest Control defines IPC as "a pest management system that, in the context of the associated environment and the population dynamics of the pest species, utilizes all suitable techniques and methods in as compatible a manner as possible and maintains the pest populations at levels below those causing economic injury". IPC requires a strong interdisciplinary approach to generate and apply pest control knowledge in an integrated fashion. Some of the operational components of an IPC program include pest surveillance, pest forecasting, crop loss assessment and economic injury level identification, all of which share many common techniques with the systems approach. It is therefore not surprising to see many proponents and applications of a systems approach to pest management (14, 15, 18). Many interdisciplinary problem-solving activities embody systems concepts, implicitly or explicitly, and IPC is definitely an interdisciplinary activity.

THE SOFT SYSTEMS APPROACH

The soft systems approach has been used to analyze both structured and non-structured systems, even though it is better for non-structured systems with no clearly defined objectives (3). In the context of plant protection, examples include identifying research needs for potato pest management (7), analyzing the status of plant protection knowledge and infrastructures in selected West and Central African countries (21), and identifying research and extension needs for rice pest management in Malaysia (11) and for millet and cotton in semiarid Africa (4). In using the approach for pest management at the field level, Teng (20) adapted the procedures of Norton (11) and specified distinctive components for applying the approach (Figure 1). A pest management system may be described by considering a biology/technology subsystem and a management/economics/sociology subsystem (Figure 1). The first represents the ecosystem and its potential, quantifiable components while the second represents the human activities affecting the ecosystem.

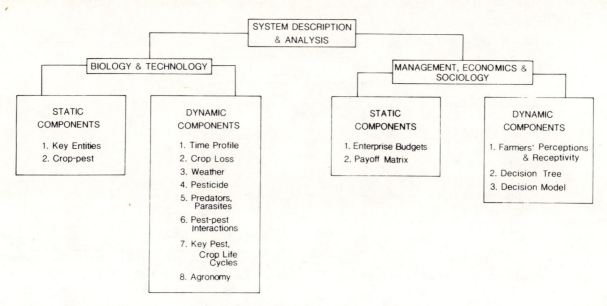

Figure 1. Components of a Soft Systems approach for Crop-pest Management System Analysis (after Norton, 11).

Biology and Technology

The ecological subsystem may be divided into static and dynamic components for analysis.   The static components are those that constitute the structure of the ecosystem, for example the key pests and key crop under consideration. This commonly involves an enumeration of all the pests considered important in the management of the crop, e.g. with the potato cropping system in north central U.S.A., Johnson et al. (7) identified eight insects and twenty-three diseases as forming the potato system structure.   The key entities are other biological organisms, such as predators and parasites, considered important for pest management.

The dynamic components analysis produces information that is represented as two-dimension rectangular matrices   (Table 1). Some matrices commonly used are those dealing with pest-time profile, crop loss, pest - weather, pesticide, predators/parasites, pest-pest interactions and pest-crop growth stage profile. An example using the pest - weather effects matrix (Table 1) illustrates the potential of this form of analysis. For each pest identified in the static system structure, life processes important for understanding the populations dynamics of the pest or for forecasting is arranged as horizontal rows in the matrix. The vertical columns are weather variables considered necessary for pest growth, development and survival. The "box" of intersection between a weather variable (e.g. temperature) and a life process of the pest (e.g. sporulation) may be used to denote a) what the effect of the weather variable is, b) how complete the knowledge is / whether a quantitative relationship exists or c) what the source of the information is. In Table 1, the boxes are filled in with either a +, -, 0, or ?, respectively denoting that a weather variable has a positive, negative, null or unknown effect on the pest life process. The kind of analysis performed using this matrix presents a summary of what is known or not known of the effect of weather on different pests. It is also capable of revealing where areas of past research emphasis have been, and from this point to areas which may be critical for pest forecasting but have not been researched. The matrix is therefore a descriptive framework of our current knowledge of different aspects of the pest ecosystem.   How many "boxes" are filled in further gives an indication of how much data is available for developing a system model for computer simulation.

Table 1. Pest-Weather effects matrix for dynamic components analysis of potato diseases.

DISEASE	WEATHER VARIABLES					
	temperature	rh	wind	rain	dew	sun
Early Blight						
Sporulation	+	+	0	-	-	-
Germination	+	+	0	+	+	-
Infection	+	+	0	-	+	-
etc.						
Late Blight						
Sporulation	+	+	0	?	+	-
Germination	+	+	-	0	+	-
Infection	+	+	0	0	+	-

Note :   + positive effect   - negative effect
         0 no effect   ? unknown effect

A series of matrices to analyze all the important dynamic components of the pest ecosystem (Figure 1) will lead to an improved understanding of the knowledge base available for improving pest management. Although these analyses appear subjective, they are a simple way of summarizing even quantitative information into a common format. Results from the analyses may form the basis for designing management procedures, for example, a matrix with different tillage practices as columns and different pests as rows will show clearly which tillage practices affect several pests and which pests are not suited for cultural control (11). As Rabb (12) noted, the practice of pest management depends on a thorough understanding of pest biology.

Management, Economics and Sociology

The static components used for this analysis are enterprise budgets and payoff matrices and the dynamic components are studies on farmers' perceptions of pests and their control, decision trees and decision models. Enterprise budgets categorize the direct and indirect costs of crop production in general and pest control specifically. These provide a basis for determining the short term economic value of improving pest management, as well as for analysis of longer term benefits to be expected from research on pest management (7).

Pest management programs designed by scientists are likely to fail unless they recognize the needs of farmers and the ability and receptivity of farmers to new technology (20). Data provided through intensive surveys are good measures of the farmers' needs (9). Farmers' decision processes further need to be analyzed and considered in the design of any pest management program. Norton (10) has developed a decision-tree procedure based on operations research, which traces the many decisions involved in producing a crop, from before planting up till harvest.

The soft systems approach enables a systematic definition of the knowledge base for improving pest management. However, there are other factors to be considered in the design and implementation of pest management programs. The infrastructure for information generation, information synthesis, information adaptation, information dissemination and information reception and evaluation, also needs to be analyzed. Often, the lack of improvement in pest management practices is not due to a lack of knowledge or available technology, but to the dissemination or adaptation processes (20).

THE HARD SYSTEMS APPROACH: SYSTEMS MODELING

Because of the availability of modern electronic computers, the systems approach and systems analysis have become almost synonymous with computer modeling (19). As the above discussion shows, practicing the approach in a soft manner does not require a computer; also, modeling per se does not require a computer. Computers do however, facilitate the mundane aspects of modeling complex systems because of their abilities to do repetitive tasks accurately (22). With computers, the complexity of the model can be increased by incorporating a large number of variables to represent as many components of the system as is desired by the scientist. In pest management, computer models have been mainly used as 1) research tools to improve understanding of the system and point out knowledge gaps, 2) as a means of synthesizing large amounts of data collected from different sources, and 3) as part of management information systems to guide pest control decision making. The models used in pest management have included crop models, pest population dynamics models, and crop-pest coupled models (8, 14, 15, 19). With nematode management, Ferris (5) divided models into spatial distribution models, phenology and population models, yield loss models, and management optimization models.

Models as Research Tools

System models attempt to represent a natural system in as complete a manner as feasible, with the prevailing state of knowledge and hypotheses on the system (19). The size of a model, and indirectly its complexity, is only limited by the amount of computer memory available and the knowledge on the system. Computer models are an excellent means to capture knowledge in a form that then enables the researcher to determine how relevant a piece of knowledge is for meeting an objective, such as reducing pesticide use in that system. The computer takes care of the number crunching tasks in modeling, and frees the researcher to explore the subtleties of the system. With developments in user friendly languages, the concepts and data underlying a system become easier to program into a model (23).

The barley leaf rust model of Teng et al. (26) exemplifies disease epidemic models which have been shown to acceptably predict the course of disease development in the field. In this model, which is analogous to an insect population model, the field epidemic is simulated by considering the effect of six weather variables (dew, maximum daily air temperature, minimum daily air temperature, minimum ground level temperature, daily rainfall and daily sunshine hours) on components of the disease infection cycle (latent period, infectious period, sporulation, spore dispersal, germination, penetration, survival). On any one day, any component of the infection cycle may have a successful outcome, depending on the weather conditions present and the equations that relate the effect of a weather variable on a specific life cycle component. The effect of disease control options, such as a fungicide application or a certain degree of rust resistance, can also be simulated. The barley leaf rust model was used to explore strategies for optimizing control, by conducting computer experiments in which different kinds of epidemics were generated (24). Computer experimentation is potentially more powerful than actual field experimentation, as more strategies can explored in a shorter time using controlled conditions. Computer experimentation with a model that has been validated (18) is also useful for providing insights into the working of the real system. The model becomes a self-teaching tool.

Many population models have remained research tools because of the size of the model, the amount of computer memory needed, and the need to monitor the weather variables required to run the model. When pest models are coupled to crop models, the requirements are increased. However, crop-pest models enable a realistic prediction of yield in the absence/presence of different pests populations, as affected by management strategies. Even though their application for real-time pest management is currently limited, they are nevertheless useful for researching strategies (13). Examples of this are found in the work on olive and alfalfa pests (15, 16).

Models as Management Tools

To use models for pest management requires that they be part of a management information system in which data are collected from and delivered to farms. The pest management model is commonly a "skeleton model" (2), containing only the basic equations and logic of the system, but still requiring location specific crop, cultural and environment data to make any predictions. The information system ensures that data needed by the model are received in a timely manner, that recommendations from the model are conveyed to the farmer for timely action, and that the effects of any management decisions are monitored and evaluated to ensure any further corrective action on the pest (25).

On-farm use of a disease management model is exemplified by the barley leaf rust model developed (25, 26) and implemented in New Zealand (27, 28). The model required cropping history and environmental data from each barley field, in order to make a prediction of disease progress and yield reduction. In addition, fields had to be regularly surveyed to determine the mean rust severity, to initialize and calibrate the model. Strategies for rust control were evaluated using economic and risk criteria, based on probabilistic simulation of rust and yield loss (27, 28). The implementation of the barley leaf rust information system was found to be limited by access to a mainframe computer needed to run the model, although the authors believed this limitation will be minimized in the future (27).

Simulation models resemble "IF-THEN" calculators and do not produce an optimal solution for any pest management problem (19). To achieve this requires the use of simulation models in conjunction with quantitative management techniques such as those found in Bayesian decision theory, mathematical programming and risk analysis.

Key activities in applying the systems approach to pest management are therefore: a) identifying and describing the key pests, crops, beneficials and environment in the system, b) defining the management unit of the system, c) developing reliable biological and environmental monitoring techniques, d) developing the pest management strategy, e) establishing economic thresholds based on sound pest-loss information, f) developing descriptive and predictive models of pests and their losses and g) developing interactive channels for the generation, synthesis, adaptation, dissemination and reception of information. These activities are a pragmatic translation of systems concepts and are by no means exclusive. However, their inclusion in the design and implementation of a pest management program will ensure a degree of "holism".

LITERATURE CITED

1. Bawden, R.J., Macadam, R.D., Packham, R.J., Valentine, I. 1984. Systems thinking and practices in the education of agriculturalists. Agric. Systems 13:205-225.
2. Blackie, M.J., Dent, J.B. 1974. The concept and application of skeleton models in farm business analysis and planning. J. Agric. Econs. 25:165-175.
3. Checkland, P.B. 1981. Rethinking a systems approach. Journal of Applied Systems Analysis 8:3-14.
4. FAO, UN. 1984. Analysis and design of integrated crop management programmes. Working Paper no. AGP:PEST/84/WP/3:4, FAO Committee of Experts on Pest Control, Third Session, Rome 24-26 October 1984. 18 pp.

5. Ferris, H. 1985. Basic modeling strategies for nematode management. In An Advanced Treatise on Meloidogyne, Vol. II Methdology, ed. K.R. Barker, C.C. Carter, J.N. Sasser, pp. 205-213. Raleigh, NC: North Carolina State University Graphics. 223 pp.

6. Geier, P.W. 1982. The concept of pest management - integrated approaches to pest management. Protection Ecology 4:247-250.

7. Johnson, S.B., Teng, P.S., Bird, G.W., Grafius, E., Nelson, D., Rouse, D.I. 1985. Analysis of potato production systems and identification of IPM research needs. Final Report of the Potato Task Force, NC-166 Technical Comm. on Integrated Pest Management. 145 pp.

8. Loomis, R.S., Adams, S.S. 1983. Integrative analyses of host-pathogen relations. Ann. Rev. Phytopathol. 21:341-362.

9. Mumford, J.D. 1981. A study of sugarbeet growers' pest control decisions. Ann. Appl. Biol. 97:243-252.

10. Norton, G.A. 1976. Analysis of decision making in crop protection. Agroecosystems 3:27-44.

11. Norton, G.A. 1982. Report on a systems analysis approach to rice pest management in Malaysia. MARDI Report. 62 pp.

12. Rabb, R.L. 1970. Introduction to the conference. In Concepts of Pest Management, ed. R.L. Rabb, F.E. Guthrie, pp. 1-5. Raleigh, NC: North Carolina State Univ. Press. 242 pp.

13. Reichelderfer, K.H., Bender, F.E. 1979. Application of a simulative approach to evaluating alternative methods for the control of agricultural pests. Am. J. Agric. Econ. 61:258-267.

14. Ruesink, W.E. 1976. Status of the systems approach to pest management. Ann. Rev. Entomology 21:27-44.

15. Shoemaker, C.A. 1980. The role of systems analysis in integrated pest management. In New Technology of Pest Control ed. C.B. Huffaker, pp. 25-49. New York: John Wiley & Sons, Inc. 500 pp.

16. Shoemaker, C.A., Huffaker, C.B., Kennett, C.E. 1979. A systems approach to the integrated management of olive pests. Environ. Entomology 8:182-189.

17. Stern, V.M., Smith, R.F., van den Bosch, R., Hagen, K.S. 1959. The integrated control concept. Hilgardia 29:81-101.

18. Teng, P.S. 1981. Validation of computer models of plant disease epidemics: A review of philosophy and methodology. Z. Pflanzenkr. Pflanzenschutz 88:49-63.

19. Teng, P.S. 1985. A comparison of simulation approaches to epidemic modeling. Ann. Rev. Phytopathol. 23:351-379.

20. Teng, P.S. 1985a. Integrating crop and pest management: the need for comprehensive management of yield constraints in cropping systems. J. Plant Prot. in the Tropics 2:24-39.

21. Teng, P.S. 1985b. Plant Protection Systems in West and Central Africa, A Situation Analysis. Final Report to FAO Plant Protection Service, University of Minnesota, St. Paul, MN 55108. 191 pp.

22. Teng, P.S., Rouse, D.I. 1984. Understanding computers - applications in plant pathology. Plant Dis. 68:539-543.

23. Teng, P.S., Zadoks, J.C. 1980. Computer simulation of plant disease epidemics. In McGraw-Hill Yearbook of Science and Technology 1980:23-31. New York: McGraw-Hill. 447 pp.

24. Teng, P.S., Blackie, M.J., Close, R.C. 1977. A simulation analysis of crop yield loss due to rust disease. Agric. Syst. 2:189-198.

25. Teng, P.S., Blackie, M.J., Close, R.C. 1978. Simulation modelling of plant diseases to rationalize fungicide use. Outlook on Agriculture 9:273-277.

26. Teng, P.S., Blackie, M.J., Close, R.C. 1980. Simulation of the barley leaf rust epidemic: structure and validation of BARSIM-I. Agric. Syst. 5:55-73.

27. Thornton, P.K., Dent, J.B. 1984. An information system for the control of Puccinia hordei: I - Design and Operation. Agric. Systems 15:209-224.

28. Thornton, P.K., Dent, J.B. 1984b. An information system for the control of Puccinia hordei - II - Implementation. Agric. Systems 15:225-243.

CHAPTER 18

THE CONCEPT OF THRESHOLDS: WARNING, ACTION
AND DAMAGE THRESHOLDS

J.C. Zadoks

Laboratory of Phytopathology,
Agricultural University, Binnenhaven 9
6709 PD Wageningen, The Netherlands

Pest and disease management couples the disciplines of biology and economics. For example, it is necessary for the farmer to know when it is economical to treat his crop and when it is not. However, when the disease or insect situation develops to a degree such that the need for treatment becomes obvious, the crop may already be unsalvageable. This situation can be avoided with foresight.

Application of knowledge of the population dynamics of insects and diseases is one such precautionary method. Scientists study population dynamics to predict the probable development of a population under a given set of conditions. Field trials and farmers' experience can help to determine, given a set of conditions, the insect/disease level at which the treatment cost equals the treatment benefit. Above this level, treatment is beneficially economical; below this level, treatment will not be beneficial. Yet, the problem of determining a scientific basis for this beneficial, bio-economical level remains.

Determining and accepting such a level at which to act implies a rejection of the idea that a harmful agent must be eradicated at any price, as well as a rejection of schedule spraying. It implies an acceptance of the idea of "managing" insects and diseases by keeping them within economically acceptable limits. It also implies an acceptance of the idea that crops and their harmful agents should be watched closely, with careful quantification of observations. Implicated as well is the idea that crops should be monitored and that insight, knowledge, and "know how" should be used to obtain the best possible economic results.

INJURY, DAMAGE AND LOSS

Any biological agent injuring the crop is called a "harmful organism" (1), be it insect, nematode, fungus, bacterium or virus. Three essentially different concepts can be distinguished by giving them three concise "names". "Yield" is the measurable produce of economic value from a crop (2). Any visible and measurable symptoms caused by a harmful organism are, collectively, called "injury", the equivalent of the obsolete "crop damage". Any reduction in quantity and/or quality of yield is called "damage", the equivalent of "crop loss". The reduction in financial returns per unit area due to harmful organisms is called "loss".

Injury may lead to damage and damage may lead to loss, though by necessity, price mechanisms may sometimes interfere. The demand for agricultural commodities usually is inelastic. Prices will fall deeply if the market is saturated and will rise steeply if there is a scarcity of the commodity. An example is given in

Table 1. Newspaper message (NRC/Handelsblad, 16 March, 1982, translated)
         indicating the difficulty of establishing a damage-loss relationship.
         Crop values added.

===

Dutch fruit growers harvested 325 million kg of apples last year; one third less
than in 1980. This was due to the frost period late in April. The smaller yield
was meanwhile favorable to price development. The mean auction price per kg in
the second part of last year was about Dfl* 1.16 against Dfl 0.46 in the same
period of 1980.

Year	Crop value
1981	325 * 10-6 kg at Dfl 1.16 = Dfl 377 * 10-6
1982	3/2 * 325 * 10-6 kg at Dfl 0.46 = Dfl 224 * 10-6

----

* At the time, 1Dfl was approximately equivalent to US$ 0.40.

Table 1.

Population densities of harmful organisms can be determined directly by
counting, e.g. with insects, or indirectly by injury assessment, e.g. with foliar
pathogens (7). The "damage function" relates damage to injury:

$$D = f(I).$$

The "loss function" relates loss to damage:

$$L = f(D).$$

The mathematical equation describing the damage function can be established with
reasonable accuracy for any pathosystem within the context of a well-defined
cropping system. The parameters will vary according to cultivar, locality, soil
type, and other factors. The loss function is far more difficult to determine; it
depends on the economic situation of the location and the time or period.

THE DEFINITION OF THRESHOLDS

The equilibrium level indicated above has been called the "economic injury
level". The classic definition of the "economic injury level" is "the lowest
population density that will cause economic damage" (14). A recent definition is
"the level of pest attack at which the benefit of control just exceeds its cost"
(8). This nomenclature, once introduced by entomologists, is somewhat confusing.
Pathologists proposed the term "damage threshold" as the phytopathological
equivalent of "economic injury level" (18).

There are good reasons not to postpone a treatment until the harmful agent
has reached the damage threshold. One reason is the risk of treating too late
because, for example, the pest developed slightly faster than expected. Another
reason may be that the disease cannot be stopped instantaneously. In such cases,
action should be taken before the damage threshold is reached. Therefore, an
"action threshold" or to use entomological terminology, an "economic threshold"
is determined.

When the insect/pathogen population has reached the action or economic
threshold, it is time for action, and large scale action requires preparation.
Machines must be ready; chemicals must be available; people must be at hand. If
all the preparations must be done at action time, it may be too late for

appropriate action; hence the need for a "warning threshold", a level of insect/pathogen population density which indicates the possibility of future danger. The warning issued is a "stand-by" warning, a signal to "be prepared". The warning threshold is a means for timely supply of pesticides to farmers and for timely readying of the spray equipment. It may also help the retail trade to be stocked in time.

Whereas t(d) and t(a), the actual dates at which the damage and action thresholds are reached, could be determined per field, per farm, or per region, warning time, t(w), could be determined on a regional or national basis. The damage, action and warning thresholds all have the physical dimension of population size. The threshold concept, corner-stone of the IPM movement, was a great innovation. However, it suggests a stability and a reliability which do not exist in real life. In addition, there is a certain degree of subjectivity in the establishment of fixed thresholds. The difficulties encountered in the implementation of thresholds, discussed in detail by Zadoks (16), will be mentioned briefly.

AGRONOMIC DIFFICULTIES IN USING THRESHOLDS

The economic implications of production level have been overlooked in earlier literature on damage thresholds. An Australian wheat farmer operating at a production level of two metric ton per ha can spend less money to control yellow stripe rust (Puccinia striiformis) than his colleague in the Netherlands expecting eight tons per ha. De Wit (3) distinguished four elementary production situations (Table 2). The physical risks, the types of harmful organisms, the economic benefit of treatment, and the farmer's perception of the economic benefit vary with the level of production. Therefore, the EPIPRE system for supervised control of pests and diseases in wheat asks the farmers to specify an "expected yield" for every field registered, and to indicate the production level (10, 15).

Soil type and fertilizer usage also influence operation decisions. The nitrogen status of plants has an effect on the development of disease. A high nitrogen content of wheat leaves favors rusts and mildew, but retards some perthotrophic fungi such as Septoria nodorum. If an epidemic of P. striiformis has passed a certain level in the spring, it may not be economical to apply the N top dressing so that the extra rust damage due to a high N status of the plants does not surpass the expected benefit of the top dressing (12).

P. striiformis is a great risk on Dutch clay soils but is nearly harmless on the sandy soils in northeast Netherlands. Glume blotch (S. nodorum) is a great risk on these sandy soils but is relatively innocuous on clay soils. As a consequence, complexes of such unexplained interactions between soil types, microclimate, plant physiology, and pathogens force one to bypass the threshold theory with its logic based on population dynamics, and, instead, force one to determine empirical damage thresholds which are independent of soil type (Figure 1).

PHYTOPATHOLOGICAL DIFFICULTIES IN USING THRESHOLDS

The threshold theory only holds if the monitoring is done within a specified time span. The sensitivity of a crop to injury varies during its growing season (5). The damage threshold must be positioned in time relative to crop growth and development. The point was admirably illustrated for aphids in wheat (10). One aphid in early spring may prelude a severe pest outbreak, whereas, one aphid at the end of the season is harmless. Damage increases not only with growing numbers of aphids and with production level, but also with crop development, until development stage DC 77, after which even severe aphid infestations can be disregarded (Figure 2).

Table 2. Four production situations sensu De Wit (3). The production situations deteriorate from situation 1 to 4. Constraints and productivity are indicated.

	Crop production situation			
	1	2	3	4
Temporary constraints	-	water	N	P
Complementary contraints	-	-	water weather	N water weather
Cropping season in days	>> 100			=< 100
Crop productivity in kg per ha per day	150-350			10-15

For harmful agents in wheat, EPIPRE applies "sliding" action thresholds, which depend primarily upon the developmental stage (Figures 1 and 2). They are also called "fluctuating threshold levels", "sliding scales", or "dynamic thresholds".

Rarely does a crop suffer from just one harmful agent. Usually, several pests and diseases occur more or less simultaneously, though at different levels. Harmful agents interact among themselves and with the host crop. The damage done by two interacting harmful agents may either exceed the sum of the damages caused by each separately at the same intensity, or fall short of that sum. This fact profoundly affects the economics of treatments.

Modern systemic fungicides usually are broad-spectrum, that is they control more than one disease. Broad action can also be improved by various tank-ready or farm-made mixes. Broad action may become problematic when beneficial insects are killed by broad-spectrum insecticides, or when beneficial fungi are killed (e.g. fungal antagonists to fungi and fungal parasites of insects).

Threshold theory tends to consider insects and diseases one by one, as if they never occur simultaneously. The costs of controlling two harmful agents decrease when their control can be performed in one machine run. The situation may arise, that two diseases occur in one crop, that both are below their respective damage thresholds, and that nevertheless simultaneous treatment pays. With concurrent diseases, damage thresholds are no longer independent from each other.

The point is reached that the decision to treat can no longer be made after a simple addition. The calculations supporting a decision to treat and the choice of chemical(s) have to be made by computer. In Europe, EPIPRE is a computerized decision-support system for wheat using a central computer and mail or telephone interaction between farmer and computer (15). SIRATAC in Australia is a comparable system for cotton, though it mainly focuses on one pest (6). In the USA several computerized decision-support systems operate in the crop protection area.

Classical damage threshold theory cannot easily cope with concurrent harmful agents. One answer is to calculate stepwise: incidence - present severity - future severity- expected damage - expected loss, to add up the expected losses of all the harmful agents monitored, and to balance the total expected loss against

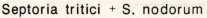

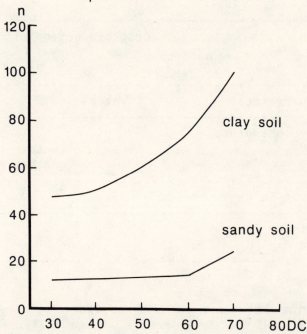

Figure 1. Sliding damage thresholds for Septoria leaf flecks on wheat, EPIPRE, Netherlands. DC indicates the developmental stage (17) of the wheat. n is the number of wheat leaves per sample with at least one fleck. Treatment is needed when n(DC) is above the curve (After 11, with kind permission of the authors).

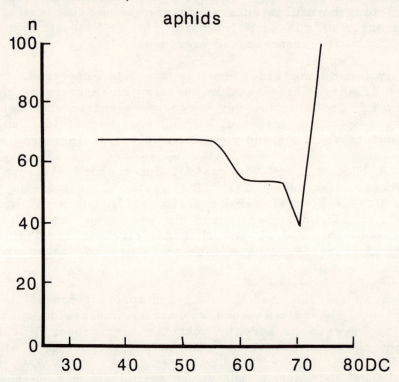

Figure 2. As Figure 1, for aphids. n is the number of wheat stems per sample with at least one aphid.

the expected treatment costs (15). In doing so, the choice of the pesticide or pesticide mix has to be matched to the harmful agents present. A point is reached where the damage threshold no longer has operational meaning. Researchers have long been trained to study one harmful agent at a time, and project funding has tended to stimulate this erroneous research attitude. Only recently is the methodology for a more comprehensive approach being developed.

ECONOMIC DIFFICULTIES IN USING THRESHOLDS

Fixed prices (annually determined ahead of the growing season) exist for some basic food commodities, as for wheat in the European Common Market. Slight variations in price due to cultivar, seed quality and cleanliness do not invalidate this point. Fixed prices of agricultural commodities however, are the exception rather than the rule. Normally, farm-gate prices at harvest time are rather unpredictable due to fluctuations in supply and demand on faraway markets. The time span between $t(a)$, the time at which the action threshold is attained, and $t(d)$, the time at which the damage threshold is reached, is relatively short in comparison to the time span between action time and sales time. What if a short-term decision has to be based on a long-term unpredictability?

In most cases, the position of field or farm relative to road and market has to be considered, as transportation costs affect production costs and/or farm-gate prices. The price expectation of the farmer is not constant over the season. It varies with his appreciation of the market. One solution of this problem is to utilize the prices of the futures market. But which date in the future? In many western countries, where information is plentiful and education is thorough, farmers play the futures market anyhow.

Thresholds, originally presented as physical values measured in terms of population density, are in fact financial values which apply standard economic costs and returns analysis. It is technically feasible to handle a variable expected price in a threshold-based decision system, but the practicality of such "flexible thresholds" is questionable (4). Practicality seems to be limited to situations where prices are more or less fixed either by individual action, e.g. contract between farmer and buyer or collective action, e.g. political decision.

SOCIAL DIFFICULTIES IN USING THRESHOLDS

So far, the threshold theory has been considered a rigid system, a magic algorithm, which produces an imperative threshold value upon receipt of the farmer's specifications. Such a procedure ignores the farmer as an "independent decision maker", a "major actor" with his own objectives, perceptions, and attitudes. Disease and pest management schemes may fail if they do not comply with the cultural elements of the human environment for which they are intended (13). The farmer lives within a social context, which guides, yet constrains him.

A discrepancy between farmers' objectives and scientists' objectives was revealed in the early days of EPIPRE. When EPIPRE began in 1977, the objective of many Dutch wheat farmers was to maximize yield, whereas, EPIPRE was intended to maximize profits.

Farmers' objectives may vary. They may be, for example, maximization of profit, or alternatively, minimization of risk. In subsistence farming food security will be the primary objective, whereas, reduction of labor may be secondary. In general, a hierarchy of objectives is expected (9). The farmer's cropping plan may dictate that hierarchy of objectives. A farmer specializing in row crops (e.g. sugar beet and potatoes) can spend his time more profitably in his row crops than in his wheat. When time is a limiting factor, he will have the wheat treated by a contractor, even when contract-treatment costs more than farmer-treatment when time is not limiting. At one stage, EPIPRE accounted for the farmer's hierarchy of objectives by incorporating the factor "cropping plan".

What can be done when there exists such variance in farmers' attitudes and perceptions? There are several options:

(a) Introduce the best action threshold and unify attitudes and perceptions by intensive instruction. EPIPRE had a unifying effect, correcting extremes in overtreatment and undertreatment.

(b) Produce "flexible thresholds" one for every type of situation and/or type of farmer. This option appears to be highly confusing for farmers and unmanageable for extension workers.

(c) Treat every case as a special case and provide instructions for the calculation of the action threshold, more or less as EPIPRE does.

Whatever the choice, the farmer is an independent decision maker, and should and will make his own decisions in crop protection.

THRESHOLDS OR ADDED VALUE?

Mumford and Norton (8) distinguish four models of decision making:

(a) the threshold,
(b) the marginal cost,
(c) the decision theory,
(d) the behavioral.

Here, the emphasis has been on the threshold model. It is intended to help move away from "schedule spraying" or "standard operating procedures" (calender or development stage bound) and to opt for "flexible response" by "monitoring-and-spraying" in an "adaptive" management mode.

In-depth analysis of the threshold theory raises questions concerning its operational use. The threshold theory gradually vanished; yet, does this mean that the theory is useless? I do not believe so, for two reasons. Firstly, though the threshold theory is not easily applicable in complex disease/insect situations, it is a useful concept from an educational point of view. It can serve as a starting point to explain economically sound integrated pest management. Secondly, not all agricultural or horticultural situations are complex, or exhibit extensive annual variation in the dominance of harmful agents, e.g. the north-west European farming scene. There still is a limited scope for the threshold theory.

The threshold theory considers one pathosystem at a time and focuses on disease or pest intensity level. The theory, therefore, fails to integrate pathosystems. Essentially, the threshold theory is a financial approach, a special case of a broader "added value theory". The latter states that every intervention, including reasoned non-intervention, in the development of a crop and of its harmful agents leads to an added value of the harvested crop, which may be positive, nil, or negative. Intervention will take place only if the expected added value is sufficiently positive. The added value approach has several advantages: (a) It adjusts to different production levels and conditions, (b) It integrates pathosystems, (c) It acknowledges broad spectrum activities of pesticides, and (d) When calculating net benefits, it "talks money". The drawback of the added value approach is that a great amount of knowledge has to be accumulated before implementation can occur.

The concept of value is the corner stone of the computerized calculations in EPIPRE. The farmer is able to see an immediate, expected monetary result of his intervention. In addition, the farmer may add or subtract a certain sum for risk coverage and the government may do so for environmental side effects.

LITERATURE CITED

1. Chiarappa, L., ed. 1971. Crop Loss Assessment Methods. FAO manual on the evaluation and prevention of losses by pest, diseases and weeds. Commonw. Agric. Bur., Farnham Royal, UK: loose leafed.

2. Chiarappa, L., ed. 1981. Crop loss assessment methods - Supplement 3. FAO/Commonw. Agric. Bur., Farnham Royal, UK. 123 pp.

3. De Wit, C.T. 1982. La productivite des pasturages saheliens. In La productivite des pasturages saheliens. Une etude des sols, des vegetations et de l'exploitation de cette ressource naturelle, ed. F.T.W. Penning de Vries, M.A. Djiteye, pp. 22-35. Pudoc, Wageningen. Agric. Res. Rep. #918. 525 pp.

4. Farrington, J. 1977. Economic thresholds of insect pest infestation in peasant agriculture: A question of applicability. PANS 23:143-148.

5. Gaunt, R.E., Lim, L.G., Thomson, W.J. 1982. The identification of disease constraints to cereal yields: crop/food loss appraisal report. FAO Plant Protection Bull. 30:3-8.

6. Ives, P.M., Wilson, L.T., Cull, P.O., Palmer, W.A., Haywood, C., et al. 1984. Field use of SIRATAC: An Australian computer-based pest management system for cotton. Prot. Ecol. 6:1-21.

7. James, W.C. 1974. Assessment of plant diseases and losses. Annu. Rev. Phytopathol. 12:27-48.

8. Norton, G.A., Mumford, J.D. 1983. Economics of decision making in pest management. Ann. Rev. Entomol. 29:157-174.

9. Norton, G.A., Mumford, J.D. 1983. Decision making in pest control. Adv. Appl. Biol. 8:87-119.

10. Rabbinge, R., Rijsdijk, F.H. 1984. Epidemiological and crop physiological foundation of EPIPRE. In Cereal Production, ed. E.J. Gallagher, pp. 227-235. London: Butterworths. 354 pp.

11. Reinink, K., Drenth, H. 1983. EPIPRE 1983. Instructieboekje. PAGV Verslag #10. 28 pp.

12. Rijsdijk F.H. 1980. Systems analysis at the cross-road of plant pathology and crop physiology. Z. Pflanzenkr. Pflanzenschutz 87:404-408.

13. Rosenfield, P., Youdeowei, A., Service, M.W. 1983. Socio - economic considerations in the management of tropical pests and disease vectors. In Pest and Vector Management in the Tropics, ed. A. Youdeowei, M.W. Service, pp.343-364. London: Longman. 399 pp.

14. Stern, V.M., Smith R.F., Van den Bosch, R., Hagen, K.S. 1957. The integrated control concept. Hilgardia 29:81-101.

15. Zadoks, J.C. 1984. EPIPRE, a computer-based scheme for pest and disease control in wheat. In Cereal Production, ed. E.J. Galagher, pp. 215-225. London: Butterworths. 354 pp.

16. Zadoks, J.C. 1985. On the conceptual basis of crop loss assessment: threshold theory. Annu. Rev. Phytopathol. 23:455-473.

17. Zadoks, J.C., Chang, T.T., Konzak, C.F. 1974. A decimal code for the growth stages of cereals. Weed Res. 14:415-421.

18. Zadoks, J.C., Schein, R.D. 1979. Epidemiology and Plant Disease Management. New York: Oxford Univ. Press. 427 pp.

CHAPTER 19

THE ROLE OF PREDICTIVE SYSTEMS IN DISEASE MANAGEMENT

Kenneth B. Johnson

Department of Plant Pathology, University of Minnesota,
St. Paul, Minnesota 55108, U.S.A.

Why include a chapter on disease predictive systems in a book about crop loss assessment? The answer to this question lies in understanding how farmers can put crop loss information to use. Very often, the data we generate with crop loss experiments is only a guide to strategic crop management. That is, the disease-intensity/yield-loss relationship, which is developed from past or current infection levels, only provides an estimate of how much yield loss has occurred. From the grower's viewpoint, this information is useful for future years, as it may change crop production strategies such as the cultivar, the frequency of rotation, or the use of pesticides. However, for current season disease control, crop loss assessment methodologies are not as useful because they only predict yield loss after the disease has occurred. To go a step further in developing comprehensive disease management plans requires a tactical tool to predict disease before it occurs, so that the grower can initiate a control program.

The tools developed for this purpose have been given various terms including disease predictive schemes, disease forecasting (7, 19), and disease warning systems (42).   Zadoks (42) has stated that disease forecasting is currently in fashion.   This is because the potential benefits of this approach include cost effective disease control, limited environmental contamination because chemical treatments are minimized, and an increased awareness by crop producers in regard to the biology of the farming system. Thus, disease predictive systems are both economically and ecologically sound and fit within the conceptual framework of integrated pest management (Chapter 17).

The purpose of this chapter is to examine the practical aspects of disease prediction, in particular, the basic concepts behind predictive systems, necessary developmental tools, and four different approaches to forecasting plant disease. Previous reviews of the topic are those of Bourke (3), Fry (7), Krause and Massie (16), Miller and O'Brien (19, 20), Shrum (26), Young et. al (40), and Zadoks (42).

CONCEPTS OF DISEASE PREDICTION

Diseases Suited for Forecasting

Three principles that determine whether or not a disease is suitable for prediction are:
1) The disease is economically significant in terms of quantity or quality.
2) Some aspect of the disease is variable between seasons.
3) Disease control measures are available and economically feasible.
Because growers are profit motivated, it is unlikely that they will adopt a tool that does not provide an economic benefit. Additionally predictive systems have their own costs of implementation, monitoring, and delivery that must be considered when going through this initial analysis. Seasonal variability in some

176

aspect of the disease is absolutely essential. If disease onset is always on the fourth week after planting and nearly always reaches the same level, there would be little advantage to go any further in developing a predictive system. For many diseases, there is variation in onset time, the rate of the epidemic and the final disease level. The variability in these characteristics of an epidemic must be attributable to some measurable disease influencing factor(s) on which the predictive system will be based.

The ability to implement control measures within the time alloted after a warning has been made, is also important. Response times can vary from hours, as in apple scab (15), to weeks, where a predictive scheme is being used to determine the need for soil fumigation (7, 24). Because relatively rapid responses are usually required, diseases controlled by chemotherapy have been the most frequent targets of predictive work. With increased intensification in agriculture, the list of diseases controlled by chemicals is continually growing. Predictive assessments are now needed for many diseases that in the past were controlled by rotation. Examples are Verticillium wilt of potato (24) and glume and leaf blotch of wheat (42).

The Disease Triangle

Disease is a function of host, pathogen, and environment; all components of the disease triangle. The disease triangle is based on VanderPlank's equivalence theorem (37, 41), which states that the effects of the environment, pathogen, and host can each be translated into the terms of the epidemic rate parameter. The result is that changes in any one of the disease triangle components - from a more to less susceptible host, from a favorable to unfavorable environment, or from a more aggressive to a less aggressive pathogen - all have an equivalent effect on the epidemic. Therefore, it is not surprising that disease management is centered around the equivalence theorem (41) and that disease predictive systems are based on one or more components of the disease triangle.

Table 1 gives examples of some simple predictive systems. By far the most common basis for predictive systems is environment. The importance of environment is reflected in both its direct and indirect effects on disease. The direct effects include the physical conditions - temperature, humidity, leaf wetness, etc. - required for germination, penetration, and infection. Environment also has effects that are less direct but, nonetheless, also influence the variability in disease. These effects include conditions for pathogen survival, pathogen dissemination, and changes in host susceptibility (e.g. the effect of nitrogen on host receptivity). Table 1 also lists several systems based on the measurement of the pathogen. Relationships between inoculum density and disease have been a major research area for many years so it is natural that it extends into disease prediction. Techniques such as spore trapping and determination of inoculum densities in soil have been used to schedule fungicide sprays and predict the potential for disease (1, 2, 9, 24). The measurement of disease itself has often been considered an indirect measurement of inoculum (6).

Forecasting systems utilizing only the host are rare because between and within season variation is usually relatively small compared to the environment and pathogen components. Many of the systems with a mixture of components incorporate host phenology, which may reflect one of several underlying causes including a change in the microclimate within a field (12), a change in host receptivity with age (23), or a differential sensitivity of the crop to damage at different times of the season (6).

It is generally not necessary for all three components of the disease triangle to be included in a predictive scheme. In many instances, disease variability can be attributed mainly to only one component, with the other components exhibiting greater stability. Predictive systems with a mix of compo-

Table 1. Examples of disease predictive systems based on one component of the disease triangle.

Disease Triangle Component	Disease	Pathogen	Disease influencing variable measured	Reference	
Environment	Late blight of potato	_Phytophthora infestans_	Rainfall and temperature	Hyre et al	(10)
			Rainfall and relative humidity	Wallin and Waggoner	(38)
			Synoptic weather charts	Bourke	(3)
	Apple scab	_Venturia inequalis_	Leaf wetness duration and temperature	Mills	(21)
	Early blight of tomato	_Alternaria solani_	Leaf wetness duration, relative humidity, rainfall and temperature	Madden et al	(34)
	Fire blight of pear	_Erwinia amylovora_	Temperature	Thompson et al	(34)
Pathogen	Early blight of celery	_Cercospora apii_	Number of trapped spores	Berger	(2)
	Root rot pea	_Aphanomyces euteiches_	Bioassy of field soil	Sherwood and Hagedorn	(27)
	Southern root rot	_Sclerotium rolfsii_	Direct assessment of inoculum density	See Fry (7) for several references	
Host	Early blight of potato	_Alternaria solani_	Host "physiologic-age"	Pscheidt and Stevenson	(23)

nents are, however, becoming more common because measurement of only one component means assumptions are made about the other two. A natural first step to improve the accuracy and sophistication of a simple system is to add a second (or third) component. This is exemplified by the work to incorporate the effect of cultivar resistance into environmentally based BLITECAST warnings for potato late blight (8). Zadoks (42) reminds us that prediction based on a few observations of the weather or pathogen goes against what is learned in epidemiology in regard to the complexity of disease.

Predictive Systems Within Their Own Context

Of the examples given in Table 1, four are schemes developed for late blight of potatoes. Two reasonable questions to ask are why would several systems be needed to predict one disease, and, is there not a universal system to predict plant disease? Zadoks (42) gives the answers to these questions by using Venn diagrams; he reasoned that the complete reality of the potato late blight pathosystem - made up of biologic, ecologic, economic, and sociologic components - is too complex to be comprehended by a single human mind. Consequently, of all the late blight predictive systems that have been developed, each only represents a small subset of the total lateblight pathosystem. Many of the predictive systems use similar variables and intersect one another but each contains elements that make it unique.

Disease Predictive Systems and Economic Thresholds

The economic threshold concept (Chapter 18, 20) involves determining the pest population level where economic injury occurs and deploying various techniques to keep the population at or below this level. Visual field scouting is commonly used to monitor the pest population level and drastic control actions such as the application of a pesticide are only taken if the population is likely to exceed the economic injury level.

Disease forecasts fit loosely within the economic threshold concept. However, plant pathologists have been somewhat reluctant to adopt the terminology (7). Implicit within every forecast is a notion that an economic threshold will be surpassed without a preventive control action but the techniques of disease prediction are much more diverse than those outlined by economic threshold theory. Although visual monitoring is a technique of disease prediction, for some diseases, such as late blight of potato, visual monitoring is difficult due to latent periods (the period after infection but before visual symptom expression (7)) and because of very low control thresholds that are at or near the perception level (usually considered 1%-5% disease for the non-specialist (17, 41)). For many predictive systems, the greater reliance on environmental monitoring, risk assessment, and infection periods as action triggers seems conceptually different from explicitly defined economic thresholds.

DEVELOPMENTAL TOOLS

The exercise of developing a disease predictive system challenges the researcher to extract essential information on disease epidemiology and synthesize it into a predictive algorithm. Towards this end, systems theory (Chapter 17) plays an integral role. The three phases of the systems approach - analysis, synthesis, and system management - are distinctly evident in disease forecasting. Moreover, the agroecosystem can be split into biological, environmental, economic, and technological subsystems, each of which must be understood independently as well as in how they relate to one another.

Role of Technology

There are two ways to view technology's relationship to disease prediction. Firstly, there are many examples of predictive systems developed largely on keen observation of the biology of a disease and its influencing factors (3, 19). To typify the early work in disease forecasting, the pathosystems selected were usually explosive, highly weather dependent diseases of moderate to high value crops (e.g. late blight of potatoes, apple scab, downy mildew of grape, blast of rice). Development was done by persons who had many years of experience obtained through observation. Meteorological conditions for infection, the first sign or symptom of the disease, and crop phenology were common bases for early systems. Warnings were often given on a regional basis by radio, telegraph, the local press, and even church bells (3). The growers responded by beginning protective fungicide sprays such as Bordeaux mixture. However, the success of what could be considered a "low" technology approach is limited by the inability to monitor and implement predictive systems on a far more localized basis (e.g. a specific field) and by the inability to incorporate the complex interrelationships between environment, pathogen, and host (16).

The second view places greater emphasis on technology to improve disease forecasts and to open new approaches to prediction. In the past decade, the development of the microprocessor has begun a revolution in how scientists and agriculturalists collect, analyze, and deliver information. Detailed environmental monitoring is now done with automated microprocessor-based dataloggers which have the capacity to continuously monitor several localized meteorological instruments at one time (34). These instruments also have the capability to compute integrations in the field and even make the disease predictions in the field (15, 17). Alternatively, field-based automated dataloggers can be coupled to larger computers by telephone or with recordings made in the field on magnetic tape (34).

Recently, there have also been improvements in field sampling procedures, laboratory assessment techniques, data management systems, and statistical analysis packages. These improvements in all aspects of technology allow for a great number of variables to be measured and analyzed. In the future, computer technology will be used more widely for information delivery.

Data Collection and Analysis

In the field, with so many interacting factors influencing disease - dew, humidity, temperature, wind, light, temporal and spatial dynamics of the host, pathogen, and disease, etc. - a tradeoff must often be made between extensive data collection on many factors and intensive observation of a few factors. For determining which variables to measure, the nature of the disease, previous field and laboratory studies, experience, and intuition play a part as does availability of automated collection equipment and the amount of time alloted for detailed observation. The choices of variables to measure will also depend on the approach taken to predict disease. In a later section, four approaches to forecasting will be discussed with many examples of the types of variables chosen.

The first step in data analysis is to examine the relationship between disease and its influencing variables. Usually the investigator has some notion of which factors are influencing the system; nonetheless, it is important to thoroughly explore all the variables with the possibility of finding more efficient or hidden relationships. With large sets of data, the techniques first employed are commonly correlation analysis, regression and multiple regression (4), or principal component analysis (Chapter 15).

Correlation is the general idea that two variables are related in some definite manner. Variables highly correlated with disease, either negatively or positively, are desirable in predictive systems. Examples are the duration of leaf wetness and degree of infection (25), and the amount of inoculum and incidence of disease (1). The correlation coefficient measures whether the correlation is real or due entirely to chance (29). Two problems associated with correlation analysis are that 1) very often disease influencing factors are intercorrelated with one another, and 2) the observational conditions do not result in enough variation for an important influencing variable to be identified as significant. With the first problem, each independent variable should be tested against the dependent variable (e.g. the amount of disease) and also against each other. If the intercorrelations between two disease influencing variables are real and not just due to chance, a decision must be made as to which variable to include in the predictive algorithm. The degree of correlation, efficiency of measurement, and how the system is to be implemented guide the choice.

To deal with the second problem, experiments or observational studies must be planned to get the widest possible range in the variables being examined. Without range or variation in the disease influencing factors, there is not a single statistical procedure that will flag a variable as important. In contrast, although a certain influencing factor may be important in other regions or in laboratory studies, the natural variability of that factor within the locality in which it was measured may be too low to justify inclusion in the predictive scheme. The best approach to obtain data sets with wide variation is to use many sites over several years possibly including treatments at the sites such as cultivar, row spacing, fertilizer, irrigation, and inoculation.

Some of the most powerful predictive variables are created as functions of the original variables. Such variables are often averages, summations, and indices of an important influencing factor governing variation in disease. For example, a relative density index of the crop canopy can be created as a function of the row width, crop width, and crop height (12), because a high density index favorably affects the microclimate required for infection. Another variable may sum the number of hours the relative humidity has exceeded 90% (38) if this is the range where spore germination occurs. The deviation of the average daily temperature below a base temperature may also be summed over the winter months if the severity of the winter influences pathogen survival (5).

The Predictive Algorithm

The final step in development of a predictive scheme is the synthesis of predictive variables into an algorithm for decision making. Although the backbone of the algorithm may range from a complex simulation model (35) to a few simple rules (e.g. the Dutch rules for potato late blight forecasting (20)), the purpose is still the same; to inform growers on the likelihood of a yield or quality reducing level of disease in the future.

FOUR APPROACHES TO PREDICTIVE SYSTEMS

The four approaches to be discussed account for differences in the types of disease influencing factors, the degree of balance between components of the disease triangle, and the degree of technology required for implementation. They are: 1) disease-prediction, 2) infection-prediction, 3) risk assessment, and 4) epidemic projection. The first two approaches are terms introduced by Krause & Massie (16). Risk assessment is a simple tool suited to diseases that are influenced by a number of balanced factors where the decision to control the disease is made only once in the growing season. The last approach, termed epidemic projection, incorporates some ideas of the first three approaches but relies heavily on disease monitoring. It is suited for prediction of polycyclic diseases because sequential estimates of disease severity are used as inputs to a mathematical model to project the probable course of an epidemic.

Infection- and Disease-Prediction: Some Terminology

The terms prediction and forecasting are often used interchangeably. Both these terms imply an aspect of the future. Nevertheless, some forecasting schemes, most notably BLITECAST (17), actually predict disease after infection has occurred. This is in contrast to other systems that predict disease before infection occurs. Krause & Massie (16) introduced the terms disease-prediction and infection-prediction to distinguish between the above two systems.

Disease-predictions are statements made indicating whether or not the biological and meteorological conditions for disease have been fulfilled. These systems have also been termed late warning (16) and post factum prediction (42). Disease-prediction is not prediction in the true sense because it does not forecast future events. Rather, disease-prediction eliminates the need for visual verification of infection, which can be a difficult task when the level of disease is below the perception threshold, or if the disease has a long latent period. Generally, these systems are based on environmental factors that influence disease.

Infection-prediction is the case where disease is forecast several time units in advance. Although these systems can be based on any component of the disease triangle, pathogen based systems predominate because of relationships between inoculum density and future disease intensity. Risk assessment and epidemic projection are actually specialized forms of infection-prediction.

Disease-Prediction Systems

Two well documented disease-prediction schemes are BLITECAST for potato late blight and the Mill's table for prediction of apple scab infection periods.

BLITECAST   The system is a combination of two systems developed by Wallin (38), and Hyre (10). Either system can independently trigger BLITECAST to recommend a fungicide application (17). The trigger based on Hyre's system examines the past ten days of weather to determine the favorability for blight. A day is considered blight-favorable if the ten day cumulative rainfall exceeds 3 cm and the mean temperature for the last five days did not exceed 25.5 C. The system forecasts the initial appearance of late blight 7 to 14 days after the first occurrence of ten consecutive blight-favorable days. The second trigger, Wallin's system, is based on the accumulation of severity units which are a function of the relative humidity and temperature (Figure 1). A spray is recommended if accumulated severity units exceed four or five per week (depending on the amount of rainfall) but only after the seasonal total of accumulated severity units has surpassed a spray threshold of 18-20 units (17).

MILL'S TABLE   This was published in graphical form in 1944 by W. D. Mills (21)(Figure 2). In essence, it is a response surface of the time required by Venturia inaequalis, the apple scab fungus, for infection of apple at varying combinations of leaf wetness duration and average temperature. By obtaining the mean temperature and cumulative leaf wetness for the last 10 to 48 hours, the degree of infection - ranging from none to severe - is determined.

The value of disease-prediction systems is twofold. First, for the early stages of an epidemic, when due to latent periods and perception thresholds, disease monitoring is difficult, these systems give information about when infection has occurred. If an assumption is made about the initial level of inoculum, the amount of disease in the field can be estimated from the accumulated severity units (17). As is done with BLITECAST, the accumulated severity units can then be related to a disease threshold where the first spray should be applied. This may save some unnecessary spray applications early in the season when the level of disease is too low to justify control.

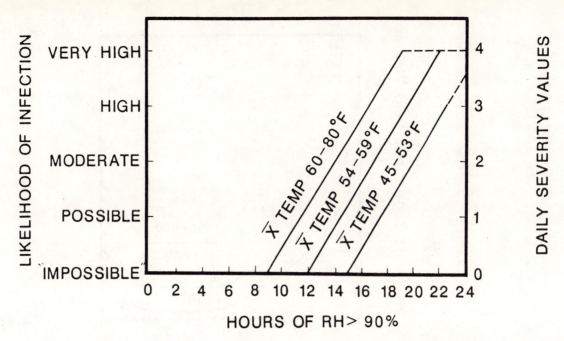

Figure 1. Relationship of the duration of high relative humidity periods and the average temperature during that period to the liklihood of infection and the corresponding severity value (source: Mackenzie (17)).

The second application involves controlling secondary infection cycles with eradicant or therapeutic fungicides. For apple scab, fungicides are rated as to the number of hours they can be used from the beginning of a wetting period and still eradicate recent infections (14). To control the disease, the grower then must act quickly to apply the fungicide and kill the newly developing infections. Using the system in this way introduces some risks for the grower. More conservatively, disease-prediction systems can also be used by growers who predominantly apply protective fungicides but have eradicant compounds available. During prolonged periods of rainy weather favorable to disease, when application of protectants is difficult due to washoff, a disease-prediction system can provide the information for a grower to make the decision to switch from protectant to eradicant chemicals (15).

Both BLITECAST and the Mill's tables have been implemented on automated within-field microprocessors (15, 17). The instruments monitor temperature, rainfall, humidity, and leaf wetness; integrate the information and give the disease-prediction in the field. The microprocessor used in predicting apple scab is currently being adapted to cherry leaf spot, black rot of grapes, and anthracnose of turf (15). BLITECAST has also been implemented on a centralized computer and on hand-held programmable calculators (17).

Other disease-prediction systems include the German negative forecast scheme, PHYTOPROG, for potato late blight (16) and FAST, a predictor of early blight infection periods on tomato (18). FAST has a dual severity unit accumulation system based on leaf wetness duration, temperature, rainfall and relative humidity similar to BLITECAST. PHYTOPROG uses a multiple regression model to arithmetically weight past weather. The system emphasizes blight-free periods, thus is termed a negative forecast.

Infection-Prediction Systems

Infection prediction schemes attempt to forecast disease severity at some point in the future based on past and/or current levels of an influencing variable. These systems are usually quite simple, and very often, pathogen based.

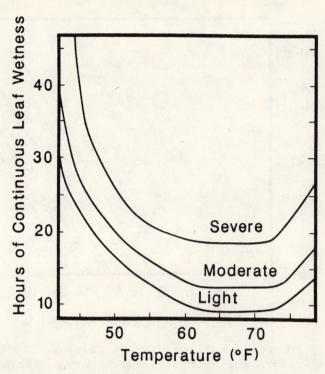

Figure 2. Relationship between intervals of leaf wetness, temperature, and probability of occurrence of apple scab (induced by <u>Venturia</u> <u>inaequalis</u>). If an apple scab infection period occurred, and apple trees were not protected by a fungicide, a grower might wish to apply a fungicide with postinfection efficacy. (Relationships were identified by Mills, 1944; reproduced with permission of Cooperative Extension Cornell University).

Direct measurement of inoculum density is the simplest form of infection prediction. For many diseases, there is good correlation between amount of disease and the level of inoculum; the so called inoculum-density/disease-severity relationship (1). Soil-borne diseases, such as Southern root rot caused by <u>Sclerotium</u> <u>rolfsii</u> (7) and Verticillium wilt caused by <u>Verticillium</u> <u>dahliae</u> have been predicted using this approach (24). Also, some foliar diseases, where there is a relationship between the amount of disease and the number of spores caught by a spore trap (2, 9, 11) may be adapted to this approach. A drawback of using just inoculum density is that the qualitative potential of the inoculum is ignored. Factors which influence the ability of the pathogen propagule to infect include propagule age, propagule distribution, and aspects of the biological and physical environment (24).

A second approach to infection-prediction is to use indirect methods to assess a component of the disease triangle. Some indirect variables include past cropping history, the disease severity the last time the crop was grown in a field, and the amount of disease on bioassay plants grown in field soil before the crop is sown (24, 27). The above variables all indirectly measure inoculum and are well suited to soil-borne diseases to determine the need for rotation and fumigation.

There are two good examples of indirect infection-prediction systems that measure pathogen survival. In the Pacific northwest of the U.S.A., stripe rust caused by <u>Puccinia</u> <u>striiformis</u> is a common disease of winter wheat. A correlation was found between the ultimate disease severity and the number of negative degree days (computed as the deviation of the daily average temperature below a base temperature of 7 C summed over the winter (7)). Because <u>P</u>. <u>striiformis</u> is an obligate rust with no telial stage, it is thought the negative degree day is a relative measure of the number of overwintering lesions. Similarly, Stewart's wilt of corn is a second example where the severity of the preceding winter negatively affects the disease in the subsequent crop (30). In this case, however, the correlation between cold weather and reduced disease is a measure of the survival of the corn flea beetle which vectors the bacterial Stewart's wilt pathogen, Erwinia stewartii.

A last indirect infection-prediction scheme is the system developed to predict fire blight of pear. In fire blight, disease is initiated when epiphytic populations of Erwinia amylovora develop on pear blossoms (34). By monitoring the air temperatures during the bloom period, the population development can be predicted and an antibiotic spray applied when temperatures are favorable for epiphytic colonization (34).

Risk Assessment

Risk assessment schemes are very broadly based systems attempting to consider many factors that influence disease. These systems have also been termed "point" systems (28, 39) because a point score is given to each disease influencing factor. Often incorporated are factors such as planting date, row spacing, cultivar, rotation practice, relative abundance of rainfall and/or irrigation management strategy.

For each factor incorporated into the scheme, a score is given over a range of possible values (zero to a maximum score) by the person making the assessment. The maximum score a factor can receive is a relative weighting of the overall importance of this factor in the disease. After all factors have been assessed, the scores are totaled and compared to a decision threshold value. If the total is below the decision threshold, no action is taken. If the totaled score exceeds the decision threshold, the disease control action is warranted.

Tables 2 and 3 give two examples of risk assessment schemes, both for fungal diseases of legumes (28, 39). The scheme used for snap beans has two decision thresholds; a high threshold to indicate a fungicide is needed and a lower threshold indicating a fungicide may not be needed if irrigation practices can be modified (40).

In general, flexibility is a desirable feature of risk assessment schemes. A system can be developed to be used as a preplanting management tool to assist in crop sequence or rotation decisions; or the assessment scheme can be based largely on information obtained during the season to time the need for a chemical spray. Simplicity, comprehensiveness, and low technology requirements are also important features. Risk assessment schemes serve as simple aids for educating growers on the important aspects of disease management.

Epidemic Projection

Among the most recent developments in disease predictive systems has been a shift toward systems that utilize disease monitoring. Disease monitoring is defined as the actual observation and quantification of the disease level in the field. This change to greater disease monitoring may be due to the trend toward widespread crop scouting and is certainly more in line with how entomologists view integrated pest management.

The main advantage of disease monitoring is knowing with reasonable precision the level of disease at a particular crop growth stage. By obtaining sequential estimates of disease over a period of time, it is possible, using some type of mathematical model, to project the future disease severity level. VanderPlank's (37) model for logistic growth might be one choice. Although one might argue that epidemic projection is implicit in all forecasts, it's the explicit use of disease

Table 2. A risk assessment scheme for determining the need for foliar fungicides to control pod and stem blight of soybeans.

Condition at R3 growth stage (pod set)[a]	Point Value
Rainfall, dew, and humidity up to R1 to R2 (early bloom) and R3 (pod set).	
Below normal	0
Normal	2
Above normal	4
Crop history	
Soybeans grown the previous year	2-3
Tillage practices	
Chisel-plow, disk, or no-till	1
Disease symptoms	
Pycinidia visible on fallen petioles or Septoria brown spot obvious on lower leaves.	2-3
Cultivar selection	
Early-maturing cultivar	1-2
Full-season cultivar[b]	0
End use	
Soybeans to be sold or used for seed	3
Yield potential	
Better than 2.354 kg/ha (35 bu/ha)	2
Less than 2.354 kg/ha	0
Seed quality (warm germination test)	
Less than 85% germination	1
Other	
The 30 day weather forecast indicates above normal rainfall expected. Or, the field has a history of disease.	1-3

a) If a condition does not apply, the point value is zero. Add the point value for each applicable condition. If the total is 12 or more for seed production fields, or 15 or more for grain production fields, foliar fungicides will probably increase yield and seed quality.

b) Foliar fungicides should not be used on full-season cultivars.

Source: Shurtleff, Jacobsen, and Sinclair (28).

Table 3. A risk assessment scheme for decision making for white mold (<u>Sclerotinia</u> <u>sclerotiorum</u>) control in snap bean.

Condition[a] at mid-bloom	Point Value

Prebloom apothecia counts (Examination of 100 m of row over two sampling dates)

Zero	0
One to two	3
Three to five	6
Greater than six	10

Moisture

More than three irrigations planned during and after early bloom	3
Cool wet weather during early bloom	3
Late afternoon or early evening irrigations planned	1

Canopy density at early bloom

Greater than 18 inches of open space between rows	0
Zero to 18 inches of open space beteewn rows	1-5

Crop history (in previous three years)

Beans 3 times	6
Beans 2 times	4
Beans once (last year)	3
Beans once (not last year)	1

Cultivar

Ramono type	2
Other	0

a) Add the point value for each applicable condition. If the total point value exceeds 10, the white mold potential is high. A fungicide and revised irrigation plans (if possible) are recommended. If the total is greater than six, the white mold potential is moderate. Revise irrigation plans if possible; a fungicide is not recommended. At point totals less than six, control is not recommended.

projections based on actual disease estimates that sets this technique apart. Diseases best suited to this approach are polycyclic, can be visually monitored (e.g. leaf spotting organisms), have a relatively slow rate of increase (at least in the initial stages), and are below the economic injury level at the time of detection. Conceptually, epidemic projection is closely tied to the methodologies of crop loss assessment because actual disease severities are projected and the disease-severity/yield-loss relationship is the basis for all decisions. With more sophisticated models, it is possible to compute the yield loss expected under several environmental scenarios using historical weather as a database. With the frequency distributions of past weather, a degree of risk or uncertainty associated with each decision can be computed. Two examples of disease projection systems are EPIPRE (42), developed in the Netherlands, and FUNGINFO (35), developed in New Zealand.

EPIPRE   The system was designed for use in winter wheat and entirely on disease monitoring and does not include any component of the weather. EPIPRE gives recommendations for six diseases and one insect - stripe and leaf rust, powdery mildew, eye spot, leaf and glume blotch, and aphids. In the Netherlands, the farmers do their own monitoring and send their findings to a central processing center.   Disease and insect levels are entered into a computer and future projections are made using modified exponential growth equations. Recommendations for all pests are sent out by mail on the same day they were received.   The recommendations are "do not treat", "treat", or "wait and see" (42).   For the disease projections, the value of the exponential equation's infection-rate parameter, "r", is computed as a function of the disease or insect, cultivar, soil type, growth stage, nitrogen fertilizer level, and the use of plant growth regulators. This function, although quantitative in form, is not distinctly different from a risk assessment approach. Using the computed "r", the disease level is then projected and coupled to a disease-intensity/economic yield-loss damage function , which serves as the basis for control decisions. The EPIPRE system has been adopted in several North European countries and the original program has also been modified into an "expert system" in the U.K.

FUNGINFO   This system uses disease and environmental monitoring to initiate a disease simulator - BARSIM (32) - to project disease development of barley leaf rust caused by Puccinia hordei (35). Disease projections are made after running the model stochastically many times, with sampling from historical frequency distributions of temperature, rainfall, and dew periods on each run. This process generates a distribution of probable disease severities that is translated to a yield loss probability distribution with an empirical multiple point model. Using the yield loss distribution, yield expectations are compared for the "spray now" and the "do not spray" strategies. A recommendation to spray is made if the yield loss expectation without spraying is greater than the break even cost. Implementation of FUNGINFO has been slowed by farmers lack of access to the appropriate computing facilities (36). As an alternative implementation strategy, Thornton & Dent (36) used FUNGINFO to generate tables of spray recommendations based primarily on the number of dew periods preceding the current date. The date of planting (reflective of crop growth stage) and date of disease onset are used as coordinates in the tables. After disease onset, the tables are designed for weekly "spray" or "do not spray" decisions to be made. The tables also allow the growers to determine their own level of risk.

CONCLUDING REMARKS AND OUTLOOK

Compared to a few decades ago, our understanding of plant disease and our ability to predict them have vastly improved. Researchers have done well to develop and adopt new technologies for assessing predictive variables. However, while there is still the need to improve existing systems and to develop new schemes for additional diseases, the challenge for the future is to obtain wider implementation of predictive systems than presently exists.

With rising energy, ecologic, and sociologic costs attributed to pesticide use, the future of pest management and disease forecasting should be bright. Yet, many predictive systems are currently limited to the literature. Some hard questions need to be asked on why predictive systems have not been more widely adopted . Have we done all we can to validate and understand the uncertainties of predictive systems? Have we developed the systems to relate smoothly to other aspects of crop management? Or, on a more positive note, have we just failed to adequately document where and how frequently predictive systems are used?

More than ever, systems thinking needs to be adopted for the development and implementation of disease forecasting systems. The bringing together of crop management and pest management is needed. Holistic approaches, such as EPIPRE, which integrate single pest forecasts into multiple pest systems have been shown to offer much benefit to growers.   Further work is necessary to unify the biology of disease prediction with the economic aspects of yield and quality loss. Additionally, our efforts must be tailored to the decision making processes of growers.

LITERATURE CITED

1. Baker, R. 1978. Inoculum potential. In Plant Disease Vol. II., ed. J.G. Horsfall, E.B. Cowling, pp. 137-157. New York: Academic Press. 436 pp.
2. Berger, R.D. 1969. A celery early blight spray program based on disease forecasting. Proc. Fla. State Hortic. Soc. 82:107-111.
3. Bourke, P.M.A. 1970. Use of weather information the prediction of plant disease epiphytotics. Annu. Rev. Phytopathol. 8:345-370.
4. Butt, D.J., Royle, D.J. 1974. Multiple regression analysis in the epidemiology of plant diseases. In Epidemics of Plant Diseases: Mathematical Analysis and Modeling, ed. J. Kranz, pp. 78-114. Berlin: Springer-Verlag. 171 pp.
5. Coakley, S.M., Line, R.F. 1981. Quantitative relationships between climatic variables and stripe rust epidemics on winter wheat. Phytopathology 71:461-467.
6. Eversmeyer, M.G., Burleigh, J.R., Roelfs, A.P. 1973. Equations for predicting wheat stem rust development. Phytopathology 63:348-351.
7. Fry, W.E. 1982. Principles of Disease Management. New York: Academic Press. 378 pp.
8. Fry, W.E., Apple, A.E. Bruhn, J.A. 1983. Evaluation of potato late blight forecasts modified to incorporate host resistance and fungicide weathering. Phytopathology 73:1054-1059.
9. Harrison, M.D., Livingston, C.H., Oshima, N. 1965. Control of potato early blight. II. Spore traps as a guide for initiating applications of fungicides. Am. Pot. J. 42:333-340.
10. Hyre, R.A., Bonde, R., Johnson, B. 1959. The relation of rainfall, relative humidity, and temperature to late blight in Maine. Plant Dis. Rep. 43:51-54.
11. Jeger, M.J. 1984. Relating disease progress to cumulative numbers of trapped spores: Apple powdery mildew and scab epidemics in sprayed and unsprayed orchards. Plant Path. 33:517-523.
12. Johnson, K.B., Powelson, M.L. 1983. Influence of prebloom disease establishment by Botrytis cinerea and environmental and host factors on gray mold pod rot of snap bean. Plant Dis. 67:1198-1202.
13. Johnson, S.B., Teng, P.S., Bird, G.W., Grafius, E., Nelson, D., Rouse, D.I. 1985. Analysis of potato production systems and identification of IPM needs. Final Report of the potato task force, NC-166 Technical Committee on Integrated Pest Management. 145 pp.
14. Jones, A.K., Fisher, P.D. 1980. Instrumentation for in-field disease prediction and fungicide timing. Prot. Ecol. 2:215-218.
15. Jones, A.L., Fisher, P.D., Seem, R.C., Kroon, J.C., Van DeMotter, P.J. 1984. Development and commercialization of an in-field microcomputer delivery system for weather driven predictive models. Plant Dis. 68:458-464.
16. Krause, R.A., Massie, L.B. 1975. Predictive systems: Modern approaches to disease control. Ann. Rev. Phytopathol. 13:31-47.
17. MacKenzie, D.R. 1981. Scheduling fungicide applications for potato late blight with BLITECAST. Plant Dis. 65:394-399.
18. Madden, L., Pennypacker, S.P., MacNab, A.A. 1978. FAST, a forecast system for Alternaria solani on tomato.   Plant Dis. 68:1354-1358.
19. Miller, P.R., O'Brien, M. 1952. Plant disease forecasting. Bot. Rev. 18:547-601.
20. Miller, P.R., O'Brien, M. 1957. Prediction of plant disease epidemics. Ann. Rev. Microbiol. 11:77-110.
21. Mills, W.D. 1944. Efficient use of sulfur dusts and sprays during rain to control apple scab. N.Y. Agric. Exper. Stn. Ithaca Bull. 630. 4 pp.
22. Mumford, J.D., Norton, G.A. 1984. Economics of decision making in pest management. Ann. Rev. Entomol. 29:157-174.

23. Pscheidt, J.W., Stevenson, W.R. 1983. Forecasting and control of potato early blight, caused by <u>Alternaria</u> <u>solani</u>, in Wisconsin. (abstract) Phytopathology 73:804.

24. Rouse, D.I. 1985. Some approaches to prediction of potato early dying disease. Am. Pot. J. 62:187-193.

25. Royle, D.J. 1972. Quantitative relationships between infection by the hop downy mildew pathogen, <u>Pseudoperonospora</u> <u>humuli</u>, and weather and inoculum factors. Ann. Appl. Biol. 73:19-30.

26. Shrum, R.D. 1978. Forecasting of epidemics. In Plant Disease, Vol. II, ed. J. G. Horsfall, E.B. Cowling, New York: Academic Press. 436 pp.

27. Sherwood, R.T., Hagedorn, D.J. 1958. Determining the common root rot potential of pear fields. Wis. Agric. Exp. Stn. Bull. 531. 12 pp.

28. Shurtleff, M.C., Jacobson, B.J., Sinclair, J.B. 1980. Pod and stem blight of soybean. Report on plant diseases, No. 509 (revised). Dept. of Plant Pathology, University of Illinois, Urbana, IL. 6 pp.

29. Steel, R.G.D., Torrie, J.H. 1980. Principles and Procedures of Statistics. New York: MacGraw-Hill, Inc. 633 pp.

30. Stevens, N.E. 1934. Stewart's disease in relation to winter temperatures. Plant Dis. Rep. 12:141-149.

31. Stuckey, R.E., Moore, W.F., Wrather, J.A. 1984. Predictive systems for scheduling fungicides on soybeans. Plant Dis. 68:743-744.

32. Teng, P.S., Blackie, M.J., Close, R.C. 1980. Simulation of the barley leaf rust epidemic: Structure and validation of BARSIM-I. Agric. Systems 5:85-103.

33. Teng, P.S., Rouse, D.I. 1984. Understanding computers: Applications in plant pathology. Plant Dis. 68:539-543.

34. Thompson, S.V., Schroth, M.N., Moller, W.J. 1982. A forecasting model for fireblight of pear. Plant Dis. 66:576-579.

35. Thornton, P.K., Dent, J.B. 1984. An information system for the control of <u>Puccinia</u> <u>hordei</u>: I. Design and operation. Agric. Systems 15:209-224.

36. Thornton, P.K., Dent, J.B. 1984. An information system for the control of <u>Puccinia</u> <u>hordei</u>: II. Implementation. Agric. Systems 15:225-243.

37. VanderPlank, J.E. 1963. Plant diseases: Epidemics and control. Academic Press, New York. 349 pp.

38. Wallin, J.R., Waggoner, P.E. 1950. The influence of climate in the development and spread of <u>Phytophthora</u> <u>infestans</u> in artificially inoculated potato plots. Plant Dis. Rep. Suppl. 190:19-23.

39. Wienzierl, R., Koepsell, P., Fisher, G., William, R. 1982. A guide for integrated pest management in western Oregon snap beans - 1982. Oregon State University Extension Service, Corvallis, Or. 30 pp.

40. Young. H.C., Prescott, J.M., Saari, E.E. 1978. Role of disease monitoring in preventing epidemics. Ann. Rev. Phytopathol. 16:263-285.

41. Zadoks, J.C., Schein, R.D. 1979. Epidemiology and plant disease management. Oxford: Oxford University Press. 427 pp.

42. Zadoks, J.C. 1984. A quarter century of disease warning, 1958-1983. Plant Dis. 68:352-355.

CHAPTER 20

ECONOMICS OF INTEGRATED PEST CONTROL

J. D. Mumford and G. A. Norton

Silwood Centre for Pest Management, Imperial College at Silwood Park,
Ascot, Berkshire SL5 7PY, United Kingdom.

The economics of pest control concerns two questions. The first question, positive or scientific, asks what outcome can be expected from a particular crop production system when different crop protection inputs are used? The second question is normative (requiring value judgments) and asks what protection inputs give the best outcome in a particular production system?

These questions can be considered on an individual, national or international level. They are asked by farmers concerned with a particular field and by international policy makers concerned with how pest management affects society as a whole. At each level problems occur which concern the relevancy and gathering of information pertinent to the questions asked, and on the policy level, there exists the further problem of deciding which portions of society will benefit from pest management and which will bear the costs. This chapter examines these questions on the individual, national and international policy level and suggests methods of solution at these varying levels.

PEST CONTROL ECONOMICS FOR THE FARM

When viewed from the standpoint of decision making two types of pest control procedures can be distinguished:
a) Standard operating procedures which are adopted regardless of current levels of attack, and
b) Adaptive (or responsive) procedures in cases where the decision on control is made on the basis of information concerning current or future pest attack.

Though different economic models for each of these control procedures exist, the basis for decision making will remain constant as the farmer will only apply control procedures when the benefits of control exceed the costs incurred. In practice, farmers are likely to consider costs and benefits in relation to a range of objectives such as safeguarding sprayer operators, reducing environmental hazards, maintaining the quality of rural life, and avoiding risk of severe yield losses (see Chapter 21). However, if the strictly economic or commercial aspects of costs and benefits are not sufficiently accounted for, the farmer is likely to put his particular farming operation in financial jeopardy. Therefore, economic objectives in pest control are likely to be the first consideration.

Looking at the economic costs and benefits of control, and ignoring the problem of risk for the present, we can say that a necessary condition for taking a control action is that

$$B > C \qquad (1)$$

in which B is the economic benefit of control and C is the economic cost.

Costs of Control

Two types of costs in pest control can be considered. The first type of cost is associated with the expenditure on extra inputs required when a particular control action is taken. It includes the costs of materials such as pesticides, the cost of farm or contracted labor involved in applying controls, the running costs and depreciation of machinery used, the costs of monitoring pests, and any extra cost of resistant seed.

The second type of cost is associated with a loss to the farmer associated with the control action taken. For instance, when pesticides are applied by tractor mounted sprayers in certain crops, the damage caused by this operation can itself cause a reduction in yield. Similarly, the phytotoxic effect of certain pesticides can reduce yield. This second category also includes the opportunity costs of pest control. These are losses incurred by farmers as the result of diverting resources from other activities. For instance, if in applying pesticides, labor has to be taken from an important activity like sowing another crop, resulting in losses due to less timely sowing, these losses should be regarded as additional costs of pest control.

Benefits of Control

It is generally easier to assess the costs of control than to determine control benefits. Since the benefit of pest control is a reduction in loss caused by pests, the obvious starting point is to determine the damage function, the relationship between pest population and crop revenue. The damage function may also be an index of pest population, such as per cent defoliation or area of lesions (20).

The damage function is determined by experiments comparing the crop yield or crop quality at different levels of pest infestation or infection (Chapters 11-13). The yield can be converted to revenue if the crop price is known, giving a direct assessment of the lost crop value associated with any pest population (or pest index). With a given level of attack and damage, the benefit from control will depend on the extent to which the control measure reduces the level of damaging attack, that is, the control function.

By applying control measures to crops with a range of pest populations the combined effect of these two functions can be determined. Converting yields to revenue and subtracting the cost of control will give curves that can be directly compared with the original damage function or the damage function with other control measures (Figure 1). The most profitable action is the one with the highest curve at the given level of pests.

Having considered the basis of pest control economics, let us now look more closely at the two types of control.

Adaptive/Responsive Pest Management

The economic threshold, developed in 1959 by Stern et al. (21), is the central concept associated with adaptive/responsive pest management. It operates on the principle that control measures should not be taken until the pest population reaches a level at which the loss prevented by control justifies the cost of control. Incorporating this principle into the economic condition in equation (1), Norton (15) produced a more formal equation expressing the economic threshold:

$$PDAK > C \hfill (2)$$

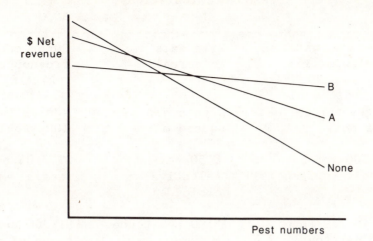

Figure 1. Damage functions, with no control, control, control A, and control B.

where A is the untreated level of pest attack; P is the price of crop per unit; D is the pest damage, expressed in terms of loss per ha per unit of pest attack; K is the effectiveness of treatment in reducing the level of pest attack; and C is the cost of treatment per hectare.

Thus, the level of attack above which it will be profitable to spray, A(t), is

$$A(t) = C/PDk. \tag{3}$$

To illustrate how this threshold can be calculated for different types of pests three examples are given in Table 1.

It is evident that formalizing the definition of an economic threshold, as in expression (3), requires a number of simplifications. First, the parameters shown in Table 1 may not be constant and, therefore, may affect the value of the threshold. For instance, P is likely to vary from season to season, and C and K will vary as different chemicals and application methods are considered (14). However, since the economic threshold is defined explicitly in equation (3), the implications of likely changes in parameters on threshold values can be directly assessed (15).

Another simplification made in expression (3) concerns the damage coefficient (d), which implies that the relationship between the level of attack and damage is linear. Investigation of the literature tends to indicate that this assumption is not correct (see Chapters 11 - 14). However, for economic purposes, the portion of the damage function that is of most interest is associated with lower levels of pest attack. In this context, Figure 2 shows how the two categories of damage relationship (20) can be linear with little, if any, loss in terms of determining the economic threshold.

While pests with a type I damage relationship can be accommodated by the economic threshold definition in equation (3), pests with a type II damage relationship clearly cannot. As indicated in Figure 2, this can be easily resolved by determining the threshold level of attack, A(d), at which damage begins to occur and then incorporating this in the appropriate economic threshold definition for type II situations. In this case, the level of attack at which it is profitable to treat, A(t) is

$$A(t) = A(d) + C/PDK. \tag{4}$$

193

Table 1. Examples of economic threshold calculations.

	Potato cyst eelworm (15)	Wild oats in cereals	Sugar cane froghopper (4, 16)
A	eggs per gm of soil	plants per square m	adults per 100 stools
C	£100 per ha (soil fumigant)	£50 per ha (selective herbicide)	$TT7.04 per ha (aerial spray)
P	£40 per tonne of potatoes	£100 per tonne of wheat	$TT200 per tonne of sugar
D	0.1 tonnes per ha per egg per gm of soil	0.75% yield loss per plant per square m (5)	.03228 kg per ha per adult-day per 100 stools
K	0.8 (80% kill)	1.0 (100% kill)	0.8 (80% kill)
A(t)	31 eggs per gm of soil	8.3 plants per square m	273 adults per 100 stools
		(at 8 tonnes yield per ha)	(assuming avg adult life of 5 days)

While parameter variability and nonlinear damage relationships can be dealt with as described above, the following features encountered in real pest problems are not so easily accounted for within the economic threshold concept (equations (3) and (4)):
(a) variable emigration and immigration
(b) effect of control on population dynamics of the pest
(c) influence of natural enemies
(d) timing of control measures
(e) other pests in the system
(f) effects of control measures for other pests.
However, defined as a breakeven concept, according to Stern et al. (21) and equations (3) and (4) above, these complications do not necessarily matter since all we are concerned with is whether control is profitable or not. Unfortunately, some confusion has been caused by economists (10, 13) who have defined the economic threshold as the optimal level of control. While one may be able to theoretically define optimal control, in practice, the optimal decision rule requires consideration of the complicating factors mentioned above, as well as others not mentioned here. Although attempts have been made to incorporate these factors in deriving decision rules (4, 6, 19), Hall and Norgaard (8) conclude that the increase in realistic variables rapidly reduces the manageability of mathematical models.

On the other hand, some of the features that need to be considered in producing an optimal decision rule undoubtedly are taken into account in practice, and should be considered when converting an economic threshold (equations (3) or (4)) to a practical decision rule or action threshold. Thus, in determining an action threshold, a good starting point is to try to derive the economic threshold (expression (3) or (4)), and test its sensitivity to a likely range of parameter

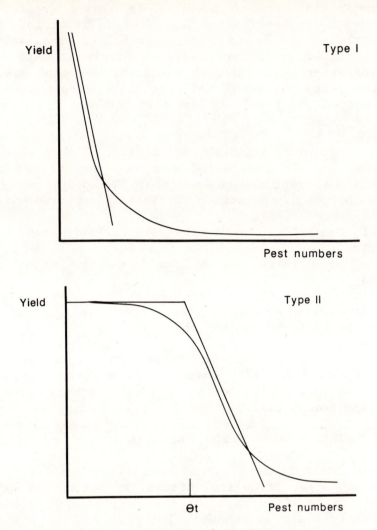

Figure 2. Type I and type II damage functions.

values. Other important considerations (such as (a) to (f) above) and experience from past years and other trials can then be used to derive a working action threshold that, while not optimal, is likely to be quite satisfactory.

Standard Operating Procedures

The decision theory concept is relevant in instances where farmers adopt control measures on a regular or calendar basis, without attempting to determine the actual level of attack in that season (13, 15). In deciding upon control procedures, the benefits need to exceed the costs of a potential control program; however, the controlling criterion is not the benefit of a specific year but rather the expected benefit of the entire program. Determining this expected benefit involves (13, 15):
(a) assessing the outcome of a strategy for a range of attack levels,
(b) weighting this outcome by the probability of that level of attack, and
(c) summing all the weighted outcomes to produce the expected outcome.
These assessments can also be used to determine the effectiveness of this strategy in avoiding the risk of undesirable or disastrous outcomes. Because of the considerable uncertainty associated with pests, farmers are not only interested in benefits and costs of control but also in the risk avoiding function of control. When the objective is risk avoidance, it may not be necessary for control action benefits to exceed costs. Pest control is often considered as a form of insurance (12, 13), and is not judged on the basis that the control procedure implemented returns as much in benefits as it costs annually. With other forms of insurance,

195

for instance car insurance, the cost of premiums is always greater than the insurance company's estimation of the expected loss (the value of loss multiplied by the chance of occurrence), but insurance holders are content because their risk is reduced. The same is true of prophylactic pest control, it may not pay for itself in reduced crop loss every year, but it remains a means of assurance for the farmer (2). Endogenous pests present a different type of problem, since the probability of attack changes progressively as pest numbers build within the field.

In most pest control situations, standard operating procedures seem to be the rule rather than the exception. In a detailed study of insect pest control on apples in the United Kingdom, Fenemore and Norton (7) identified four major constraints to supervised control (using action thresholds) compared with calendar spraying (a common control procedure):

(a) calendar spraying of fungicides means that insecticide can be applied cheaply at the same time,
(b) economic threshold decision rules are difficult to design and employ in English apple orchards,
(c) trained labor for monitoring pests is unavailable or expensive, and
(d) the risks of a reduced proportion of top class fruit associated with supervised control is likely to be unacceptable.

At the farm level, pest control economics is primarily a matter of ensuring that relatively short term benefits exceed costs. The complexity of factors when considering optimal control strategies often prevent farmers from adopting suggested standard operating procedures.

PEST CONTROL ECONOMICS FOR PUBLIC POLICY

What is good for individual farmers is not necessarily good for farmers in aggregate or for society at large. Price inelasticity of demand for agricultural products, external costs and benefits associated with pesticide production and use, and societal opportunity costs associated with pest control practices make regional or national aspects of pest control economics different from those on the farm.

Because conflicts of interest occur between different segments of society, it is difficult to maintain a positive or objective view of pest management economics beyond the farm level. Value judgments must be made about the relative claims of consumers and farmers, of farmers growing different crops, and of farmers growing the same crop in different places or under different conditions. While an economist concerned with pest control at this level can indicate the likely outcomes of a pest management policy, the subsequent policy decision is based ultimately on political judgments rather than economic facts alone.

Aggregate Effects of Changes in Pest Control Practice

Agriculture viewed as an industry differs from other industries in two important respects: the demand for most agricultural products is inelastic and as an industry made up of millions of small firms, concerted action is difficult (9).

Inelastic demand means the consumer's desire or need for a commodity is less flexible than the commodity's price. So if the amount of an inelastic product available falls 25% due to crop failure, for instance, the price will then increase more than 25%. If supply increases 25%, as it could with the widespread adoption of an improved pest control technique, the price would fall more than 25%. Generally, products with elastic demand are less essential items; the amount consumers demand stretches greatly with price changes. The quantity of inelastic products which consumers desire, however, does not stretch much relative to price.

196

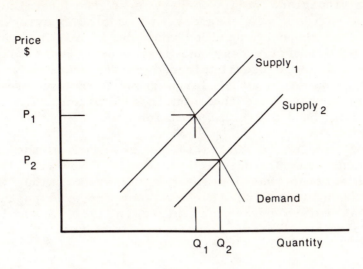

Figure 3. Demand curves for elastic and inelastic commodities.

Because of the price inelasticity of most crops, an economic spiral develops forcing individual farmers to adopt yield increasing technologies, such as pest control even when it is not in their best interest collectively. In Figure 3, the total income for farmers as a group is the product of the equilibrium price and quantity. With more effective pest control, the supply curve moves to the right (assuming farmers can produce more crop for the same effort). The new product price and quantity is less than before, and the group as a whole suffers, unless production costs have been reduced enough to offset the fall in value of the crop. Yet, the pressure remains for all farmers to adopt new techniques, for any individual who did not would suffer more extensively by receiving the old, low yields at the new, low prices (17).

Public Pest Control

Both inelasticity and conflicts of interest present problems for public pest control policy. Consider the case of cotton and the control of the cotton boll weevil (Anthonomus grandis). Nearly $200 million is spent annually on control of the cotton boll weevil, a major pest of cotton in most of the southeastern US cotton belt (1). In addition, boll weevil control measures reduce the natural enemies of other cotton pests, resulting in increased costs for control of pests otherwise naturally regulated. In the late 1970's, the USDA conducted a pilot program to test the feasibility of eradicating the cotton boll weevil using sterile male release in North Carolina. This technique would need to be used throughout a region to be effective, and be organized by government agencies. Eradication appeared to be technically possible, but what would the benefits be?

The USDA determined that the price elasticity of cotton was -0.8. This means that if the amount of cotton produced increased 8%, the price would fall 10%. As a result, the USDA estimated that increased production associated with eradicating the boll weevil would result in a $48 million dollar annual cost to farmers as an aggregate.

The loss, in addition, would not be evenly distributed. Cotton growers who normally suffer from boll weevil attack would benefit considerably because their yields would improve and their costs would be lower. It was estimated that this group would benefit by $69 million annually. By contrast, those cotton growers whose farms are outside the geographic range of the weevil and consequently do not normally suffer weevil damage, would not increase their yield but would suffer from the falling cotton prices due to the increased yields in the rest of the cotton belt. These farmers, as estimated, would lose $117 million a year.

In this type of situation, policy makers need to determine that the net benefit is enough to warrant some segments of society paying a high price. Consumers would benefit as the cotton prices would be lower. The environment would be less contaminated with insecticides, and less energy would be used in making and applying insecticides. Boll weevil plagued cotton growers would benefit by about $25 per hectare. However, another large group of cotton growers might lose considerable income, or perhaps lose their entire operation. The resolution of a conflict of this sort must be a political decision.

A similar type of problem in Australia has resulted in the introduction of legislation in Parliament, the Biological Control Bill 1984 (3). The plant Echium plantagineum, Paterson's Curse, has long been a declared noxious weed of pasture in many parts of southern Australia. It chokes out improved pasture species and contains alkaloids that poison horses and pigs. In 1982, Australian government entomologists began releasing insects from Europe in a trial to control the weed biologically.

Soon after the release project started, two graziers and two beekeepers obtained a court injunction to stop the releases because of the threat to their livelihoods. In the drier regions where the graziers lived, E. plantagineum (known there as Salvation Jane) provides important green feed for stock during droughts. Beekeepers claimed they relied on E. plantagineum pollen for honey production. As the law stood, the courts could only consider the potential harm to the people complaining, the benefits to others were not legally relevant. The Biological Control Bill aims to ensure that when biological control organisms are to be released both sides are considered by public inquiries which will determine whether there is a net benefit to the nation.

Public or regional pest control can, however, have clear economic advantages. In addition to improved control through thorough overall treatment in a large area another major benefit is in economies of scale. For example, Lewis and Norton (11) found that aerial application of baits to control leaf cutting ants on citrus and cocoa in Trinidad (which can only be done on a large scale) reduced the cost of control considerably compared to on-farm ground application of insecticides. On-farm control costs were between $TT10.46 and $TT28.28/ha (depending on ant density), while by air the cost was only $TT6.18/ha. Along with these lower costs, the level of control would be maintained or improved, and the amount of insecticide used would be greatly reduced.

External Effects

As we have seen in the cotton and Echium examples above, pest control benefits and costs extend outside those persons directly involved in the pest control procedure. There are also smaller scale examples, for instance, the loss incurred to the beekeeper when his bees are poisoned during visits to neighboring crops treated with insecticides is an external cost of the farmer's control decisions. An external benefit not directly relevant to the decision maker might be the elimination of mosquito breeding sites, as the reduction of potential diseases is extended to neighboring residents.

External effects of a pest control decision can distort economic assessments and prevent benefits to society as a whole. A pest control action may seem very cost effective viewed on the farm itself, but neighbors may suffer losses that would make it unjustifiable. Similarly if everyone in a village considered only the benefits to themselves when deciding to eliminate mosquitoes they may not bother. Adding in the advantages to others would make it worthwhile. Regulations are often needed to "internalize" the costs and benefits of pest control actions. In the case of bee poisoning, regulations can prohibit spraying flowering crops or ensure that beekeepers can claim damages which force the pesticide user to consider bees in his cost benefit accounting. Regular mosquito control can be

198

carried out by government agencies or householders can be required by law to control mosquitoes on their own property to ensure that actions that give wide benefit are taken.

From the discussion of aggregate effects and externalities above it should be clear that policy makers face difficult decisions on what constitutes success in a pest management policy. Traditionally, it has been felt that what is good for individual farmers should be the objective for policy makers (18). This is based on the assumption that the long term interest of the farmer and society are the same. However, as concern for the environment and occupational health increases, more conflicts are likely to arise. Resolution of these conflicts, through the development of acceptable integrated pest management strategies, for example, is only likely to occur if policy makers understand the objectives of each group and work towards compromise policies that attempt to balance costs and benefits among all their various constituents.

LITERATURE CITED

1. Anon. 1981. Overall evaluation: beltwide boll weevil/cotton insect management programs. Natural Resource Economics Division, Economic Research Service, United States Department of Agriculture, Washington, DC. Staff Report No. AGES810721. 45 pp.
2. Carlson, G.A., Main, C.E. 1976. Economics of disease-loss management. Ann. Rev. Phytopathol. 14:381-403.
3. Carne, P.B., Whitten, M.J. 1984. The context of biological control research in Australia. XVII Int. Congr. Entomology, Hamburg, Fed. Rep. Germany, 20-26 August, 1984. Abstract Vol. pp. 801.
4. Conway, G.R., Norton, G.A., Small, N.J., Small, A.B.S. 1975. A systems approach to the control of the sugar cane froghopper. In Study of Agricultural Systems, ed. G. E. Dalton, pp. 139-229. London: Applied Science. 441 pp.
5. Cousens, R., Peters, N.C.B., Marshall, C.J. 1984. Models of yield loss - weed density relationships. Proc. 7th Int. Symp. on Weed Biology, Ecology and Systematics. Paris, France. pp. 367-374.
6. Croft, B.A. 1975. Integrated control of apple mites. Michigan State Univ. Ext. Ser. Bull. E825. 12 pp.
7. Fenemore, P.G., Norton, G.A. 1985. Problems of implementing improvements in pest control: The case of apples in the U. K. Crop Protection 4:51-70.
8. Hall, D.C., Norgaard, R.B. 1973. On the timing and application of pesticides. Am. J. Agric. Econ. 55:198-201.
9. Hallett, G. 1981. The Economics of Agricultural Policy. 2nd ed. Oxford: Basil Blackwell. 365 pp.
10. Headley, J.C. 1972. Economics of agricultural pest control. Ann. Rev. Entomol. 17:273-286.
11. Lewis, T., Norton, G.A. 1973. Aerial baiting to control leaf-cutting ants (Formicidae, Attini) in Trinidad. III. Economic implications. Bull. Ent. Res. 63:289-303.
12. Mumford, J.D. 1981. A study of sugar beet growers' pest control decisions. Ann. Appl. Biol. 97:243-252.
13. Mumford, J.D., Norton, G.A. 1984. Economics of decision making in pest management. Ann. Rev. Entomol. 29:157-174.
14. Norgaard, R.B. 1976. The economics of improving pesticide use. Ann. Rev. Entomol. 21:45-60.
15. Norton, G.A. 1976. Analysis of decision making in crop protection. Agroecosystems 3:27-44.
16. Norton, G.A. 1985. Economics of pest control. In Pesticide Application: Principles and Practice, ed. P. T. Haskell, pp.175-189. Oxford: Oxford University Press. 450 pp.
17. Ordish, G., Dufour, D. 1969. Economic bases for protection against plant diseases. Ann. Rev. Phytopathol. 7:31-50.

18. Perkins, J.H. 1982. Insects, Experts and the Insecticide Crisis. New York: Plenum Press. 304 pp.

19. Shoemaker, C.A. 1979. Optimal management of an alfalfa ecosystem. In Proceedings of a Conference on Pest Management, ed. G. A. Norton, C. S. Holling, pp. 301-315. Oxford: Pergamon Press. 352 pp.

20. Southwood, T.R.E., Norton, G.A. 1973. Economic aspects of pest management strategies and decisions. In Insects: Studies in Population Management, ed. P.W. Geier, L.R. Clark, D.J. Anderson, H.A. Nix, pp. 168-184. Canberra: Ecological Society of Australia. 295 pp.

21. Stern, V.M., Smith, R.F., van den Bosch, R., Hagen, K.S. 1959. The integrated control concept. Hilgardia 29:81-101.

CHAPTER 21

ANALYSIS OF DECISION MAKING IN PEST MANAGEMENT

J. D. Mumford

Silwood Center for Pest Management, Imperial College at Silwood Park
Silwood Park, Ascot, Berkshire SL5 7PY, United Kingdom

The process of decision making combines pertinent information with individual preference. Decision making in pest management requires information concerning the extent and likelihood of the pest problem and the expected performance of available control measures so that the decision maker can choose, according to his or her objectives, the most appropriate pest control action. In this chapter I will examine briefly how pest management decisions are made and how decision making can be improved.

HOW DECISIONS ARE MADE

All pest management decisions can be described by the simple decision model (11, 15) illustrated in Figure 1. Each component of the model is a separate process of decision making and requires individual analysis.

The Problem

The pest and its relationship to the crop is the origin of a pest management decision. However, the pest does not need to be physically present in the field to be a problem; the mere threat of pest attack may also constitute a problem.

The Options

The actual control options available to the decision maker establish the range of choices to be considered. However, these may be limited by internal or external constraints; for instance, lack of suitable application equipment on a farm, or pesticide restrictions imposed by government regulatory agencies. It should also be noted that a decision to not act upon a pest problem is always a possible option.

The Perceptions

Though the basis of the decision process is the pest problem and control procedures, both the problem and options must be interpreted and understood by the decision maker. Farmers may lack specific knowledge, such as the number of insects in the field, or more general knowledge, such as the difference in effect of systemic and surface contact pesticides.

Investigative studies on how farmers perceive pest problems (7, 8, 17, 19) reveal that while some aspects of pest problems and control are well known, others are not. The decision maker's perceptions are formed on the basis of direct experience with a pest or control technique, association with other pest management situations, agrichemical advertising, advisory programs and many other sources (2, 4, 5, 6, 9, 12). Poor or incorrect perceptions and options are primary causes of bad pest management decisions.

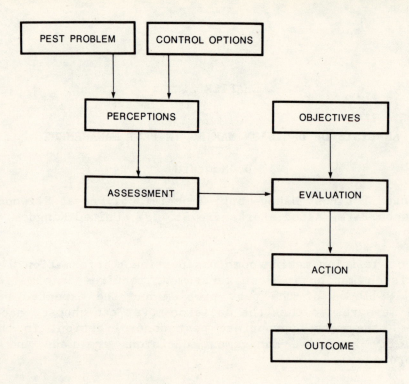

```
        ┌──────────────┐   ┌────────────────┐
        │ PEST PROBLEM │   │ CONTROL OPTIONS│
        └──────┬───────┘   └───────┬────────┘
               │                   │
               ▼                   ▼
        ┌──────────────┐     ┌──────────────┐
        │ PERCEPTIONS  │     │  OBJECTIVES  │
        └──────┬───────┘     └──────┬───────┘
               │                    │
               ▼                    ▼
        ┌──────────────┐     ┌──────────────┐
        │  ASSESSMENT  │────▶│  EVALUATION  │
        └──────────────┘     └──────┬───────┘
                                    │
                                    ▼
                             ┌──────────────┐
                             │    ACTION    │
                             └──────┬───────┘
                                    │
                                    ▼
                             ┌──────────────┐
                             │   OUTCOME    │
                             └──────────────┘
```

Figure 1. The basic decision model (15).

Assessment

The decision maker must determine how each control option can be expected to perform in relation to the perceived pest problem. Assessment of each possible action may not necessarily be exact; for instance, a farmer may anticipate a range of performance from a pesticide without being sure of precisely what will happen upon application. Value judgments do not enter into the decision making process at this point. At this level of the decision making process, one assessment is not considered better than another.

Objectives

The farmer's objectives are the standard by which perceived options are judged. Objectives are determined by the decision maker's own personal value judgments, for example, the emphasis placed upon effectiveness and environmental safety. Decisions can only be judged as being "good" or "bad" in relation to the decision maker's objectives. The objective of pest management has traditionally been treated as being profit maximization (11). However, insurance against catastrophic losses is likely to be a much more prevalent overall objective (11). In addition, each manager is likely to establish a hierarchy of other objectives, including human and farm animal safety, environmental protection, "clean" field appearance, etc., all of which may also influence the "best" action to be taken.

Evaluation

The crucial stage of the decision process is determining how well each option, based on individual assessment, meets the decision maker's objectives. If the objectives are relatively simple (such as maximization of the current year's profit) and the perceived problem and options are quite narrowly defined, then the evaluation should be very simple. However, as the complement of objectives becomes more complex, and the degree of uncertainty about the extent of the problem and the performance of the control options increases, the choice becomes much more difficult.

Action

The choice is implemented at this stage.

Outcome

The outcome of the pest management action may not be what was assumed at the assessment stage, since only the anticipated conditions could be considered at that point.

A "bad" outcome should be distinguished from a "bad" decision, since a decision is made on the basis of the best available information at the time. An action may give the most favorable outcome over the range of conditions expected and be, therefore, a "good" choice. However, unexpected natural events could occur in which an alternative action would have been better in those particular circumstances. In this case the "good" choice would give a "bad" outcome.

This decision model has been described in terms of an individual farmer making a tactical decision about a relatively immediate local pest problem. It could also be used to describe longer term strategic decisions at the farm level, or the model could be extended to describe regional, national, or international decision making. The evaluation process becomes increasingly difficult in the latter cases as objectives become more complicated and the planning horizon gets further into the future (making accurate assessment harder).

HOW DECISIONS CAN BE IMPROVED

Since decision making is dependent on the two inputs, information and choice, there are two general ways of improving decisions: better information can be provided for decision makers, and/or their ability to make choices can be improved.

Better Information on Pests and Controls

There are four categories in which pest management information could be made better:
(a) more accessible (or more available)
(b) more specific to the problem/options (i.e. to the field, pest species, control technique, etc)
(c) more reliable
(d) earlier.

Extension and commercial advisers should present pest problems and control options in a complete, easily available form. This is often not the case in practice, particularly in the information supplied by agrichemical firms. Lawson (5), for example, showed that most of the commercial literature available to farmers on oilseed rape pests in the United Kingdom left out some information needed to make a sensible control decision (such as how to recognize the pests, what the economic threshold is, or the degree of plant compensation for attack).

While public extension information is often very complete (as Lawson (5) found), it may not be possible for the grower to get it when he wants it due to shortage of staff or other resources. Telephone advisory tapes and direct on-farm computer connections via telephone (21) can help give farmers immediate access to pest management information very efficiently. While these systems are quick and easy (provided, in the case of computer links, that they are "user friendly") they may have the drawback of not being specific enough for the farmer's conditions. This can be overcome to some extent by systems such as the Dutch EPIPRE computerized cereal disease advisory program (16) in which there is an interactive relationship between the farmer and the computer. This allows the computer to get the necessary individual inputs for the farmer's specific conditions.

Information that is both reliable and timely may be impossible to achieve in many cases. The earlier that predictions of pest development and loss are made, the less accurate they are likely to be. However, some sacrifice in reliability may be worthwhile if earlier information allows the decision maker to consider more options, or provides more time to make a good decision.

Better information should primarily improve the perception and assessment stages of the decision process.

Better Choice Process

There are three general approaches that can be taken to improve the choice process: management training, deferring to a specialist manager, and adopting an expert system. These tend to improve the assessment and evaluation stages of the decision process.

Pest managers can learn to be better decision makers. There are two principal factors that can be improved by establishing a clearer understanding of the problem, objectives, and options. Understanding pest problems and controls can be improved by basic education on pest management principles, and also by training managers to present the problem in a complete and logical form. For instance, listing the options and their expected benefits and costs, drawing decision trees, practicing with management games (10), etc. will contribute to a logically presented problem. Decision trees are a particularly useful way to present a decision maker, adviser, or researcher with the heart of a pest problem and its options for control.

The decision tree shown in Figure 2, for example, describes the insect and disease control choices that an English sugar beet grower has. The tree illustrates what the options are, and in what sequence they occur, and can be used to help a grower plan his information needs. A decision tree can be made more useful by including outcomes and probabilities if these were known for the various branches of the tree.

Such information may be difficult to obtain empirically, especially for a complex series of options, but it can sometimes be done subjectively, as by Valentine et al. (20), in describing the value of a scouting program for gypsy moths in New England forests (Figure 3). The decision maker has a choice of scouting or not, and then spraying insecticide, applying a bacterial spray, or doing nothing. The probabilities and outcomes for each action have been determined subjectively by canvassing opinions of forest pest control experts. This form of decision tree shows not only the options and sequences of decisions, but also the probabilities of pest outbreaks of various levels if the scouting is positive or negative, and the likely outcomes if particular actions are taken. A user can then determine the degree of risk associated with each strategy, and can choose according to his own objectives.

If the pest problem and control options are clearly understood by the decision maker, then the "best" choice is likely to be seen more readily. Similarly, if the objectives of the decision maker are clearly defined it is much easier to rank options during the evaluation phase of decision making.

Rather than putting effort directly into becoming better decision makers, many pest managers in North America, Europe, and elsewhere have deferred this role to specialist consultants (1, 3, 12). The consultant is aware of the farmer's objectives and uses specialized knowledge such as field scouting, contacts with pesticide manufacturers and researchers, etc., to determine the best course of ac-

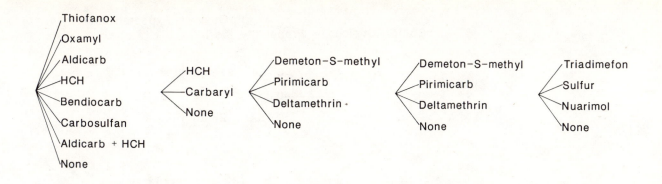

Thiofanox				
Oxamyl				
Aldicarb		Demeton-S-methyl	Demeton-S-methyl	Triadimefon
HCH	HCH	Pirimicarb	Pirimicarb	Sulfur
Bendiocarb	Carbaryl	Deltamethrin ·	Deltamethrin	Nuarimol
Carbosulfan	None	None	None	None
Aldicarb + HCH				
None				

| **Soil Pests** | **Flea beetles** | **Aphids** | **Aphids** | **Powdery mildew** |
| March | May | June | July | August |

Figure 2. A decision tree for sugar beet insect and disease control in England.

tion for the farm. The consultant may either decide directly what action is required, or clearly present the problem and options to the farmer so that the farmer can make a sensible choice based on these informed assessments. The decision process through the assessment or evaluation stage is carried out by a well informed expert, rather than by the farmer himself, whose knowledge of pest management practices is often deficient in some aspects (as shown by a number of surveys (2, 4, 8, 17, 18)).

Another way of improving decision making in pest management is to rely on an "expert system" to help define the problem and options and to then choose an action according to predetermined criteria. There are many examples of these based on computer or calculator programs (21), but even something as simple as a wall chart or sliding card can be an expert system. In principle they are all similar, helping to guide the decision maker through the relevant aspects of the problem to an appropriate solution.

The pocket computer/calculator type of "expert system" can serve as an example of how such systems work. A program is built into the machine which asks the user a series of questions about the situation (i.e. the crop, soil type, sowing date, pest presence, weather, expected crop price, etc.). Using mathematical relationships based on previous field observations on pest development and control performance, the program then computes the expected loss and degree of control associated with the various options available. It will display the choice that best fits the predetermined objective, generally a maximization of expected revenue. The pest manager benefits from using this system because the computer program makes sure that all of the relevant information is obtained and considered, and automatically presents the choice.

In theory, such expert systems are very good. In practice, however, there may be considerable doubt about how "expert" they are. If the system is used over a wide area the pest development and control relationships built into the program may not be applicable to crops a great distance from the sites where the models were developed. A pocket computer supplied by a pesticide manufacturer to aid decision making for cereal diseases may choose well, but only from the limited range of sprays sold by that company. Furthermore, the information given by the farmer in response to the questions the machine asks may not be correct (particularly estimates of pest abundance), so the computer may have incorrect perceptions as well since it relies on nonexpert inputs.

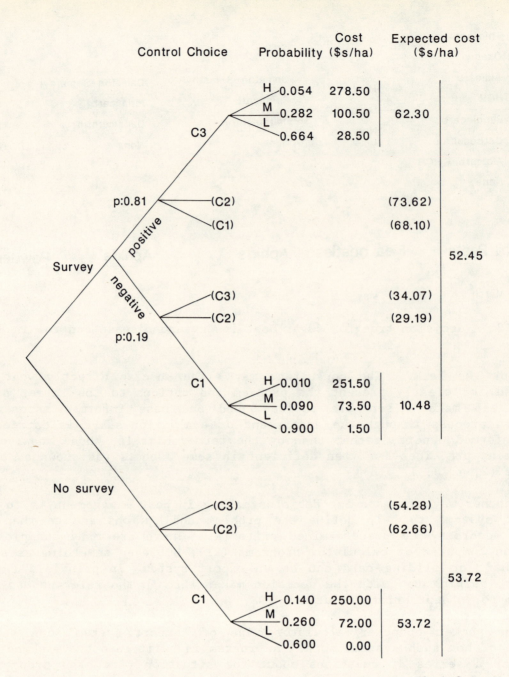

Control Choice		Probability	Cost ($s/ha)	Expected cost ($s/ha)

C3 — H 0.054 278.50
M 0.282 100.50 — 62.30
L 0.664 28.50

p:0.81 — (C2) (73.62)
(C1) (68.10)

positive — 52.45

Survey

negative — (C3) (34.07)
(C2) (29.19)

p:0.19

C1 — H 0.010 251.50
M 0.090 73.50 — 10.48
L 0.900 1.50

No survey

(C3) (54.28)
(C2) (62.66)

— 53.72

C1 — H 0.140 250.00
M 0.260 72.00 — 53.72
L 0.600 0.00

Figure 3. Decision tree for gypsy moth egg scouting and control (after Valentine et al., 20). C1 = no spray, C2 = bacterial spray, C3 = chemical spray. L = light defoliation, M = moderate defoliation, H = heavy defoliation. The expected costs are weighted by the probabilities of each outcome. Control choices in parentheses are ones that would not be chosen because they are always worse than one of the other choices at that branch (as can be seen from their high relative cost).

Another potential drawback is that the decision criteria in the program may not be the same as the farmer's (profit maximization versus risk minimization, for instance). If the expert system has crop/pest/control relationships that do not apply to the farm, a more limited range of options than is truly open to the farmer, inputs that are no better than the farmer's own perceptions, and objectives that differ from those of the user it will be worse than useless.

Mainframe computer systems overcome some of the shortcomings of portable computers because the programs can be updated very quickly to reflect new information on the problem or changes in options (such as introduction or removal of pesticides on the approved list) and can hold a range of programs to cover different regional conditions. They also have the advantage that many farmers or advisers are providing inputs and the system could compare inputs from one user with the rest and either question unusual entries or alert advisers of changing pest patterns. However, mainframe systems can delay decisions if they are not "on-line".

All expert systems are also valuable learning tools for farmers until they become familiar with the sort of decisions the model will make. The farmer may then have enough confidence in his ability to recognize the symptoms of the problem and know enough about the available options that he may no longer need the expert system, having become an "expert" himself.

CONCLUDING REMARKS

The most likely improvements that can be made in pest management decision making will come first from making already available research information more accessible to decision makers and secondly, from experts or expert systems that perform at least part of the decision process for the farmer or adviser. The latter is becoming increasingly important as pest management decisions become more complicated, requiring more specialized knowledge in what is only one of many components of overall farm management.

LITERATURE CITED

1. Addison, S.J. 1982. Independent crop consultants in agriculture: Why do farmers employ them and how do they benefit from doing so? M.Sc. thesis, Imperial College, University of London.
2. Beal, G.M., Bohlen, J.M., Lingren, H.G.. 1966. Agricultural chemicals and Iowa farmers. Iowa State University. Special Report 49. 24 pp.
3. Bristow, C.M. 1983. Preliminary report on independent crop consulting in the United Kingdom. Unpublished Report, Silwood Centre for Pest Management, Imperial College, London.
4. Brown, J.D. 1968. Adoption and purchasing of agricultural pesticides. University of Georgia, College of Agriculture Experiment Stations. Research Bulletin 39. 42 pp.
5. Lawson, T.J. 1982. Information flow and crop protection decision making. In Decision Making in the Practice of Crop Protection, ed. R. B. Austin, pp. 21-32. Monograph No. 25. Croydon, U.K.: British Crop Protection Council. 238 pp.
6. Leeks, D.E., Mumford, J.D. 1982. Advertising and pesticide use in Britain, 1960-1979. Protection Ecology 4:59-65.
7. Mumford, J.D. 1980. A survey of pests and pesticide use in Canterbury and Southland. Research Report No. 107, Agricultural Economics Research Unit, Lincoln College, New Zealand. 39 pp.
8. Mumford, J.D. 1981. A study of sugar beet growers' pest control decisions. Ann. Appl. Biol. 97:243-252.
9. Mumford, J.D. 1982. Farmers' perceptions and crop protection decision making. In Decision Making in the Practice of Crop Protection, ed. R.B. Austin, pp. 13-19. Monograph No. 25. Croydon U.K. British Crop Protection Council. 238 pp.
10. Mumford, J.D. 1984. A computerized decision model of sugar beet insect and disease control. pp. 13-19. Proc. 1984 British Crop Protection Conference, Brighton, U. K. 1207 pp.
11. Mumford, J.D., Norton, G.A. 1984. Economics of decision making in pest management. Ann. Rev. Entomol. 29:157-174.
12. Norgaard, R.B. 1976. The economics of improving pesticide use. Ann. Rev. Entomol. 21:45-60.
13. Norton, G.A. 1976. Analysis of decision making in crop protection. Agroecosystems 3:27-44.

14. Norton, G.A., Mumford, J.D. 1982. Information gaps in pest management. In Proc. International Conference on Plant Protection in the Tropics, ed. K.L. Heong, B.S. Lee, T.M. Lim, C.H. Teoh, Y. Ibrahim, pp. 589-587. Kuala Lumpur, Malaysia: Malaysian Plant Protection Society 743 pp.
15. Norton, G.A., Mumford, J.D. 1983. Decision making in pest control. Adv. Applied Biology 8:87-119.
16. Rijsdijk, F. 1982. The EPIPRE system. In Decision Making in the Practice of Crop Protection, ed. R. B. Austin, pp. 65-76. Monograph No. 25. Croydon, U.K.: British Crop Protection Council. 238 pp.
17. Tait, E.J. 1978. Factors affecting the usage of insecticides and fungicides on fruit and vegetable crops in Great Britain: II. Farmer-specific factors. J. Environ. Management 6:143-151.
18. Turim, J., Reese, C.D., Kempter, J., Muir, W. 1974. Farmer's pesticide use decisions and attitudes on alternate crop protection methods. U.S. Environmental Protection Agency, Washington, D.C. EPA-540/1-74-002.
19. Turpin, F.T., Maxwell, J.D. 1976. Decision making related to use of soil insecticides by Indiana corn farmers. J. Econ. Entomol. 69:359-362.
20. Valentine, W.J., Newton, C.M., Talevio, R.L. 1976. Compatible systems and decision models for pest management. Environ. Entomol. 5:891-900.
21. Welch, S.M. 1984. Developments in computer-based IPM extension delivery systems. Ann. Rev. Entomol. 29:359-381.

Chapter 22

PEST SURVEILLANCE SYSTEMS IN THE USA - A CASE STUDY USING THE
MICHIGAN CROP MONITORING SYSTEM (CCMS)

Stuart H. Gage and Howard L. Russell

Department of Entomology, Michigan State University,
East Lansing, Michigan 48824, U.S.A.

The Cooperative Crop Monitoring System (CCMS) was developed to provide a standardized, inter-agency information collection and retrieval system for agricultural pests, the crops they infest and the actions taken to control these pests.   Initiated because of the economic necessity to share resources and expertise, CCMS gathers, manages and preserves crop and pest observations made by a cooperative network of observers and then makes these data available for a wide array of applications. Quality data collected for one purpose which are recorded in a standardized format are shared by cooperating local, state and federal agencies for many additional purposes.

Since the early 1970's the concepts of Integrated Pest Management (IPM) have flourished in the U.S.A. and many states have developed IPM programs.   The principal components of on-line IPM (21) - environmental monitoring, biological monitoring, modeling and information delivery - contribute to a systems approach to pest control through their integration.   Considerable effort has been made in Michigan and other states to initiate operational information management and dissemination systems to aid pest managers (8).   Welch (22) reviewed the development of the computer-based extension IPM delivery system, and, earlier, Haynes et al. (10) addressed the needs of environmental (weather) information necessary to time biological models dependent on real-time weather input.

IPM has contributed much to our understanding of the extremely complex interactions of host crops with insects, plant pathogens, nematodes, viruses, vertebrate pests and environmental parameters. IPM advocates recognized that reasonable estimates of pest and pathogen levels can be made only through field observation (biological monitoring).   This was a challenge because of lack of knowledge of sampling methods for even some common pests and diseases causing loss to crop production.   Pest management programs contributed significantly to developing such sampling methods and to training of the field observers who have responsibility for making reliable assessment of pest and disease severity. This need has led to the development of field manuals and sources of information useful to aid the field observer.

RATIONALE FOR DEVELOPING A CROP MONITORING SYSTEM

Ecologists, epidemiologists and students of the dynamics of biological organisms know that regular quantitative estimates of the occurrence and the abundance or incidence of organisms in space and time are essential if we are ever to predict future pest and crop events.   Without some form of information gathering, storing and retrieval of comprehensive agricultural information, knowledge of the factors which affect the dynamics of pests in agricultural production systems will remain poorly understood. Only if we record our actions and document the subsequent results, will we understand whether the strategy we apply to the system has been positive or negative. The lack of quantitative historical information on biological events is probably the most serious obstacle affecting our ability to make reliable predictions and forecasts of pest potential and to estimate crop loss associated with pests and plant pathogens. The value of a crop surveillance system that can provide these information needs, not only for government agencies and those responsible for solving immediate pest and disease problems, but also for scientists determined to understand the function of the agroecosystem has long been recognized. The presence of such surveillance systems in Saskatchewan, Canada, allowed grasshopper populations to be examined over a 32 year period with reference to weather and crop production to develop infestation probabilities and economic consequences of grasshoppers (3). Without regular and consistent field observations recorded using standard observation methods, the opportunity for such analysis is not possible.

Unfortunately, prior to 1982, there were very few sources of long-term quantitative information on factors influencing plant health in the U.S.A. One of the primary reasons for development of the Michigan CCMS was to begin a scientific, systems approach toward pest and disease survey and to develop a survey system that would serve the pest information needs for many users as efficiently and as cheaply as possible.

In Michigan, as in many other states, the results of scientific pest surveys were stored in a file cabinet belonging to the person or agency that conducted the survey, and to a large degree, were inaccessible to others interested in the data. A survey system had existed in Michigan for the cereal leaf beetle (Oulema melanopus), to quantify the changes in distribution and abundance of this newly introduced pest. The Michigan Department of Agriculture had conducted surveys of the Gypsy Moth since 1960 to delimit the distribution of this introduced pest. However, the volume of information generated by these two programs alone made management and analysis difficult without the aid of high speed computers. Also, there were scattered and unstructured records of outbreaks of other pests which caused severe crop damage, but there were no records describing these populations when they were not in an outbreak state.

With the number of IPM scouts increasing, the opportunity to collect and preserve pest information in a standardized format, and from a wide array of cropping systems, was at hand.

Michigan's IPM program was gaining momentum in the late 1970's. The Pest Management Executive System (PMEX), a computerized information delivery system, was in operation to facilitate the exchange of information among university specialists, Cooperative Extension field staff and IPM scouts. These IPM scouts, hired by private and state organizations to monitor individual fields, orchards and forest stands, often used different monitoring techniques within the same cropping system. Clearly, there was a need for the Cooperative Crop Monitoring System in which observers employ standard monitoring techniques and data recording formats to build a common accessible database for current and historical applications.

Need for a common set of sampling methodologies for specific insect pests was recognized and developed by a group called the North Central States Survey Entomologists (12). The objective of the coordinated effort on plant pest surveillance is to provide pertinent information about the current status of plant pests and to provide the ability to extract specific information about factors affecting plant health within a time frame so that some action can be taken or so that a plan can be formulated and crop protection programs can be developed. The records collected can also provide the information necessary to develop models to forecast pest occurrence.

The Michigan Pest Monitoring Coordinating Committee was formed in 1978, to address many of these same issues at the state level. Comprised of representatives from the major state and federal agencies involved in pest survey, this committee sought to develop a system to standardize and manage biological observations made by employees of these agencies. This committee provided a forum in which common pest information needs were discussed and several key issues were identified. The most significant of these were: recognition of the need for cooperation between and within agencies in the collection and management of pest information; a need for data collection and recording standards; recognition that a labor base already existed within agencies to collect pest information; and a recognition of the need for an integrated (multi-crop multi-pest) program to address pest information needs.

The Michigan Pest Survey Committee agreed that a biological monitoring system should be developed. A mandate was given to the State Survey Entomologist (senior author) to provide leadership of this multi-agency, multi-disciplinary pest and disease surveillance system. Design of the overall system structure including data management and computer applications would reside within the university, while responsibility of data collection was designated to the state and federal agencies (Michigan Department of Agriculture, Michigan Department of Natural Resources, the Michigan Cooperative Extension Service and USDA/APHIS) because of the stable observer base available through such agencies.

DESIGN AND DEVELOPMENT OF CCMS

Gage and Mispagel (4) described the basic philosophy and structure of CCMS and outlined its potential as a statewide data collection and information retrieval system for factors affecting agriculture in Michigan. At the same time, codes were developed for Michigan crops and pests so that a coordinated field manual could be produced which could be distributed to cooperators.

The structure and operation of the Cooperative Crop Monitoring System has changed considerably since the development of the basic components necessary to collect, record, manage and disseminate field observations. The basic components include:
1. a network of cooperative field observers,
2. mark-sense forms for recording biological observations,
3. a manual detailing techniques and codes to assist in form completion,
4. computer programs to summarize data at the county level, and
5. a newsletter to provide current information on the status crops and pests in Michigan.
In 1978, the first cooperative data were collected by personnel of the Michigan Department of Agriculture. In this pilot program, six crops were monitored and observers were instructed to rate the abundance of two selected pests in each crop. Almost 1000 observations were made and the data were summarized manually by crop and pest for each county. Analysis and evaluation of the information indicated that more than two pests could be recorded per crop and observers should not be restricted to two crops.

Time taken for the summary of the information also pointed out the need for an automated process to enable reports to be summarized more rapidly. A general system was envisioned where any pest could be observed on any crop at any location in the state. The primary focus of the monitoring system was to be the crop and the availability of field personnel to make rapid assessment of crop health. Collection of crop parameters (i.e. variety, growth stage) as well as factors affecting the health of the crop, had to be monitored to provide more complete information.

By 1980, the demands on the system resulted in significant modifications to CCMS. The system evolved from a general county level pest information reporting system to also include a specific field level data collection and summary system. These modification requests came primarily from the Michigan fruit IPM programs. In addition to crop and pest data, IPM programs required detailed data quantifying pesticide applications and crop damage on field level basis in order to assess the site specific importance of beneficial and pest organisms and to evaluate the various levels of IPM in practice in Michigan. In response to these needs, CCMS personnel developed new reporting forms, training materials and software required to summarize these additional data. Because of the amount of information generated by a single IPM scout, CCMS personnel also modified existing crop and pest data report forms to reduce the completion time and to provide for recording data on 32 separate pests on a single form. It was at this time that the boundaries separating IPM programs and CCMS became less distinct.

DATABASE MANAGEMENT AND CCMS

During the 1982 growing season, over 1,200 sites were monitored in Michigan, resulting in the contribution of over 120,000 crop and pest records to CCMS. In 1983 the number of records increased to 147,000. These data represented observations from 40 different crop species from about one half of the counties in Michigan. Machine and resource limitations precluded use of a commercial Data Base Management System (DBMS) for management and summary of the Michigan CCMS data until 1983, although the need for a relational DBMS had been recognized for some time.

The structuring and management of the information contained in the growing Michigan CCMS data base caused numerous modifications to data processing. Continuous development of new software to satisfy specific information summaries needed by observers, extension specialists, researchers and administrators was necessary. Specific requests could only be satisfied through writing user specific software. The number and diversity of requests and the time needed to develop software caused delays in use of the Michigan CCMS information. The problem and its solution was recognized by Brown et al.(2).

The data processing component of CCMS went through a major transition during 1983-1984. In 1983, a relational DBMS (INGRES from Relational Technology) was installed on a VAX 11/730 minicomputer and the CCMS data were restructured for management via INGRES. The transition from a Pascal software based data processing system to a relational DBMS is being completed.

In addition to crop/pest observations recorded by IPM scouts and other cooperators, gypsy moth pheromone trap catch data is also managed under the umbrella of CCMS. Since 1981, information from over 60,000 pheromone traps have been input into the system via mark sense forms (17).

Changes in data storage efficiency and access to specific CCMS records managed under DBMS have been significant. In 1983 under the Pascal based system, CCMS data required 14 million bytes of disk storage, while the same data under DBMS requires only 3.5 million bytes (16). User specific requests for reports which previously required three weeks of computer program development may now be developed in less than 30 minutes for the more straight-forward requests.

Under the INGRES DBMS, CCMS data are organized into tables or relations of similar data, which together make up the CCMS database, now called the Michigan Agricultural and Natural Resources Information System (MANRIS). Figure 1 shows the general relationship among the different tables in the MANRIS database. Site, crop, pest, pesticide use and weather tables exist for each year the data were collected. Static tables contain information (e.g. scientific names, pesticide active ingredients) about individual crops, pests, pesticides and other parameters that remain relatively unchanged from year to year. By linking or joining the various tables together, one can examine the complex interactions that exist among

MICHIGAN AGRICULTURAL and NATURAL RESOURCES INFORMATION SYSTEM

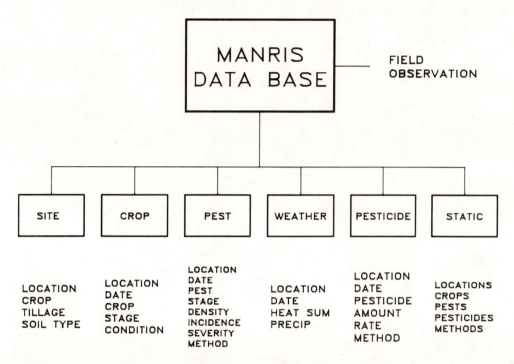

Figure 1. Logical schema of the MANRIS relational database which organizes CCMS data into tables of similar information. This provides users a consistent and orderly view of pest surveillance data. Tables can be queried separately or joined together based on location or data attributes.

crops, pests, pesticides and weather without a vast knowledge of computer languages and complex operating systems. Our experience with a relational DBMS has convinced us that it is an exceptional tool for pest surveillance systems. For the first time, we are limited in our analyses by our understanding of the data, not our understanding of the computer. However, the transition is not simple and there must be a working knowledge of the concepts and philosophy of a relational DBMS.

Weather information was recognized as a vital element in the crop surveillance program. A major effort has been made to deliver real time weather information to IPM participants so they can access degree day accumulations for timing pest specific sampling activities (7). Computer software was developed (5) to link the field observations collected at specific locations with degree-day and precipitation accumulations calculated for nearest neighbor weather stations. This satisfied the requirement to link the weather information with the crop, weed and pest growth stages.

ACCESS TO TYPES OF INFORMATION

Figure 2 illustrates the uses of the information collected by CCMS. Once assessment methods and monitoring standards are developed and the information access system is structured, the use of the information ranges from near real-time decision-making to long-term policy planning. Pest forecasting systems and loss assessment estimates will require both detailed research information from designed experiments as well as the pest surveillance information collected over a larger area.

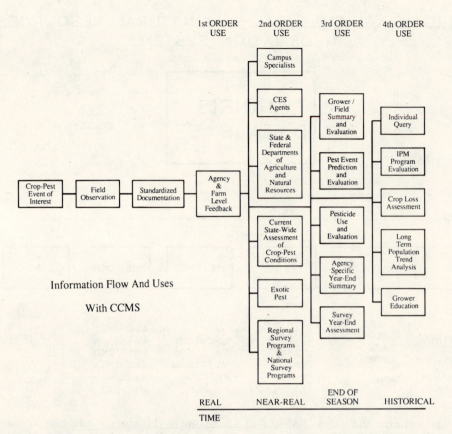

```
              1st ORDER    2nd ORDER    3rd ORDER    4th ORDER
                 USE          USE          USE          USE
```

Figure 2. Information flow and uses of CCMS information. Standardized collection and reporting methods allow continual use and benefit of CCMS data.

CCMS information is regularly disseminated to the public via the weekly Michigan State University Cooperative Extension Service (MSU/CES) Pest Alerts Newsletter; the Pest Management Executive computer system and pest management field workshops; annual CCMS reports to participating agencies; IPM year-end field summaries; and CCMS training sessions.

Field summaries prepared for each grower participating in MSU/CES affiliated IPM programs are one of the more important uses of CCMS data. These summaries, which may include 20 pages for an apple orchard, are important not only because they provide growers and CES agents with a comprehensive report of crop development, pest populations and pesticide use, but also, they serve as the motivation for scouts to complete the CCMS report forms. Without these field summaries, IPM scouts and consultants would not realize any short-term benefit of investing the time and effort in additional CCMS training and paperwork. This "what's-in-it-for-me" rationale for participating in CCMS is not peculiar to IPM programs, it seems that busy people everywhere are primarily interested in their immediate information needs and not so much interested in the needs of regional pest forecasting and crop loss assessment programs. In order to acquire the quality data needed to perform these long-term and "to-the-greater-benefit-of-society" applications, CCMS has to satisfy the immediate, often mundane information needs of participating agencies and individuals.

A single IPM year-end field summary comprises the following tabular and graphical components:

1. Pest Abundance Table - summarizes crop and pest observations by date: includes crop stage; degree-day accumulations; observed pests, life stages, density and damage levels, and sample method.

2. Pesticide Use Table - summarizes pesticide applications by date: includes crop stage; degree-day accumulations; variety and number of acres treated; application method; pesticide brand name, formulation, concentration, and amount applied.

3. Pesticide Cost Analysis Table - seasonal summary of the pesticides used: includes pesticide brand name, total amount applied, actual and recommended rate, estimated seasonal cost of each pesticide; and total seasonal cost of all pesticides applied per acre and total seasonal pesticide cost for the entire site.

4. Damage Survey Summary - bar chart summary of special damage surveys conducted in fruit sites; includes the source of damage and the proportion of damaged fruit associated with the source.

5. Field Summary Plot - x, y graphical summary illustrating the changes in pest density and damage in relation to date, degree-day accumulation, crop stage, pesticide applications, and levels of beneficial organisms.

This set of summaries allows growers and their pest management agents to review, both in biological and economic terms, the factors which influenced the productivity and profitability of the crop during the growing season. Shown in Figure 3 is an example of a CCMS Field Summary Plot which integrates the data collected by the scout (i.e.,crop, pest and pesticide) with environmental data furnished by the MSU Agricultural Weather Service. Assessment of crop loss is facilitated by the Damage Survey Summary as illustrated in Figure 4. These summaries allow growers to evaluate the timing, rate and method of pesticide applications in terms of pest population response and year-end damage estimates. By comparing CCMS year-end field summaries from growers who use various levels of chemical, cultural and biological control, CES agents and campus specialists can present evidence that supports the concepts of IPM.

CCMS also prepares annual state-wide and regional summaries which provide a seasonal overview of pest importance, pesticide use and crop development. These are usually included in departmental reports which can be reviewed by any interested party or they are specially prepared at the request of participating agencies. These "canned" packages include the following applications:

1. Seasonal Pest Importance - seasonal summary of the relative importance of Pests by crop which presents the proportion of monitored fields that were threatened by individual pests;

2. Gypsy Moth Pheromone Trapping Program Summaries - a set of tabular and graphical reports which summarize trapping results by state, county and township; and

3. State Forest Pest Summaries - a set of tabular and graphical reports which summarize forest pest populations in state forest management areas.

University researchers and extension specialists use and re-use CCMS data for purposes other than those for which the data were originally collected. These applications are facilitated by the INGRES DBMS which organizes and structures the data into a logical and efficient information retrieval system. INGRES also provides a common user interface to the CCMS data. This common interface is the INGRES query language which unlike PASCAL or other computer languages, allows users to simply tell INGRES what they want without having to specify the exact series of steps necessary to accomplish the task. For example, consider the following query which requests the average abundance rating of European Corn Borer in corn in Tuscola County, Michigan, between July 1 and July 15, 1983:

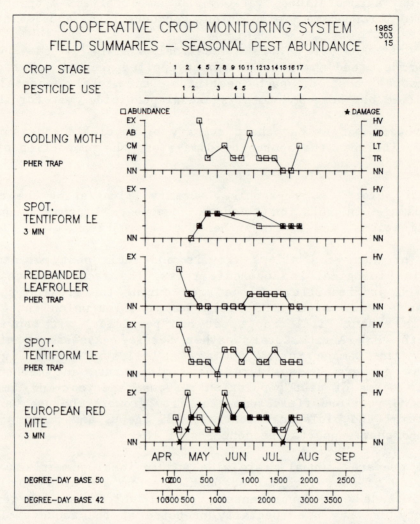

Figure 3. CCMS year-end field summary plot showing the changes in pest abundance and damage severity levels in relation to date, degree-day accumulations crop stage and pesticide applications. These summaries provide growers and CES agents the ability to evaluate pest importance, pesticide performance and pest management strategies.

```
/* Step 1: identify which records will be queried. */
/* Comments which are bounded by /*..*/ are not    */
/* executed, but serve to document query.          */

range of c is cropobs83
range of p is pestobs83

/* Step 2: retrieve selected crop variables and pest variables */
/* into a temporary table where the criteria are satisfied.    */

retrieve into mydata (c.form, c.date, c.crop, c.county, p.pest, p.abund)

where (c.crop = "corn") and        /* select crop          */
      (c.county = "157") and       /* select county        */
      (c.date >= "07/01/83") and   /* select beginning date */
      (c.date <= "07/15/83") and   /* select ending date    */
      (p.pest = "1229") and        /* select ECB pestobs    */
      (c.batform = p.batform)      /* join cropobs and pestobs */
```

216

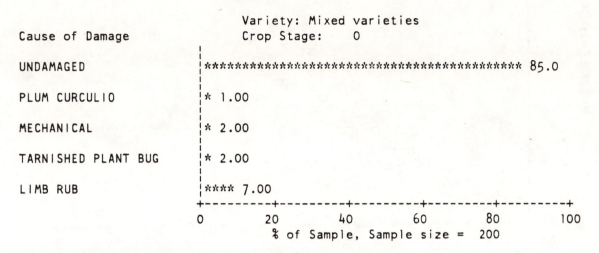

Damage Survey Date: 8/16/85

Figure 4. CCMS year-end damage survey summary showing the causes and extent of crop loss attributed to pests and other sources of damage in an individual apple orchard. Using these summaries CES specialists are able to compare damage levels among orchards in which varying levels of pesticides are applied.

```
/* Step 3: use the average function to calculate the    */
/* average abundance rating and join the county, crop    */
/* and pest codes with the appropriate static to display */
/* the name rather than the code                         */

range of m is mydata
range of cnty is countynames
range of crops is cropnames
range of pests is pestnames

retrieve (cnty.county, crops.crop, pests.pest, average_rating =
          avg(m.abundance by m.crop, m.county))
where (cnty.code = m.county) and
      (crops.code = m.crop) and
      (pests.code = m.pest)

execute query ......
```

COUNTY	CROP	PEST	MEAN RATING
TUSCOLA	CORN	EUROPEAN CORN BORER	2.7

Accomplishing the same query using a PASCAL program would require several hours of programmer time and considerable computer time to debug and test to ensure proper execution.

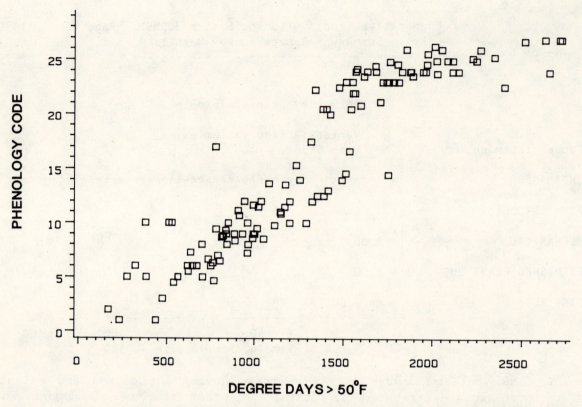

Figure 5. Corn phenology model developed from CCMS data in which growth stages are plotted against degree-day accumulations. CCMS observers are urged to collect standard crop phenology data to facilitate this type of long-term investigative application of the CCMS database.

The ability to request information using the above format is essential if the information is to be used effectively for analysis or for the development of decision and forecast models of pest and pathogen populations. A relational DBMS is one solution to the provision of user accessible data. Management of CCMS information under this type of system has opened many areas of potential analysis and investigation. For example, the 2000 observations annually on corn phenology throughout the state provide the opportunity to quantify the relationship between crop development, pest incidence and degree-day accumulation.

Figure 5 shows some preliminary results of this investigation in which corn phenology is plotted against degree-day accumulations. Similarly, Figure 6 shows the relationship of peak corn rootworm adult observations to degree-day accumulations. By combining these two types of analyses, researchers can build predictive models of crop development and key pest events. This information was extracted from the CCMS database via INGRES and displayed using an on-line statistical plotting package. Other current investigative applications of CCMS data include analysis of pesticide use in apple orchards and corn fields; and assessment of corn tillage practices in relation to pest abundance or incidence, and weed phenology.

CONTRIBUTIONS TO THE DEVELOPMENT OF THE U.S. NATIONAL SURVEY PROGRAM

The U.S. Department of Agriculture's Animal and Plant Health Inspection Service (USDA/APHIS) requested the Intersocietal Consortium of Crop Protection to form a committee in the late 1970's to address national needs for a Pest and Plant Pathogen Surveillance System. Concepts began to crystalize (11) and Michigan had already begun to develop a structured approach to pest survey (4).

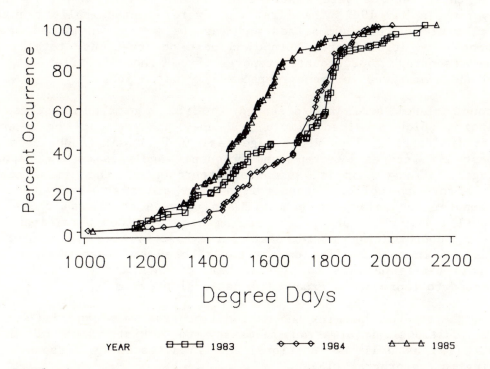

YEAR ▭—▭—▭ 1983   ◇—◇—◇ 1984   △—△—△ 1985

Figure 6. Cumulative percent of the number of Michigan corn fields in which the adult corn rootworm populations were reported at the maximum level plotted against degree-day accumulations. Such studies are used to develop pest event prediction models to assist Michigan pest management programs.

By 1981 CCMS was being used regularly by the IPM program in Michigan as its data collection and summary system. Several thousand observations were being entered and managed under CCMS, and weekly reports were being produced which summarized current pest levels based on observations made throughout the state. It was also during this period that a National Pest Survey Program was evolving.

In the spring of 1982, Michigan CCMS personnel assisted USDA/APHIS in initiating a national pest survey program (13). Using CCMS as a model, representatives from 16 pilot states agreed to a standard coding system for transmitting field observations to the USDA/APHIS Fort Collins Computer Center (FCCC) in Colorado (15). As USDA/APHIS was unable to develop computer software needed to summarize these data due to other program objectives, Michigan assisted in this development under a cooperative agreement with USDA/APHIS, by developing the necessary software and was responsible for weekly summarization of the national program data during the pilot year of 1982.

The basic information collected from the field and sent to the FCCC by sixteen cooperating states consisted of two record types: a crop record and one or more pest records. The data records contained the following data elements:

Crop record:
    Record type, record number, year, month, day, state, county, crop, crop stage.

Pest record:
    Record type, record number, pest code, predominant pest life stage, other pest life stages, abundance or incidence rating, damage or severity rating, sample method.

Variables in these records were a subset of records developed for the multi-crop Michigan system. In May 1982, the first data were electronically transferred to the Michigan minicomputer, summarized and reports were transmitted back to the FCCC for access by Pilot States. Crop and pest observations, totalling 170,000 records, representing 2500 county reports from the 16 Pilot states, were transmitted and summarized in twenty-three weekly batch jobs.

The county level summaries prepared in 1982 consisted of three separate tabular reports which provided simple frequency analyses of the reported pest abundance or incidence ratings and damage ratings by crop; frequency analysis of reported pest life stages by crop; and frequency analysis of the reported crop growth stage. These simple summaries represented the first cooperative multi-state effort to collect standardized crop and pest observations, and to utilized computer technology to transmit and provide access to these data on a national level (14). The 1982 pilot program also provided participants valuable experience in coordinating state pest survey activities to capture quality observations and in constructing data records based on these observations. The result of this experience was to provide participants a better understanding of what was needed to improve and expand the national program.

In 1983 the record types for the national program were expanded significantly to include additional information necessary to satisfy some of the Federal requirements (19) as well as additional needs suggested by personnel in states participating in the program (18).

The following list provides the major record types used in the National Cooperative Plant Pest Survey and Detection Program in 1983 :

TYPE	INFORMATION
10	Crop
11	Location
21	General pest
22	Pest presence or absence
23	Pest diagnostic information
24	Export certification
25	Trap pest catch
31	Unknown crop damage (biological)
32	Unknown crop damage (physical)
41	Indirect beneficial organisms
42	Direct beneficial organisms

The new record types were developed to reflect the diversity in which crop and pest observations were collected. These expanded record types provide flexibility because they contain specific data elements which are peculiar to the different sources of crop and pest data that exist within each state. Another important objective in organizing the national data into record types was that it facilitated data processing and access in the national program.

Illustrated in Figure 7 is the conceptual schema of the national pest survey database structured on the record type concept. By grouping data into tables of similar data and utilizing modern relational database management systems, users can access data in specific tables without having to consider thousands of other records which are inappropriate for the present application.

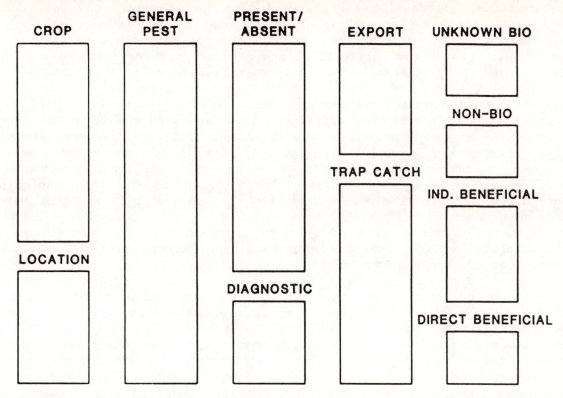

CROP   GENERAL   PRESENT/   EXPORT   UNKNOWN BIO
        PEST   ABSENT

                                              NON-BIO

                               TRAP CATCH

                                           IND. BENEFICIAL

LOCATION

              DIAGNOSTIC

                                           DIRECT BENEFICIAL

Figure 7. The proposed relational schema of the U.S. National Plant Pest Survey and Detection Program database developed by CCMS in 1983 for USDA/APHIS. Separate tables for each of the national program record types serve to organize the database into logical units based on data content and objective. Under this schema, the different record types can be accessed independently without having to sort through thousands of other nonpertinent data records.

INTERNATIONAL CROP INFORMATION SYSTEMS

The concepts and implementation of crop survey systems were developed further at an international level during 1983-1984 when the senior author spent a year at the Victoria Department of Agriculture's Plant Research Institute (PRI) in Melbourne, Australia.

A team of research, extension and computer specialists designed the Victoria Crop Information Service (1). The entire service was designed, developed and executed within a six month period and was an excellent example of system development based on needs assessment, inter-disciplinary cooperation, desire, support and a great deal of effort on behalf of the PRI personnel. Special focus was placed on small grain systems with emphasis on stripe rust, locusts, cereal cyst nematodes, cape weed, armyworms and mice.

The potential for international cooperation in the area of crop monitoring systems is considerable as the need and desire to improve crop production efficiency appears to be great.

BENEFITS AND FUTURE DIRECTION OF CCMS

CCMS has played an important role in Michigan in the development of an information system focused on plant health assessment. It has encouraged and facilitated inter-agency cooperation, encouraged the development of standard sampling methods for pests and diseases, coordinated the assessment of economic injury levels of pests, developed standardized plant and insect phenology observations, and developed a standard data coding system. Together these have provided a tool to assess low pest population levels as well as assessing populations in an outbreak phase. CCMS made use of modern relational database management systems for agricultural and natural resource applications, encouraged interdisciplinary interaction and attracted resources for plant pest survey.

Pest surveillance systems are only in their infancy in the U.S.A. An organized system has been in existence for three years and is in the developmental stages. The advances made in state and national programs have been significant and major changes are occurring regularly. A major thrust toward database management systems is underway. The recognition of the value of field observations in the development of models to forecast pest potential is just beginning to be recognized. Utilization of the data, now that access to it is being developed, is the next important step. We envision that private industry will be a major contributor of field data as well as a major user for assessment of market potential for agricultural products to assist farmers to facilitate the optimal management of cropping systems.

State and local governments need quality crop and pest information to assess their economic base and to determine if crop production projections are meeting targets. Increased concerns about toxic substances require state governments with regulatory responsibilities to assess chemical inputs to agricultural and forest enterprises as well as to document certification of pest free products which may be shipped across state lines or exported to other countries. These agencies must have good knowledge about pest situations and their potential impact on crop or forest production.

Field observations collected over wide areas using closely correlated observation methods and standard observation codes can assist in determining regional trends in pest population change and enable the detection of newly introduced pests much earlier than in the past. Development of regional pest distribution and dispersal is much more feasible than it was when no communication existed between states and state surveillance programs. It is often difficult to anticipate the types of information necessary to describe and quantify a given pest situation. It is most important to evaluate past successful single pest or pathogen surveys and to glean the best methods and types of variables collected to answer the question pertaining to the situation. We must have a long range vision to collect variables we may not need immediately but may find essential in the future as we attempt to develop regional models to forecast pests. For example, a few years ago one might question the value of collecting tillage method whereas today most would insist that it be collected.

The future success of crop surveillance systems depends on the quality of the observations so that analytical interpretation of the information contained in the database can be made. The burden of quality observations lies on our educational institutions and the training sessions conducted on behalf of the field observers. Proper identification materials, detailed pest damage assessment methods and quantitative population assessment is essential for observers to provide quality observations. Historical information on pest and cropping systems is essential for development of systems to forecast pest dynamics. Research and development of these components of pest surveillance as well as investigation of new monitoring and data collection techniques can be conducted in universities and colleges. Expertise in modeling and data analysis will emanate as well from students who apply their talents to quality field observations.

CCMS has helped to stimulate and contribute to development of state, national and international crop monitoring programs. Information contained in the database will need to be continually evaluated and improved to meet the needs of users of the information. The information about factors influencing plant health is valuable because of the expense of employing people to collect it. If such observations are not recorded and stored then they can never be used. Such information, carefully collected, recorded and managed for future use can provide a valuable resource to the plant protection community in assessing and improving plant health.

LITERATURE CITED

1. Amos, T.G., Perry, M.R., Gage, S.H. 1983. A crop information service for Victoria. Research Rep. 185. Vict. Dept. Agric. 21 pp.
2. Brown, C.G., Lutgardo, A., Gage, S.H. 1980. Data base management systems for IPM programs. Environ. Entomol. (Forum) 9:475-842.
3. Gage, S.H., Mukerji, M.K. 1978. Crop losses associated with grasshoppers in relation to economics of crop production. J. Econ. Entomol.71:487-498.
4. Gage, S.H., Mispagel, M.E. 1979. Design and Development of a Cooperative Crop Monitoring System. Automated Data Processing for IPM. pp 157-78. Ft. Collins Co. Lab. Infomation. Sci. Agric., Co. State Univ. 289 pp.
5. Gage, S.H., Pieronek, J.V. 1980. A microcomputer application for integrated pest management. AgComp Bull. 1:29.
6. Gage, S.H., Russell, H.L., Prokrym, T.H., Miller, D.J., Pieronek, J.V., Rosenbaum, R.R. 1982. Development and implementation of the Cooperative National Plant Pest Survey and Detection Program: Final report to the Federal Cooperative Extension Service. Ent. Dept. Report, MSU.
7. Gage, S.H., Whalon, M.E., Miller, D.J. 1982. Pest event scheduling system for biological monitoring and pest management. Environ. Entomol. (Forum). 11:1127-1133.
8. Gage, S.H. 1983. Development of processor based hierarchial system for communication of integrated information for pest management. Final. Tech. Rep. EPA Grant R805730. 390 pp.
9. Gage, S.H., Miller, D.J., Antosiak, R.A., Rosenbaum, R. 1984. Pest Management Executive System (PMEX) user's guide. Tech. Rep. No.28. Dept. Entomol., Mich. State Univ. 89 pp.
10. Haynes, D.L., Brandenburg, R.K., Fisher, P.D. 1973. Environmental monitoring network for pest management systems. Environ. Entomol. 2:889-99.
11. Kim, K.C. 1979. Cooperative pest detection and monitoring system: a realistic approach. ESA Bull. 25:223-230.
12. Lovitt, L. ed. 1981. Insect survey manual for the North Central States. Wisc. Dept. Ag., Madison, Wisc.p.115.
13. Russell, H.L., Pieronek, J.V., Miller, D.J., Gage, S.H., Prokrym, T.H. 1982. Instructions for constructing and transmitting data records for the Cooperative National Plant Pest Survey and Detection Program Pilot Program. Ent. Dept. Report, MSU.
14. Russell, H.L., Miller, D.J., Prokrym, T.H., Gage, S.H. 1982. Final report on the computer processing of Cooperative National Plant Pest Survey and Detection data. Ent. Dept. Report, MSU.
15. Russell, H.L., Gage, S.H., Miller, D.J., Prokrym, T.H., Rosenbaum, R.R. 1982. The Cooperative National Plant Pest Survey Program: Overview of the 1982 Pilot Program. Ent. Dept. Report, MSU.
16. Russell, H.L., Prokrym, T.H., Miller, D.J., Gage, S.H. 1982. Annual report of the activities of the Cooperative Crop Monitoring System. Ent. Dept. Report, MSU.
17. Russell, H.L., Prokrym, T.H., Miller, D.J., Gage, S.H., Rosenbaum, R.R. 1982. State summary of the Gypsy Moth Pheromone Trapping Progam for 1982 with individual regional summaries. Ent. Dept. Report, MSU.
18. Prokrym, T.H., Miller, D.J., Russell, H.L., Gage, S.H. 1982. 1982 CCMS Comprehensive pest summaries. Ent. Dept. Report, MSU.
19. Russell, H.L., Miller, D.J., Gage, S.H., Prokrym, T.H. 1983. Enhancements for the Cooperative National Plant Pest Survey and Detection Program: A proposal to participating states. Ent. Dept. Report, MSU.

20. Russell, H.L., Pieronek, J.V., Miller, D.J., Prokrym, T.H., Rosenbaum, R.R., Gage, S.H. 1984. Status of the Michigan Cooperative Pest Survey Program and the applicability of relational database management systems to plant pest survey data. FY 1983 Final Report to the Cooperative National Plant Pest Survey and Detection Program. Ent. Dept. Report, MSU.

21. Tummala, R.L., Haynes, D.L. 1977. On-line pest management systems. Environ. Entomol. 6:339-49.

22. Welch, S.M. 1984. Developments in computer based IPM extension delivery systems. Ann. Rev. Entomol. 29:359-391.

23. Whalon, M.E., Gage, S.H., Dover, M.J. 1984. Estimation of pesticide use through the cooperative crop monitoring system in Michigan apple production. J. Econ. Ent. 77:559-564.

CHAPTER 23

CROP LOSS ASSESSMENT IN A PRACTICAL INTEGRATED PEST CONTROL PROGRAM FOR TROPICAL ASIAN RICE

Peter E. Kenmore

FAO Inter-Country Program for Integrated Pest Control
in Rice in South and Southeast Asia, P.O. Box 1864
Manila, The Philippines

Crop loss assessment (CLA) for practical integrated pest control (IPC) is "caught between a rock and a hard place" (Figure 1). On one hand the clients of IPC programs, including farmers, their advisers, and agricultural policy makers, want more management options to respond to pest threats they perceive. They want to know what to do for the crop at hand, and possibly for one crop in the future. They have been managing pests for years, using their own perceptions of crop losses and developing attitudes to crop losses from pests. A practical IPC program must stay close to the clients and provide advice that they can understand and use.

On the other hand, there is a large and growing body of scientific literature containing more and more refined approaches to CLA, most of which seems to be written only for other scientists and not for clients with problems. When these scientists speak to clients they present formulations full of internal rigor and elegance, but with little evidence of the clients' expressed needs. Practical IPC programs must often borrow from the literature to answer some of the questions asked by clients. Occasionally, the attitudes and questions of clients can be changed. This can be part of a training, to help them solve problems more effectively for themselves. The training may also bring the clients closer to what analytical or reductionist scientists can offer. As with most problem solving in the real world, there has to be compromise. I believe IPC programs must resolve this compromise in favor of the clients rather than the scientists.

Beginning with the clients - farmers and administrators (Figure 2) - surveys of farmers and statements of administrators show that both groups usually perceive pest losses to be intolerable. They regard chemicals, especially insecticides, as essential to rice production. In contrast, field data suggest that insecticides are not essential in every rice crop. No more than half the crops we have studied in farmers' fields exhibit yield increases with insecticides. Where do these large discrepancies between clients' perceptions and IPC field results come from?

When farmers misidentify the causes of damage observed in their fields, they may apply an ineffective chemical, like an insecticide for a fungal infection. Without objective standards for evaluating the impact on yield and profits of each chemical treatment, they may conclude that their yield was saved even when there was no effect. Administrators who are several steps removed from the field, and must rely on reports from overworked or unreliable field scouts are in even weaker positions to judge the benefits derived from chemical treatment. As they often prefer to act quickly and appear decisive, they may order treatments and never evaluate because the action alone serves their political purpose and a negative evaluation can only weaken that purpose (20). Clients continue to perceive pesti-

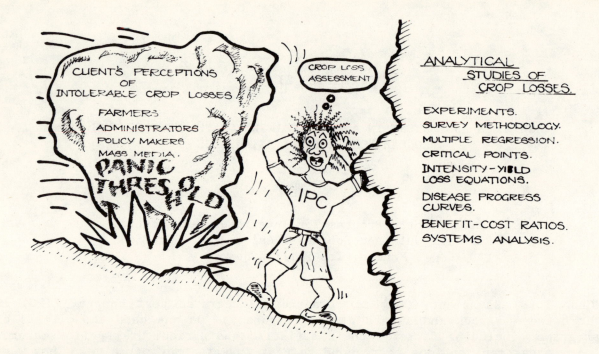

Figure 1. Pictorial representation of crop loss information: clientele dilemma.

Figure 2. Pictorial representation of crop loss information: methodology dilemma.

cides as essential and act accordingly without testing that perception.

Part of the clients' uncertainty is shared by researchers. There is little data on quantitative interactions of pests. We know that rats and weeds interact positively in rice to make crop losses more severe. But we do not know the magnitude of the interaction or even the cause of the higher loss. Do rats prefer weedy fields and feed more or does rat damage in the vegetative stage reduce the competitive ability of rice? While interactions of defoliating insects with stemboring insects also seem to be positive, the interactions among defoliators themselves are unclear. Wounds from defoliators increase the observed levels of bacterial leaf streak, presumably by opening infection courts, but the eventual impact on yield has not been estimated.

For the moment, multiple pest infestations are still outside the knowledge base of reductionistic analyses of rice crop losses. As some of these interactions may take place at low densities, the farmer has reasons to feel insecure about ignoring low density populations, even though IPC demonstrations show that yields will not be affected by low populations. If such farmers apply insecticides, and yields are high because pest populations never reached yield reducing levels, they will be reinforced in this behavior, all because practical field assessments of crop losses have not been done.

Additional ubiquitous reinforcement of insurance applications of pesticides comes from the chemical industry. As well as multi-media campaigns and extensive networks of energetic agents, companies often enlist government agencies to market their products. The Bayer company has had a multi-year high-level agreement with the National Irrigation Administration of the Philippines where it pays the national agency (which is not mandated by the government to give recommendations on crop protection) and distributes chemicals and signs. More than 900 field employees of the agency are used to set up demonstration farms where the company's products are displayed and supposedly used regularly. The additional contact time with farmers, and the legitimization implied by the government agency circumvent the rigors of free enterprise competition and improve the performance of money invested in marketing.

The chemical industry is very aware of the importance of yield and yield losses to clients. Benlate has been successfully marketed in the Philippines with advertisements that do not use the word fungicide. The entire campaign used the term yield booster instead to describe its product, without explaining how yields might have been boosted. Ambiguity about the differences between pesticides and fertilizers appears to have been maintained by the chemical industry in the Philippines.

The response of the FAO IPC program to these discrepancies has been to try and change the questions asked by clients. Instead of "what chemical do I need to save my crop?", clients are taught to ask "how much will I get in return for spending on chemicals?" This is done differently with farmers and with administrators. Neither of these questions is a central question asked by researchers working in CLA (39). They tend to ask "how much of the estimated production of this field/season will be/has been lost to the observed pest populations?" While this question is asked by some policy makers and planners, it is not asked by practicing rice crop managers or their advisers.

CLIENTS' PERCEPTIONS OF RICE CROP LOSSES TO PESTS

Nearly all irrigated rice farmers in the Philippines regard pest damage as intolerable and unavoidable if no action is taken (25). This is similar to farmers throughout the world, who usually treat their crops for insurance rather than investment, because they always perceive the pest threat as quite serious (27, 28).

When asked if they expected a lower yield from <u>not</u> treating, 94% of the rice farmers surveyed in Iloilo, Philippines said "yes" (4). The average estimate of the percentage yield farmers expected to lose was 49%. Over 88% of farmers said that treating only after pests were noticed in the field would lead to significant losses, while over 90% said that preventive calendar applications were the ideal use of insecticides. Missing one application risked the whole season's spray program, according to 90% of the farmers, and over 88% felt that calendar-based treatment allowed them to sleep easier at night.

A similar study of rice farmers in Tanjong Karang, Malaysia revealed that while the average estimated loss to pests was 25-30% of yield, of 43.5% reporting serious pest attacks during the 1980-81 season, only 27.5% felt they had controlled their problem successfully (13).

A 1984 study of rice farmers from two villages in Bicol, Philippines found that "farmers' first recourse when confronted with infestation problems is to spray with chemicals regardless of the extent of damage" (29). In this study, 83% of farmers in the richer village that was more frequently visited by extension workers, and 60% farmers in the poorer said they would spray at once when any pests were observed. On a 7-point Likert scale of attitude with 7 the highest agreement, the two villages' average reactions to the statement "Using chemical crop protection methods would increase harvest" were 5.97 and 6.47; high scores-all over 6 - were also registered in responses to statements that chemicals including pesticides and fertilizers are essential and that chemical crop protection means a farmer will avoid pest attack. The influence of income was expressed in the major problem cited by farmers: 56% of farmers in the richer village cited <u>prevention</u> of pests and diseases as their biggest problem, while none in the poorer village did so. Instead, the poorer village respondent named particular pests that had already invaded the field. Similar results were obtained by van Huis et al. (16) in Nicaragua where large commercial farmers <u>overestimated</u> the threat from insects to their maize production (as compared with results from field trials) far more often than did small-scale subsistence maize farmers. Clients' perceptions of crop losses from pests may change depending on income level, possibly because more money is at risk or because there are more response options open to richer farmers. Oliva et al. (29) also suggest that more socially prominent farmers may perceive greater potential losses and treat more often in order to maintain both their status as progressive farmers and their good relations with agricultural technicians who advise them to treat more frequently.

Two recent surveys of 300 and 658 Philippine rice farmers respectively, confirmed that crop losses were considered intolerable and requiring treatment (3, 6). In the first study, from Leyte, 65% of farmers said calendar spraying was the best method, 67% said the more you spray, the higher will be your profit, 71% said a few butterflies in the field means a farmer should spray, 49% said the presence of one moth in the field requires spraying, and 62% said they sometimes sprayed even if insects were not observed in the field.

In the second study, conducted in 9 rice growing areas of the country, 47% of farmers said that whenever high yielding, modern cultivars are planted, there will be significant numbers of pests. Seventy-nine percent said the use of pesticides will increase yields, 67% said that although pesticides are expensive, using them more often will ensure higher profits, 76% said you should spray pesticides the moment you see pests in the field even if damage looks small, and 65% said a farmer should spray his crop regularly to prevent pests even when he does not see pests in the field.

Rice farmers believe intolerable pest damage is inevitable unless action is taken. Pesticides are seen to be as essential to rice production as fertilizers, and in the Philippines, more rice farmers use pesticides in irrigated rice than fertilizer every season (21).

The attitude of agricultural administrators to crop losses is, as with farmers, a _qualitative_ rather than _quantitative_ response. A lead quotation in a national newspaper shows the near hysteria that greeted a rise in the population of rice pests in Malaysia:

"RED MENACE THREATENS PADI CROP...This disease kills and kills fast and if we don't act now, all may be lost"
   --New Straits Times (Kuala Lumpur) front page lead headline, 11 May 1983.

A subsequent study showed that only 8% of the area discussed was infested during the epidemic and only 2% was severely damaged by the "penyakit merah" ("red disease") virus or PMV (14). Over a four year period (1981-1984), the average loss to PMV in that area of Malaysia was estimated at less than 1%.

Similarly, in the Philippines during 1975-76, a new phenotype of brown planthopper (biotype 2) evolved and attacked previously resistant cultivars. Newspapers and officials of the Department of Agriculture were in a near panic. Yet internal reports from the pest monitoring system to the Bureau of Plant Industry (BPI) revealed that in no municipality was the percentage of rice area infested over 10%, and the national level of infestation far below 1% (BPI, unpublished data). When 12,000 of 103,000 hectares in North Sumatra, Indonesia, were infested by rice brown planthoppers, an emergency was declared and aerial applications of insecticide carried out (38). Similar reactions have been observed in Sri Lanka to brown planthopper and in India to brown planthopper and tungro.

In nearly all cases, there is little evidence of any yield actually being lost. The rule of thumb seems to be that when any sized political unit experiences an infestation, regardless of severity (one tungro-infested plant means the whole farm of 2-4 hectares is infested, even if the actual incidence is 10%), an outbreak is declared and extraordinary measures invoked. This may be called the "Political Pest Outbreak Panic Threshold," (cf Zadoks 1985 and Chapter 18) and further comparative work should be done to refine it to aid policy makers.

Even when the panic threshold is not exceeded, policy makers routinely order insecticides without evaluating their large-scale impact on production. There is no regular program to collect data on rice crop losses in Asia, so pesticide policies are set without comparisons to a field standard. Most policy makers identify and treat pesticides as some fixed fraction of fertilizers (generally 10-15% by value) and conceive of them acting the same way to raise production (cf Chapter 1). Atwal (1) exemplified this attitude as follows - "farmers are not the only beneficiaries of pesticide usage. The pesticide industry also make profits on the chemicals they sell. The fact is that the government and the general public are big winners. The quality of food offered today to the average consumer is undoubtedly better than ever before." Most important from the perspective of policy makers, "the government is assured of certain minimum food production in the country, which is essential for political stability and to plan future development programs."

Rice is the staple food in much of Asia, cheap food means political stability, and pesticides are thought essential for production, even though accurate crop loss assessments are not available. The identification of pesticides with fertilizers is common; it is more convenient to plan for, import, store and distribute them together rather than adapt supply policies to the particular ecological or agronomic roles served by each.

Rice farmers perceive losses from pests as intolerable and inevitable unless regular preventive actions are taken. They estimate losses between 20% and 100%. They perceive pesticides as their only way to ensure yield, and are prevented from regular treatments only by the high cost of pesticides. They perceive pest damage as potentially within human control but do not perceive pest management as following rules of investment and economic return, but as an unavoidable part of modern rice production.

FIELD EVIDENCE OF CROP LOSSES

There are few large scale studies of yield losses in rice fields. Litsinger (24) summarized eight years of trials in farmers' fields in different parts of the Philippines. In the 337 crops, only 50% showed significant yield differences between parcels receiving the maximum level of insect control (9-11 applications per 10-12 week season) and parcels in the same farm with no insecticide applied.

Sumangil (33) summarized a different series of trials from the Philippine national IPC program. In 105 crops, again only 50% showed significant yield losses when treated parcels (6-9 applications of insecticides) were compared with untreated. These results can be attributed to two major factors. First, there were large populations of natural enemies of rice insect pests in all of the crops studied. Not only did these control most pest populations in the untreated parcels but if treated with insecticides, their pest regulating role was disrupted. Second, the rice cultivars planted by most farmers during these studies expressed phenotypes resistant to some of the major insect and disease pests, so that pest population growth was limited without chemicals.

A recent study in Sri Lanka (26) showed that in 77% of farmers' fields, there were no significant yield differences between biweekly insecticide treatments (5-6 per season) and no insecticide. This was attributed to low pest pressure. Natural enemies are found in abundance in Sri Lanka rice fields and may have contributed to the low pressure.

Small scale studies (from one to five farmers' crops) have been carried out during IPC training courses in South India (Karnataka and Kerala), Malaysia (Penang, Selangor, Kedah, and Perlis), and Indonesia (six sites in Java and Sumatra). In each study the results have been similar. In no more than half the fields, and often in none of the fields, have there been significant yield losses when untreated parcels were compared with maximum levels of insecticide. Our working hypothesis is that the perceptions of our clients, that insect pests losses in tropical rice are intolerable in every season and that insecticides are at least as essential as fertilizers for high yields, are not supported by field evidence.

We do not have a complete explanation for this discrepancy. At least one study has shown that educated farmers in a very advanced rice-growing province of the Philippines applied insecticides not in correlation with insect damage levels but in significant correlation with fungal and viral diseases symptoms (25). Misidentification of pests causing damage is common, although it may be corrected through better extension and farmer training (see below). More difficult is to change farmers' decision rules. Bandong et al. (2) found that farmers without special training tended to treat at first sight of an insect either in their rice fields or in their neighbors' fields or along the dikes separating paddies. Part of this response is due to an attitude of wanting to insure the crop through insecticides, since rice farmers perceive insects and diseases as controllable natural hazards unlike floods, droughts, or typhoons (7, 8, 9). Some of it is ignorance of the interaction of crop growth stage and multiple pests. When differences in planting date of less than one month can, through fluctuations in solar radiation, result in 30% changes in yield (10), and the average range of planting dates in most irrigated doubled cropped rice areas is more than 2 months, the impact of pests on yield can be very difficult to estimate. When pests interact (37), the impact on yield may be synergistic (low levels of stem borer seem to favor whorl maggot and defoliator damage). With these uncertainties, it may be reasonable for farmers to conclude that pest losses are unavoidable and treatment always needed. The relationship between low levels of pests and yield, and the role of natural enemies in keeping pest populations below damaging levels, are not yet well understood.

A further source of the discrepancy is large-scale vested political and economic interests. Pesticides are not only a source of profits for multinational chemical companies, but are also important commodities in input subsidy programs in many countries. This means that, when governments undertake to guarantee the delivery and availability of pesticides along with fertilizers to farmers, they also guarantee intermediate income to people in the distribution channels. Both large companies and small private dealerships, including rural credit institutions, are interested in the rapid cycling, of pesticides as commodities through the agricultural system. Even if the compounds never reach the fields, once farmers are forced to accept the debts for pesticides as part of production loan packages; the government covers those debts to the lending and/or dealing institutions.

Availability begets routine use. By 1966, following 10 years of free treatments by government technicians, Philippine rice farmers had adopted insecticides widely in irrigated villages. However, it was not until the Masagana 99 program, which began in the early 1970's and included insecticides as part of the loan package, that average dosages of insecticides rose high enough to be ecologically potent (21). The aggressive publicity campaign for Masagana 99, designed by the J. Walter Thompson advertising firm and including multiple media as well as personal visits by technicians, certainly pushed the concept of insecticides as essential to rice production.

Government campaigns to promote pesticide subsidies determine the extent of insecticide use by farmers (30). In Bangladesh, a 50% reduction in the subsidy resulted in a 50% reduction in insecticide use in the late 1970's. By 1981, insecticide subsidies were completely withdrawn leading to almost no use of insecticides. During the first quarter of 1986, the subsidy on insecticides in Indonesia was reduced from 85% to 75%, causing a price rise of 66% to the user. As a result, farmers in Yogjakarta were motivated to learn IPC.

RESPONSE OF THE FAO IPC PROGRAM

Because results of field trials showed that measurable crop losses were far lower than clients' estimates, it became a major objective of the FAO IPC Program to convince clients that they could maintain production with much less insecticide. Four methods were used to accomplish this objective:
  i.   demonstrations;
 ii.   training;
iii.   reassurance by follow-up; and, for administrators,
 iv.   accumulating data from farmers' fields.

Demonstrations

These are "results" demonstrations designed to compare the yields and profits from different practices after the crop season. The first round of demonstrations in the Philippines compared 4 treatments: farmers' practice, no insecticides, the nationally recommended calendar of applications, and IPC decision rules. Each treatment was applied in at least 100 meters^2 and replicated using four farmers per village. This design allowed us to compare the return on added investment for each of the treatments with the no investment level, and showed that IPC decision rules were most profitable in 80% of the crops (33). In one set of trials, the return on investment (ROI) for IPC was higher than any other treatment in 6 of 7 main season crops and 2 out of 3 off season crops. The returns on investment for IPC ranged from 2 to over 4, whereas neither farmers' nor calendar treatment ever exceeded an ROI of 1 (32).

Table 1. Farmers' pest control investments compared to costs for IPC need-based use of insecticides in the same farm.

Type of Pest Control	Cost of Insect Control (P)	Total Cost of Production (P)	% Total Cost of Insect Control	Average Yield (Tons/Ha)
IPC	176.11	3962.62	5.63	5.01
Farmers' Practice	387.58	3978.13	10.69	4.92

Data from 93 farmers in 8 regions of the Philippines, wet (main) season 1984. Cost of IPC includes P100 imputed labor cost of time spent monitoring the crop to a whole season so actual cash cost of insecticides in IPC is P76.

A related analysis using data from regional experiment stations and researcher-managed trials in farmers' fields showed that the probability of making a profit from investment in pest control was higher when IPC decision rules were followed than for either no application or for any treatment needing a larger investment (15). Whether in farmer-managed, extension supervised or researcher-managed trials, the IPC decision rules produced far lower investments in insecticides and, even when an opportunity cost of labor for monitoring fields regularly was included, a much higher return on investment (Table 1).

In the second round of demonstrations IPC decision rules were compared only with farmers' practices in farmer-managed, extension-supervised trials.

In the preliminary Sri Lankan study reported by Matteson (26), different decision rules were tried in different sites, but all of them showed higher profits and lower insecticide use compared both to farmers' practices and to earlier decision rules that caused more frequent applications through lower action thresholds.

Training

This includes five kinds of activities:
- critical consciousness raising;
- explanations;
- field methods demonstrations;
- problem sharing and discussion; and
- review and practice including tests.

The first task is to stimulate a more critical and demanding attitude to pest control. This begins by public enumeration of problems experienced with conventional pest control, such as human poisoning, elimination of fish, frogs, snails and other important protein sources in rice paddies, increasing cash expense of pest control, and the breakdown of resistance in rice cultivars. We do this at special meetings open to everyone in a village, held in the local political hall, church or Muslim prayer house. When they list their problems farmers get more objective about pest control and begin to see it as an investment, with good and bad consequences that can be compared, rather than as a necessity of modern cultivation. This is a fundamental change in their approach to pest control.

One activity arising from this kind of meeting is an informal contract between farmers in a village and trainers to hold a series of farmers' classes on

IPC. These classes are always held when crops are in the ground so that pests and symptoms can be investigated in the field. If a results demonstration is also taking place in the village, this is a good field to use for practice. The explanations offered in the classes center on why IPC works better than conventional control. This includes:

o   factors exerting natural control on pest populations
      - natural enemies
      - weather
      - availability of host plants
o   choice of the most appropriate cultivars that express resistant phenotypes
o   crop management practices that control pests
      - fertilizer applications
      - land preparation
      - crop rotation
      - imposing a break in the availability of rice by simultaneous harvest and synchronized planting
o   proper selection and safe handling of pesticides
o   simple budgeting or accounting methods to make it easier to calculate profits from the crop.

Group discussion of explanations allows the trainers to adapt their messages to local terms and counting systems. They also reinforce socially the principle that pest control is an investment to be compared with estimated, graduated crop losses.

Field work is the only essential part of training. Extension specialists call this "methods" demonstrations, where skills are practiced by imitation of the trainers who correct on the spot. Identification is the core of these field skills. Once farmers can identify most of their common problems they can begin to compare different levels of pests and pest damage symptoms. Decision rules are learned in reference to observed field levels wherever possible; this is far more effective than learning in relation to an abstract extrapolated level. More repetitions of the standard identification-assessment-decision (treat or not) pattern throughout a season mean more confidence in the new techniques and a greater chance of adoption. When farmers compare their diagnosis and decision with that of the trainers, the IPC system gets better and better.

Problem sharing, both in discussions and through field walks, reinforces the IPC message by showing that it works over a realistic range of local conditions. Farmers like social reinforcement, and strength of numbers helps them confront the IPC trainers when a message does not make sense. While we have induced the formation of village IPC groups de novo, it is easier and faster to build the IPC group within the framework of a existing farmers' organization such as a village cooperative. We aim in either case to have groups that not only continue for the seasons after training but also are empowered to initiate useful, even critical exchange with extension services and other public and private sources of technical information. We are just beginning to develop a method of impact evaluation that uses frequency of farmer-initiated contacts as a measure of training success.

Field tests are used in the FAO IPC Program to sharpen, review and reinforce skills. These are multiple choice tests that cover identification, assessment of pest threat, choice of control methods, and decision making. They are given only in the field, with as fresh specimens as possible, and with the spirit of a game. Farmers in the Philippines respond very positively to the competitive challenge although this may not be the same in other cultures. For national extension programs, these tests are quite valuable to assess the impact of training and the performances of different trainers.

Reassurance through Follow-up

It takes more than one class and one demonstration to deliver IPC concepts to villages. Not only do pest problems change from season to season and field to field, but management choices, from cultivars to chemicals to fertilizers to machinery also change. A dynamic two-way channel of information with local agricultural advisers is the ideal, though it is very difficult to achieve. There is some success when exceptionally motivated extension workers not only keep following up their previous seasons' IPC groups but also search out new knowledge from specialists to deliver to farmers.

Accumulation of Evidence

This means not only tabulating results from demonstrations and evaluation of training but understanding the needs of administrators in relation to national agricultural policies. Production is the first concern of administrators; therefore divorcing pesticides from their concept of production is the first step. The FAO program has compiled results to show that lower pesticide use does not reduce production. This is a continual process, as new administrators come to office and as the IPC message filters down farther through the agricultural hierarchy so that more local administrators need local evidence. We welcome this because it establishes a stricter standard for evaluation of any alternative technology, not just IPC.

At the same time, by providing and insisting on economic comparisons as well as production comparisons, policy considerations of pest control and of crop loss assessment are moved towards a more specific and practical basis. Pest control is an investment for a nation just as for a farmer. By showing that losses can be reduced while foreign exchange is saved, administrators make more effective decisions based on facts from the field.

The FAO IPC Program is also attempting to develop objective criteria for extension evaluation, using data on training inputs and changes in field skills among trainees. Administrators use these data to compare the return on investment in extension with other uses of scarce funds. For example, our calculations show that the overall rate of return for investment in training costs is 4.48 over a four year period (1983-86). During this time, farmers saved approximately 14,279,200 Pesos using IPC compared to program training costs of 3,186,980 Pesos. The number of extension staff and farmers trained during the same period were 4,893 and 17,987 respectively.

We also make the point that introducing new management technology to farmers can and should be planned, monitored and evaluated in unbiased terms just as introduction of new seeds or machines. This increases administrators' interest in and use of crop loss assessment because it is in comparison with estimated losses that program benefits are best measured.

RESPONSES OF PROGRAM CLIENTELE

Farmers who have had IPC training and demonstrations do better than farmers who have not. We compared farmers in Nueva Ecija, Philippines five years after their IPC group was set up with untrained farmers in the same town (Table 2).

An intensive survey conducted in two municipalities of the same province, Jaen and Talavera, Nueva Ecija, three to four years after training and demonstration revealed differences between the knowledge of trained and of untrained farmers. When compared with untrained, trained farmers knew more about the sources of pest problems, could name and describe correctly more pests of the previous three seasons, knew and used IPC decision rules more often, applied insecticides less often, and knew more about the identities and function of natu-

Table 2.   Production comparisons of IPC-trained farmers with untrained
          farmers in the Philippines five or more years after training
          (averages of three crop seasons 1984-85)

	IPC Trained	Untrained
Yield (tonnes/hectare)	5.52	5.03
Insecticide applications per season	1.8	2.9
Costs of insect control (pesos/hectare)	295.75	486.23

ral enemies of insect pests and the effects of insecticides on those natural
enemies (21).

Significant differences in the attitudes of trained farmers were also found.
They said they thought that new resistant cultivars needed less insecticides
(untrained farmers said more were needed); they treated their fields significantly
less often when their field neighbors treated (untrained farmers treated routinely
after a neighbor's treatment to forestall anticipated immigration); they felt
their crops did not need to be treated immediately after fertilizing (untrained
farmers did feel this and did so treat); and significantly more of them perceived
IPC decision rules as being more profitable than alternative application
strategies (21).   Farmers with IPC exposure not only produced more rice with
higher profits, they also changed their attitudes to crop losses and the need for
insecticides.

Although changes in knowledge and attitudes are important responses, it is
also useful to know if farmers who have been shown and taught IPC actually behave
differently from other farmers.   While the intensive studies needed for
anthropological confirmation have not yet been done, the FAO Program has been able
to verify changes in farmers' skills using field tests.   Pretraining scores have
been compared with the posttraining scores of 10,167 farmers during 1985.   The
average pretraining score was 51% correct, the average posttraining score was 72%
correct, a highly significant difference.   This score change means higher levels
of skills in identification, damage assessment, choice of chemical, and deciding
if a treatment is necessary.   Areas where skills were still weak after training
included natural enemy identification, fungal disease identification, and control
methods for virus diseases (i.e. resistant cultivars).

A large-scale follow-up, with a more tightly controlled set of test questions
and standard test personnel and procedures, is planned for 1986-87.   It is
anticipated that this will provide evidence of longer term retention of skills
after training.   The same test will be given to extension agents to assess their
skills.   Good farmers should have good agents informing them.   If not, then deeper
study of how the good farmers got good will be very important.

Assessing the responses of administrators is a far less quantitative task.
Since the beginning of the FAO IPC rice program, significant national policy
changes have happened in five of the participating countries:

- In India, the Minister of Agriculture announced that IPM would be the
official crop protection policy for the country for the next five-year plan,
1985-1990.

235

- In Malaysia, the Minister of Agriculture announced that IPM would be expanded from rice to other crops as the national crop protection policy of the Fifth Malaysia Plan, 1985-1990, and especially tailored for the mini-estate group farming thrust planned for the rice sector.

- In the Philippines, IPM has become the official Cost Reducing Technology for pest control under the National Food and Agriculture Council's policy of making rice farming more profitable; the Deputy Minister of Agriculture has also called for IPM to be the national crop protection policy for crops other than rice as well as rice.

- In Sri Lanka, IPC has been adopted as the core of all crop protection messages for the national Training and Visit extension system in every rice-growing district.

- In Indonesia, IPM has been declared the official crop protection strategy for over one million hectares most threatened by brown planthopper and tungro virus (38).

In each case, even though accurate national estimates of crop losses are not available, the decision to adopt IPM or IPC is predicated on a new attitude to the threat of crop losses, because IPC is inseparable from the precept that low densities of pests are not threats to production. At the outset, IPC is implemented only on a pilot basis, and close attention paid to its performance. It has continued to satisfy policy makers.

CONCLUDING REMARKS: NEEDS AND PROSPECTS

What do clients still need?

Farmers still need better crop loss assessment to help them rank and select their crop management options before, during, and after each crop season (Figure 2, left-hand side) (12). They need local predictions of crop losses. Because it is still difficult to determine accurate severity-loss relations that can work in different environments, an empirical approach is the next best alternative. Farmers and technicians should begin to collect local experiences with infestations and crop losses in simple but reliable ways.

These should not be refined reductionist analyses of a small number of plots, nor should they be organic and haphazard conglomerates of local traditions. Observations on pests and crop losses should be collected using robust theory and recorded for the community, preferably by the community. The community is the logical choice of unit for this recording because it is ecologically large enough that the experiences of different fields in one season give some indications of normal variation across seasons, and yet it is socially small enough that experiences can be shared without too much abstraction. Farmers can check each others' fields, daily if desired, and they know by how much to inflate or deflate each others' reports. The greatest difficulty in teaching IPC is to make it sufficiently concrete without losing all predictive generality. Eliciting and recording the perceptions of pests and their eventual impact on yield for a given season from a group of farmers in one community is a challenge for a village-level worker, in collaboration with older or younger farmers with the time to serve as archivists.

Building community knowledge bases in this way can be an integral part of the standard IPC demonstration and training. It certainly keeps the extension agent scouting the fields and seeking out the farmers, and it can be checked by a supervisor so long as the supervisor gets to the fields during the season. Special skills that must be taught to extension agents are identifying, motivating and, hardest of all, listening to their collaborators in the community. Tully

(36) was an early master of the problem census approach. Chambers and Jiggins (5) have several suggestions about focus groups and innovator workshops that are starting points. Ledesma (23) summarized intensive but effective methods for participatory research by members of rural communities who gathered basic demographic data to evaluate the quality of health care available to them.

On the technical side, the core of this approach should be simple questions, simple methods, adequate replication, and regular consultations with knowledgeable specialists. The questions to be answered are of the form "What pests are in your field this week?" "Are there more or less than last week?" "What if anything did you do about them?" "How did you decide to do that?" If resources permit interventions, they should also be very simple. Treating a small parcel with insecticide or fungicide regularly and comparing pest numbers and yields with untreated paired plots is arduous enough for any extension worker. The point of adequate replication is not to satisfy the editors of an international journal. It is to capture a realistic range of pest variation for the sake of prediction for that community alone (which may well be recorded in a village journal!). About 10 farmers is a reasonable minimum for pest infestation records; at least 20 are needed to compare statistically two practices across the range of management styles found in a typical village.

The final requirement is regular consultation with a specialist, who may be a trained farmer, an extension agent, a chemical dealer, or a money-lender. What is important is that the exchange take place repeatedly and cover the same simple questions outlined above. Only through interaction will the goal of crop loss recording, local prediction for the community, be reached. Someone from outside the community knowledge base must interrogate that base and challenge it, or it will collapse into story telling. Someone must connect the observations of the season to the yields at season's end or prediction will never begin. Of course farmers will also construct prediction systems, and these should be compared with the specialists' systems in a thorough way. The best predictors for the community will be a combination of the two. If the farmers participating in assembling the knowledge base come to seek out and initiate the interaction, so much the better. They will be much further ahead in forging and forcing productive relationships with specialists in the years after the initial IPC project is set up.

Administrator clients need help in explaining what just happened more than in predicting what might happen (Figure 2, right-hand side). At a national level they need reasonably accurate records of the causes of crop losses no more than two months after median harvest date. Why? Because they need to respond quickly to incipient outbreaks by mobilizing stocks of resistant seeds or turning off the water in an irrigation system or scheduling large-scale chemical intervention. In the longer term, they need to plan investment in research and development using accurate figures to rank problems in order of importance. I know of no case where the costs of breeding a resistant variety have been estimated in advance to compare with the costs of controlling the pest some other way.

If the network of community knowledge bases proposed above exists, then the problem of the administrators is small. Simply survey enough communities to estimate the overall losses. In the interim before such networks proliferate, national administrators must rely on the reports of extension and small numbers of crop protection scouts. Most of their reports are still limited to incidence data. The first priority for national crop loss assessment systems is to develop simple severity scouting procedures that anyone can use. These do not have to estimate potential yield losses; all that is needed is improved recording of what proportion of plants or plant parts in a representative sample are infected. In the difficult case of virus diseases, this may require special methods for differential diagnosis in the field.

What should practical CLA provide?

Moving from severity data to yield loss prediction is where most of the papers in this book begin; it is where this one will end, because it is so far from the immediate needs of clients and so likely to lead to reductionistic elegance without impact in the field. Severity-yield loss relationships for rice should be explored <u>not</u> for their own sake but in a logical manner that maximizes the qualitative insights that can be passed on to field workers immediately.

Simulation models are a delightful teaching tool that help keep an inhumanly large number of factors in good order in a simple-minded way (34, 35). Pest damage has been linked to a simulation model of rice (18, 19): phloem-feeding rates of different ages of brown planthoppers were measured and the appropriate mass of photosynthetically produced carbohydrates was subtracted from the daily increment of growth of a simple biomass accumulation model driven by a vector of historical daily solar radiation with all other inputs held optimal. The resulting percentage yield losses fell within the standard errors of the validation trials. Yet the important outputs of the exercise were not the quantitative ones but the qualitative ones. First, early season damage appeared more important than middle season damage. This flew in the face of standard rice crop physiology which holds that panicle initiation is the most vulnerable period. The photoperiod insensitive trajectory of biomass accumulation in modern rice cultivars makes those cultivars more susceptible to early season distortions of that trajectory than were older cultivars that grew luxuriantly and indeterminately until photoperiod stimulated flowering. Modern cultivars are less able to tolerate this sort of stress because they cannot recover enough mass, in the limited time before they flower, to attain high yield.

The next point was more quantitative in origin, that the overall relationship of brown planthoppers to rice plants could be conceptualized as "negative sunlight". What solar energy put in, the BPH sucked out. There was no need to invoke toxins or other mysterious mechanisms to explain the range of results. The practical implication of this was simply that BPH was more dangerous when solar radiation was low, as in the main rainy seasons of the humid tropics. This is because the plants are already laboring near a carbohydrate stress point and the additional stress of BPH would be like extra shade at the most vulnerable times. This explained much of the confusing observations from field anecdotes, that even though BPH populations were often higher in dry sunny seasons, yield losses seemed higher in wet seasons. So we learned to look up at the sky nearly as much as down under the leaves of the rice to predict BPH damage, <u>qualitatively</u>.

It was through this qualitative looking up that the final insight came to us (18). We still had not had to explain the most sensational aspect of BPH damage; this was a drying and death of the plants characterized as hopperburn. In the final dry, sunny season of field work, we created a very localized outbreak of over 1000 BPH per plant through misuse of insecticides. While plants got their sheaths covered with sooty mold and a bit of chlorosis was observed, so long as at least 500 gm-cal/cm^2-day fell on the crop there was no hopperburn. There was more than enough photosynthate for all.

Then a tropical depression generated a dense cloud cover and reduced the radiation to below 200 langleys for two days before the sun came back. <u>In</u> those two days over 35% of the field hopperburned, and <u>after</u> those two days no more than an additional 5% of the field hopperburned. The total yield lost compared with the adjacent unsprayed uninfested field was about 35%, so only those plants killed by the interaction of low sunlight and BPH were unable to compensate for BPH damage. So we thought hopperburn was something different from the quantitative accumulation of "negative sunlight".

After further studies, we concluded that hopperburn involved not only BPH and darkness, but also accumulations of high nitrogen amino acids, ultimately including ammonia. If shade closed down enough stomata, then ammonia could not escape, and well known toxic reactions began. This taught us not only to look at the sky and under the leaves, but in the soil as well for the nitrogen levels to understand the crop losses from BPH.

The above work taught us that the exact severity-yield loss relationship is a very complicated and frighteningly local affair. Pursuing that quantitative relationship through many validation trials would have led me away from the fundamental qualitative relationships that actually determined yield. It would appear that simulations and precise quantitative relationships cannot replace basic crop physiological principles and a prepared mind open to observation and interpretation.

Practical IPC programs have to answer to clients who shout. We need qualitative rigor, and local predictions, and holism if you will, and we need to stay close to the clients.

ACKNOWLEDGEMENTS - Thank you to Tonie Putter, Pat Matteson, Au Santiago, and Bernadette Florendo of FAO; Jess Sumangil, Mike Salac and the rest of the Philippine IPC program; K.L. Heong and the rest of the Malaysian IPC program; Jim Litsinger, Jovy Bandong, Kevin Gallegher, S.K. DeDatta and the rest of IRRI; Grace Goodell, Lloyd Evans, and Jan Zadoks for guidance and great arguments; and mostly Paul Teng for the patience that passeth understanding.

LITERATURE CITED

(Editor's Note: Many of the papers cited in this chapter are not in standard journals, and exemplify a problem facing our scientific colleagues in the developing world. However, the information contained in them have been verified by us and copies may be obtained, gratis from the author.)

1. Atwal, A.S. 1983. The role and use of pesticides in agriculture. Pest. Information Oct.-Dec. 1983:25-37.
2. Bandong, J.B., Litsinger, J.A., Kenmore, P.E. 1985. Farmers' management of rice pests: Its implications to IPM implementation. Paper presented at 15th Annual Meeting, Pest Control Council of the Philippines, May 1985. 7 pp.
3. Binamira, J.S. 1985. Parameters for a strategic marketing plan for integrated pest control. Report submitted to the FAO Intercountry Program for Integrated Pest Control in Rice. 72 pp.
4. Brunold, S. 1981. A survey of pest control practices of small rice farmers in Iloilo Province. (mimeo.) 20 pp.
5. Chambers, R., Jiggins, J. 1986. Agricultural research for resource-poor farmers: A parsimonious paradigm. (mimeo.) 47 pp.
6. Escalada, M. 1985. Rice farmers' knowledge, attitudes and practice of integrated pest control (Western Leyte): Final report submitted to the FAO Intercountry Program for Integrated Pest Control in Rice. 96 pp.
7. Espina, M.L.P. 1983a. Rice farmers' perceptions of and responses to pest hazards. Report submitted to the FAO Intercountry Program for Integrated Pest Control in Rice. 129 pp.
8. Espina, M.L.P. 1983b. Some methodological guides in conducting interviews with women: A brief glimpse on the participation of women technicians in extension work. In Role and Potential of the Filipina in Rice Crop Protection, pp. 77-81. Los Banos, Philippines: SEARCA, Univ. Philippines.
9. Espina, M.L.P. 1984. Rice farmers' perceptions of and responses to natural hazards. M.A. thesis, Ateneo de Manila University, Philippines. 198 pp.
10. Evans, L.T., de Datta, S.K. 1979. The relation between irradiance and grain yield of irrigated rice in the tropics, as influenced by cultivar, nitrogen fertilizer application and month of planting. Field Crops Res. 2:1-17.

11. Felton, E.L., Sorenson, R.Z. 1967. Seed Corporation of the Philippines. Inter-University Program for Graduate Business Education in the Philippines. 44 pp.

12. Goodell, G. 1984. Challenges to international pest management research and extension in the third world: Do we really want IPM to work? Bull. Entom. Soc. America 30(3), Fall 1984:18-26.

13. Heong, K.L. 1984. Pest control practices of rice farmers in Tanjong Karang, Malaysia. Insec. Sci. Applic. 5:221-226.

14. Heong, K.L., Ho, N.K. 1986. Farmers' perceptions of the rice tungro virus problem in the MUDA irrigation scheme, Malaysia. In Integrated Pest Management: Farmers' perceptions and practices, ed., J. Tait, D. Bottrell, and B. Napompeth. In press.

15. Herdt, R.W., Castillo, L., Jayasuria, S. 1984. The economics of insect control on rice in the Philippines. In Proceedings of the FAO/IRRI workshop on judicious and efficient use of insecticides on rice, pp. 41-56. Los Banos, Philipines: IRRI.

16. Huis, A. van, Nauta, R.S., Vulto, R.E. 1982. Traditional pest management in maize in Nicaragua: A survey. Meded. Landbouw. Wageningen, Nederland. 43 pp.

17. James, W.C. 1980. The structure of disease-loss appraisal programs. In Assessment of crop losses due to pests and diseases, eds., H.C. Govindu, G. K. Veeresh, P.T. Walker, J.F. Jenkyn, pp. 49-51. Hebbal, Bangalore: Univ. of Agric. Sc.

18. Kenmore, P.E. 1980. Ecology and pest outbreaks of a tropical insect pest of the Green Revolution: The rice brown planthopper _Nilaparvata_ _lugens_ (Stal). Ph.D. diss., Div. Biol. Pest Control, Entomol. Sc., Univ. Calif., Berkeley. 226 pp.

19. Kenmore, P.E., Angus, J. F. 1979. Tropical rice and brown planthoppers: Quantitative analysis of yield losses in the field. Paper presented at the IX Int. Congr. of Plant Prot. and 71st Ann. Meeting of the Amer. Phytopathol. Soc., August 5-11, 1979, Washington, D.C., USA. 8 pp.

20. Kenmore, P.E., Heong, K.L., Putter, C.A.J. 1985. Political, social and perceptual aspects of integrated pest management programs. In Integrated Pest Management in Malaysia, eds. B.S. Lee, W.H. Loke, K.L. Heong, pp. 47-66. Kuala Lumpur, Malaysia: MAPPS.

21. Kenmore, P.E., Litsinger, J.A., Bandong, J.P., Santiago, A.C., Salac, M. M. 1986. Philippine rice farmers' insect control: Thirty years of growing dependency and new options for change through training. In Ref. 14, In press.

22. Khosla, R.K. 1980. Methodology of assessing losses due to pests and diseases of rice in India. In Ref. 17, pp. 240-248.

23. Ledesma, A.J. 1981. Participatory research for community-based agrarian reform: the SIAY BCC experience. (mimeo.) 15 pp.

24. Litsinger, J.A. 1984. Assessment of need-based insecticide application for rice. Paper presented at MA-IRRI Technology Transfer Workshop, March 15, 1984. 13 pp.

25. Marciano, V.P., Mandac, A.M., Flinn, J.C. 1981. Insect management and practices of rice farmers in Laguna. Paper presented at Crop Science Society of the Philippines Annual Meeting, April 1981. 23 pp.

26. Matteson, P.C. 1986. First rice IPC demonstration results from Sri Lanka: Guidance for the national implementation program. In Extended Abstracts, 2nd International Conference on Plant Protection in the Tropics, 17-20 March 1986, Genting Highlands, Malaysia, 269 pp.

27. Mumford, J.D. 1981. A study of sugar beet growers' pest control decisions. Ann. Appl. Biol. 97:243-252.

28. Mumford, J.D. 1982. Perceptions of losses from pests of arable crops by some farmers in England and New Zealand. Crop Protection 1:283-288.

29. Oliva, A., Imperial, S.S., Rivera, T. 1984. Farmers pest control attitudes and practices in two Bikol villages. A report of the Research and Service Center, Ateneo de Naga. 120 pp.

30. Repetto, R. 1985. Paying the price: Pesticide subsidies in developing countries, Research Report #2. World Resources Institute, Washington, D.C. 27 pp.

31. Salas, R.M. 1985. More than the Grains. Tokyo, Japan: Simul Press, Inc. 245 pp.

32. Santiago, A.C., Salac, M.M., Dayao, J., Salenga, E., Mangahas, A., Cajanding, H. 1984. Comparison of rice insect controls in five villages with farmers' integrated pest control groups in Central Luzon. Paper presented at the 1984 Conference of the Pest Control Council of the Philippines. 12 pp.

33. Sumangil, J.P. 1984. IPM for cost reduction in the Masagna-99 rice production program. Paper presented at the Agritech 84 Conference, 9 September 1984, Manila, Philippines. 6 pp.

34. Teng, P.S. 1985. A comparison of simulation approaches to epidemic modeling. Annu. Rev. Phytopathol. 23:351-79.

35. Teng, P.S., Oshima, R.J. 1983. Identification and assessment of losses. In Challenging Problems in Plant Health, eds., T. Kommedahl, P.H. Williams, pp. 69-81. St. Paul, MN: American Phytopathological Society.

36. Tully, J. 1968. Farmers'problems of behavioral change. Human Relations 21: 373-382.

37. Viajante, V.D., Heinrichs, E.A. 1985. Yield losses of rice variety IR36 damaged by green hairy caterpillar, Rivula atimeta. Paper presented at 15th Annual Meeting, Pest Control Council of the Philippines, May 1985. 9 pp.

38. Ward, W.B. 1985. The continuing battle against insects and diseases. In Science and rice in Indonesia, U.S. A.I.D. Science and Technology in Development Series, pp. 49-62.

39. Zadoks, J.C. 1985. On the conceptual basis of crop loss assessment: Threshold theory. Annu. Rev. Phytopathol. 23: 455-473.

CHAPTER 24

A COMPUTER-BASED DECISION AID FOR MANAGING BEAN RUST

Richard A. Meronuck

Department of Plant Pathology, University of Minnesota,
St. Paul, Minnesota 55108, U.S.A.

Bean rust is caused by <u>Uromyces</u> <u>phaseoli</u> and has a worldwide distribution. It is an important production problem in many areas of the world that grow dry edible beans. Yield losses are most severe when plants are infected during the pre-flowering and flowering stages of development, approximately 35 to 45 days after planting (6). Disease losses have been estimated from 18% to 100% of the potential yield. Losses of 30%-95% have been documented in commercial Pinto fields in Minnesota (4). Work done at Minnesota has shown that both Bravo and Dithane M-22 are very effective fungicides against the spread of rust (4). Results from these trials have been used to estimate yield loss and economic return of a particular spray program (5).

James and Teng (1) have suggested that, in the light of progress made in developing models for predicting yield loss due to specific levels of disease, research should now be conducted to evaluate the need for each fungicide spray in a control program. This evaluation should be based on the extra amount of benefit that any additional spray can generate (7). With potatoes, for example models are available for estimating yield loss caused by late blight (2, 3). These models predict loss from late blight severity assessed either singly or over the tuber bulking period. Such loss estimates, when compared with pesticide use on the farm, can enable a rational cost/benefit evaluation of the need for pesticides. It was with this in mind that this research project on Pinto beans was initiated.

Scouting activities during the growing seasons of 1977-1980 revealed that rust was present in dry edible bean fields in Minnesota. Bean rust was especially destructive in fields that had Pinto beans following a previous crop of Pinto beans. Crop rotation continues to be a very important practice in the prevention on bean rust, but effective and more efficient chemical control would contribute to greater flexibility in the control of this disease and result in higher yields in fields that become infected during a particular growing season with high rust severity. Experience shows that more disease control and greater returns per acre will result when a fungicide spray program is started just after rust appears in the field or in the particular growing area. Because of the experience with previous losses due to rust, certain growers have initiated preventative spray programs without concern as to whether or not rust will be a potential problem in any year. Although this control practice is very effective, often more fungicide than necessary is applied, resulting in an overuse of chemical and a lower return per acre for the overall spray program.

An alternative approach to the above would be an astute scouting procedure to identify the very early stages of rust in a field, followed by an immediate initiation of the spray program. This management technique would eliminate spray application during periods when rust was not a threat. During certain growing seasons, spray applications could be eliminated altogether by scouting fields weekly to determine that the rust infection has not started.

Certain fields of Pinto beans which are not under either a preventative spray program or an adequate scouting program may develop significant rust and sometimes are not discovered until the disease becomes advanced enough to present a potential threat to maximum yield. When rust at various stages of advancement is discovered, the question of whether a spray program will pay is always of concern. It is this question that this study is designed to help answer. The objectives of the study reported in this chapter are:

1. To provide information as to the amount of yield loss that would be caused by various levels of rust infection at various growth stages of the plant.
2. To provide information as to the amount of yield recovery that would result from various spray programs, using Bravo 500, to control rust.
3. To provide information as to the return per acre of spray programs initiated at various stages of disease development and plant growth stage.

MODEL DEVELOPMENT

Data Collection

Plots 10 feet x 30 feet having 30 inch row spacings were planted each year during the first week of June at a rate of 6 to 7 seeds per foot of row using recommended herbicides and fertilizer rates. The plots were arranged in a randomized complete block design. Each plot consisted of four rows; the outside two rows were used as border rows and the two inside rows were used as the treatment rows. The plots were deployed with sufficient alleyway area to permit tractor and sprayer operation between the plots without driving on any of the plants in the plot.

Each year the plots were inoculated with rust spores to ensure a uniform rust disease epidemic. Spores suspended in a light oil mixture were distributed throughout the canopy of the center two rows of each plot using a Mini ULVA centrifugal sprayer. These inoculations were made on different dates so as to establish different levels of rust epidemics.

A series of treatments were set up which: 1) established different levels of rust infection so as to enable determination of the amount of yield loss which would result from the unchecked growth of any given level of rust infection at any given stage of plant growth and 2) compared the effectiveness of one, two, and three spray treatments initiated at different levels of rust infection thus enabling the determination of the amount of yield loss which would be recovered due to treatment.

Treatments were initiated within one week from the time that rust was first seen in the plots (Table 1). The spray applications were made with a field sprayer calibrated to deliver 60 gallons per acre at 180-200 lb pressure.

To inhibit the development of white mold in 1983 and 1984, each plot was treated with Benlate at the rate of 2 lb product/Acre. Assessments were done on the same day that spray applications were made. In the center row of each plot, the percentage incidence of leaves with each scale of rust severity was estimated. A disease index was then calculated for each plot by summing the individual (incidence X severity scale) readings. Plots could therefore potentially have disease index values from 0 (healthy) to 500 (dead from rust).

Plots were harvested during the second or third week of September when 95% or more of the leaves were yellow, brown or dry and the pods were dry. The plants in ten feet of each of the center two rows of each plot were pulled by hand and allowed to air dry in the field for several days. While drying, the plants were left in burlap bags to prevent seed loss. The plants were then threshed and the

TABLE 1. 1983 Treatment Yields and Disease Readings, Staples, MN.

		TREATMENT SCHEDULE			YIELD
					(lb/Acre)
S	S	S	S	S	2187 a
I	-	-	S	-	1955 ab
I	-	S	S	S	1933 abc
-	I	S	S	-	1845 abcd
-	I	-	S	S	1818 bcde
I	-	-	S	S	1809 bcde
I	-	S	S	-	1788 bcde
-	I	S	S	S	1704 bcdef
-	I	-	S	-	1583 cdefg
-	I	S	-	-	1553 defg
I	-	-	-	S	1483 efg
-	I	-	-	-	1386 fgh
I	-	S	-	-	1365 gh
I	-	-	-	-	1079 h

7/12	7/19	7/26	8/2	8/9	: Date of Treatment
V8	R1	R3	R5	R6	: Growth Stage

I = inoculation , S = spray treatment

$LSD_{0.05} = 361.3$

Rust severity was assessed using a 0-5 scale, in which:

0 = No Rust,	1/2 = 1-2 spots/leaf,	1 = 10 spots/leaf,
2 = 40 spots/leaf	3 = 200 spots/leaf,	4 = 400 spots/leaf,
5 = most leaves dead from rust.		

seeds were dried for three days in a 140 F oven, weighed and yields (adjusted to 13% moisture) were calculated.

Data from experiments conducted in 1981-1984 were used in the regression analysis of the plots located in Staples, Minnesota. Readings taken each year at the same growth stages were pooled in the regression analysis to represent a possible range of disease readings which might occur over a period of years with different weather conditions. Both 2 pts/Acre and 4 pts/Acre of Bravo were used in 1981 and 1982. However subsequent analyses showed that there was no significant difference between the two rates. Because of this, these data were treated as comparable in the regression analysis.

The Rosemount experiments were done only in 1983 and 1984 and readings taken at the same growth stage each year were again pooled in the regression analysis. The percentage loss (dependent variable) was calculated as:

% Loss = Yield of top yield plot - Yield of lowest yielding plot x 100

Yield of top yielding plot

and regressed against the disease readings (0 - 500) as the independent variable. The regression analysis for the % yield recovery was done with the percentage recovery as the independent variable, calculated as:

$$\% \text{ Recovery} = \frac{\text{Plot yield - yield of lowest yielding plot} \times 100}{\text{Yield of top yielding plot - yield of lowest yielding plot}}$$

against the disease reading (0 - 500) as the independent variable.

Data Analysis

STAPLES   These experiments were done under irrigation to approximate field conditions using overhead pivot irrigation. Inoculations and spray schedules were designed to create a series of epidemics which would have different severities throughout the growing season. Regression analysis of yield losses at different growth stages with various amounts of rust infection gave single point models which predicted the amount of yield loss in any given field of Pinto beans (cultivar UI114) at certain growth stages. Although inoculations were made early enough to establish early epidemics, weather conditions were not favorable for an increase in severity. As a result, regression analysis of data collected at growth stage R_3 and R_5 did not give significant relationships (Table 2, 3). Data were consistently significant after the R_6 growth stage and were incorporated into the existing bean rust spray decision model.

ROSEMOUNT   These experiments were conducted under dryland conditions using no supplementary irrigation. Under these conditions the rust epidemics developed slower, which minimized the effect of the different inoculation and spray application dates on the attempt to establish a gradient of epidemics. There was a severe lack of moisture in 1983, which affected yield. Growing conditions were enhanced by adequate moisture in 1984 (Rainfall June - August, 12.35", departure from normal - 0.29") and increased N_2 fertilizer which resulted in substantially higher yields.

Regression analysis of yield losses at different growth stages with various amounts of rust infection provided data for single point models. These predicted the amount of yield loss in any given field (cultivar UI114) at specific growth stages under dryland conditions (Table 2).

The significance of the regression analysis of the yield recovery data due to a spray program was not as consistent, again, as in Staples, data collected at growth stages R_4, R_5 and R_6 (Table 2) explained lower proportions of variation in yield than later growth stages. Significant relationships were incorporated into the spray decision model.

STRUCTURE OF THE DECISION AID MODEL

The decision aid model estimates the return/acre of a particular spray program initiated at a given growth stage and a particular level of the rust disease. In addition to rust index and growth stage, input for the model are:
1) the expected price of the pinto beans,
2) the potential yield of the field based on production records, and
3) the spray cost/acre, which includes the cost of the chemical and the application cost.
These inputs are use in a formula to calculate return/acre as follows:

Return/Acre = Price (Potential Yield x %Loss x %Recovery x CF) - Spray Costs.

The %Loss and %Recovery are derived from the regression formulas in Tables 2 - 4. The correction factor CF equals 1 in the current version of BEANRUST and acts as a modifier for any difference in apparent infection rate between cultivars. As more information becomes available on the relationship between weather and infection rate on different cultivars, the value of this modifier will become a variable.

Table 2. Regression Analysis Summary of % Yield Recovery, Staples, MN.

(1 Spray)	Bean Growth Stage							
	R4	R5	R6	late R6	R7	late R7	R8	R9
# of obs.	18	29	61	31	52	25	0	0
R^2	15.50	13.58	23.68	44.86	37.91	51.18		
y int.(a)	32.24	50.14	64.96	85.57	80.48	122.10		
slope (b)	-0.08	-1.37	-0.16	-0.21	-0.22	-0.28		
F value	2.9	4.2*	18.3**	23.6**	30.5**	24.1**		
years	81,82	83,84	81-84	82	81,82,84	81,84		

(2 Sprays)	Bean Growth Stage							
	R4	R5	R6	late R6	R7	late R7	R8	R9
# of obs.	20	31	59	0	23	0	0	0
R^2	23.25	8.12	39.20		37.61			
y int.(a)	50.89	52.18	84.37		49.39			
slope (b)	-0.15	-0.55	-0.24		-0.17			
F value	5.4*	2.6	36.7**		12.7**			
years	81,82	83,84	81,82,84		84			

(3 Sprays)	Bean Growth Stage							
	R4	R5	R6	late R6	R7	late R7	R8	R9
# # of obs.	18	16	16	0	0	0	0	0
R^2	13.96	48.16	59.76					
y int.(a)	72.36	55.35	50.76					
slope (b)	-0.11	-1.79	-0.24					
F value	2.6	13.0**	20.8**					
years	81,82	84	84					

Table 3. Regression Analysis Summary of % Yield Loss.

(Staples)	Bean Growth Stage							
	R4	R5	R6	late R6	R7	late R7	R8	R9
#of obs.	9	36	89	81	184	189	0	50
R^2	37.91	2.06	64.15	52.73	56.16	72.16		39.23
y int.(a)	67.72	45.10	17.40	-1.03	-5.32	-41.32		-66.05
slope (b)	0.09	0.10	0.18	0.19	0.18	0.24		0.22
F value	4.3	0.7	155.7**	88.1**	233.2**	484.6**		31.0**
years	81,82	83,84	81-84	82,83	81-84	81,82,84		83

(Rosemount)	Bean Growth Stage								
	R4	R5	R6	late R6	R7	Late R8	R8	R8	R9
# of obs.	8	14	29	0	50	44	100	60	88
R^2	53.14	42.18	32.45		48.12	34.49	12.62	12.66	17.74
y int.(a)	15.09	19.20	7.28		-6.78	-11.76	-2.18	-5.19	-6.81
slope (b)	0.87	0.18	0.13		0.17	0.17	0.08	0.07	0.06
F value	6.80*	8.7*	13.0**		44.5**	22.1**	14.2**	8.4**	18.5**
years	83	83	83		83	83	83,84	84	84

*P = 0.05
**P = 0.01

Table 4. Regression Analysis Summary of % Yield Recovery, Rosemount, MN.

(1 Spray)				Bean Growth Stage				
	R4	R5	R6	late R6	R7	late R7	R8	R9
# of obs.	7	15	20	0	0	0	28	21
R^2	3.69	21.78	31.88				21.00	23.94
y int.(a)	-67.88	47.82	97.81				999.38	3526.60
slope (b)	1.56	-1.11	-0.52				-5.06	-12.47
F value	0.2	3.6	8.4**				6.9*	6.0*
years	83	83	83				84	84

(2 Sprays)				Bean Growth Stage				
	R4	R5	R6	late R6	R7	late R7	R8	R9
# of obs.	8	15	0	0	0	0	21	0
R^2	1.96	37.96					8.37	
y int.(a)	-23.23	77.19					1954.98	
slope (b)	-1.94	-0.82					-7.81	
F value	0.1	8.0*					1.7	
years	83	83					84	

(3 Sprays)				Bean Growth Stage				
	R4	R5	R6	late R6	R7	late R7	R8	R9
# of obs.	8	0	0	0	0	0	0	0
R^2	40.71							
y int.(a)	73.28							
slope (b)	-1.93							
F value	4.1							
years	83							

    The decision model has been programmed for use on a hand-held computer, the Hewlett Packard 41CV, and for a microcomputer, the IBM PC. A sample printout using the IBM PC version is shown in Appendix I. The hand-held computer version is capable of printing the same output using a portable printer, the HP 82143A, and may be used in the field to discuss rust control strategies with growers.

MODEL EVALUATION

    A field study using the dryland spray decision aid developed with data collected in 1983, was made on 5 fields of dryland Pinto beans grown near Crookston, Minnesota during 1984. The fields were divided into 3 strips. One strip used the spray decision aid to manage rust infestation. The second was to be managed by the farmer and the third strip was to be left unsprayed regardless of rust infestation. Plot areas, (10 ft x 50 ft) unrandomized but replicated 4 times and chosen to represent the field strip were selected. Rust readings were taken in these plots at various growth stages, and the plots were harvested at the end of the trial to establish yield differences.

    Field 1 was cultivar Orauy, fields 2 and 3 were cultivar UI111 and fields 4 and 5 were cultivar Fiesta. Orauy is a bush type bean of medium maturing ability with no rust tolerance. UI111 is a vine type bean of medium to early maturity with no rust tolerance. Fiesta is a semi-vine type of bean with early maturity and reported rust tolerance.

    Disease readings were taken at several dates (Growth Stages R_5-R_8) and used in the spray decision aid to predict economic feasibility of a spray program for rust control. None of the readings predicted a reasonable return for a spray program given the dry conditions of the year (July, August, Rainfall 7.28").

Table 5. Results of Model Evaluation Trials at Crookston, Minnesota.

Field #[1]/	Treatment	Average Yield	# Sprays
1	M	2473	0
1	FS	2614	1
1	VS	2494	0
2	M	2164	0
2	FS	2067	2
2	VS	2223	0
3	M	2063	0
3	FS	2095	2
3	VS	2046	0
4	M	1393	0
4	FS	1409	0
4	VS	1352	0
5	M	1781	0
5	FS	1844	0
5	VS	2291	0

[1] M = Model used as management tool
FS = Farmer sprayed/managed
VS = Unsprayed, no management

The farmer managing field 1 sprayed once on August 10 when the disease reading was 2 (range 0-500). This constitutes a trace of rust in the field. The farmers managing fields 2 and 3 sprayed twice before any rust was detected. Fields 4 and 5 were not sprayed (Table 5).

The program initiated by the farmer in fields 2 and 3 substantially reduced the disease severity at the last reading date. A single spray in field 1 did not reduce the disease severity. The disease severity readings in all treatments of fields 4 and 5 followed the same trend. Spray programs incorporating 2 sprays did decrease the disease severity rating at the end of the growing season, however, they did not affect the yield except in field 1 where average yields were slightly higher. This suggests that the disease severity was not at a high enough level during the critical periods of pod set and fill to affect yield.

The spray decision aid in this trial was an effective management tool in predicting that a spray program was not economically feasible during that particular growing season. This suggests that the model has the potential to reduce a number of unneeded chemical applications.

CONCLUDING REMARKS

Although the validation data is only tentative, we believe the BEANRUST decision aid could be used in addition to other judgmental factors such as grower experience and disease resistance of the variety. Its goal is to reduce some of the guesswork in spray decisions, so that only economical spraying is done. BEANRUST exemplifies the group of decision aids that attempt to include both quantitative and qualitative information (see Chapters 19, 21) and with more data, will develop into a disease management "expert system".

LITERATURE CITED

1. James, W.C., Teng, P.S. 1979. The quantification of production constraints associated with plant diseases. Applied Biol. IV:201-267.
2. James, W.C., Shih, C.S., Hodgson, W.A., Callbeck, L.C. 1972. The quantitative relationship between late blight of potatoes and loss in tuber yield. Phytopathology 62:92-96.
3. Large, E.C. 1952. The interpretation of progress curves for potato blight and other plant diseases. Plant Pathol. 1:109-117.
4. Meronuck, R.A. 1982. Rust on Pinto beans. Univ. of Minn. Agric. Ext. Service, Special Report 12. pp. 39-42.
5. Hardman, L.L., Meronuck, R.A. 1982. A guide to dry edible bean production and pest management in Minnesota. Univ. of Minn. Agric. Ext. Service Special Report 103. 72 pp.
6. Schwartz, F., G'alvez, G.E. 1980. Bean production problems. CIAT 420 pp.
7. Teng, P.S., Blackie, M.J., Close, R.C. 1978. Simulation modeling of plant diseases to rationalize fungicide use. Outlook on Agric. 9:273-277.

Appendix 1. Printout of sample run with rust decision model

---

1)	Crop growth stage	R7	This program is based on two year's data
2)	Disease severity	102	developed under irrigation using ground
3)	Price per pound	0.18	spray application. Application was 2-4
4)	Spray cost (per acre)	8.00	pints Bravo in 60 gals/acre of water.
5)	Potential yield (lbs/acre)	2000	Field trials indicate that Bravo and

Dithane M-22 are equally effective in
rust control.

Number of sprays: 1   2   3

Return per Acre $33.78   *   *
Projected Yield: *No data available which
   With Treatment: 1828   *   *   would recommend this
   Without Treatment: 1596   number of sprays.

BEAN RUST SPRAY DECISION AID

1)	Crop growth stage	R7	This program is based on two year's data
2)	Disease severity	22	developed under irrigation using ground
3)	Price per pound	$0.18	spray application. Application was 2-4
4)	Spray cost (per acre)	8.00	pints Bravo in 60 gals/acre of water.
5)	Potential yield (lbs/acre)	2000	Field trials indicate that Bravo and

Dithane M-22 are equally effective in
rust control.

Number of sprays: 1   2   3

Return per Acre $7.72   *   *
Projected Yield: *No data available which
   With Treatment: 1963   *   *   would recommend this
   Without Treatment: 1876   number of sprays.

CHAPTER 25

THE SIRATAC SYSTEM FOR COTTON PEST MANAGEMENT IN AUSTRALIA

P. M. Ives and A. B. Hearn

Department of Entomology, University of Minnesota, St. Paul,
MN 55108, U.S.A. (formerly of CSIRO Cotton Research Unit), and
CSIRO Division of Plant Industry, Cotton Research Unit,
Narrabri, NSW, Australia 2400

Cotton is of great economic importance to Australia, ranking 4th in agricultural export earnings with a current estimated value of A$300 million. Although accounting for little over 1% of total world cotton production, Australian yields per hectare are high, ranking from 1st to 5th in the world since 1962 (1). In Australia, cotton production involves heavy investment in land, irrigation facilities and machinery, as well as heavy use of fertilizer, insecticides and herbicides (37, 40). High yields are accompanied by high production costs and high risks of pesticide resistance, environmental pollution and human health hazards. Except where action is taken to reduce these risks, the phenomenon sometimes referred to as "the pesticide treadmill" tends to occur. This is a spiral of increasing costs and pest problems, requiring further increases in pesticide application to maintain yields. In 1966 47% of total farm pesticide use in the U.S.A. was on cotton (11); even with lower than normal pest populations and drought in some areas, the estimate had fallen only to 41% a decade or more later (10). By 1977 at least 24 spp. of arthropod pests of cotton world wide were resistant to one or more pesticides (4).

Resistance to several pesticides has been found since the early 1970's in Australian populations of both the bollworm, Heliothis armigera (15, 16, 26, 38), and Tetranychus spp. mites (V. Edge pers. comm.). Cotton production in the Ord River irrigation area of Northwestern Australia collapsed by the mid 70's, exemplifying the progression from exploitation to crisis to disaster described by Smith (34). By 1974, the last year of commercial production in the Ord, an average of 25-30 insecticide sprays were applied per season, primarily against H. armigera. Commercial cotton production in Australia is now limited to the eastern states New South Wales and Queensland. If the industry is to persist in the long term, there is a clear need for pest management that minimizes the biological costs, i.e., selection for pesticide resistance, induction of secondary pests, health hazards and environmental contamination, while maintaining profitability of cotton production.

DEVELOPMENT OF THE SIRATAC SYSTEM

Origins of SIRATAC

In the early 1970's some cotton researchers and growers began to recognize this need. Growers formed the Australian Cotton Growers' Research Association (ACGRA) and paid a levy to support cotton research. In Queensland an integrated pest management unit (IPMU) was formed at the University, to collaborate with the state Department of Primary Industries (QDPI), and in New South Wales (NSW) the CSIRO (a Federal research organization) set up a Cotton Research Unit to collabo-

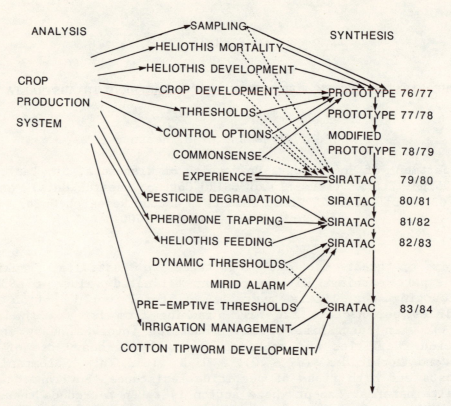

ANALYSIS   SYNTHESIS

CROP
PRODUCTION
SYSTEM

SAMPLING
HELIOTHIS MORTALITY
HELIOTHIS DEVELOPMENT
CROP DEVELOPMENT
THRESHOLDS
CONTROL OPTIONS
COMMONSENSE
EXPERIENCE
PESTICIDE DEGRADATION
PHEROMONE TRAPPING
HELIOTHIS FEEDING
DYNAMIC THRESHOLDS
MIRID ALARM
PRE-EMPTIVE THRESHOLDS
IRRIGATION MANAGEMENT
COTTON TIPWORM DEVELOPMENT

PROTOTYPE 76/77
PROTOTYPE 77/78
MODIFIED
PROTOTYPE 78/79
SIRATAC   79/80
SIRATAC   80/81
SIRATAC   81/82
SIRATAC   82/83
SIRATAC   83/84

Figure 1. Origin and development of SIRATAC.

rate with the NSW Department of Agriculture to develop ecologically and economically viable systems of cotton production. The result was the computer-based, on-line pest management system SIRATAC (an acronym for CSIRO and NSW Dept. of Agriculture tactics for growing cotton). After its initial development in NSW, SIRATAC was adapted for use in Queensland, in the light of QDPI and IPMU experience. All research groups have collaborated in subsequent development, refinement, and implementation of the system.

System development has been a two stage process -- analysis and synthesis (Figure 1). Components of the production system have initially been studied separately for a few years and then integrated to form an integrated pest management package. The first major synthesis was the construction in 1976 of a prototype on-line system (33), which has been progressively updated (2, 19, 23) as results of further research on the components have become available. Gaps in knowledge were filled at first by commonsense. This has either been replaced by research results, or confirmed or modified by experience.

Form of the Pest Management System

Any management system can be thought of as having three levels: its goals, the strategies (philosophy or general approach) used to achieve those goals, and the tactics by which the strategies are implemented. Integrated pest management (IPM) strategies may be either prophylactic or responsive (38), with prophylactic strategies being further subdivided into damage avoidance and pest reduction. The latter is aimed at lowering the general equilibrium position of the pest population, whereas responsive pest management acknowledges that for many insect pests, the injury they inflict causes damage i.e., reduces the yield value of the crop (29) (whether directly or by delaying it) only when they are above certain levels of abundance.

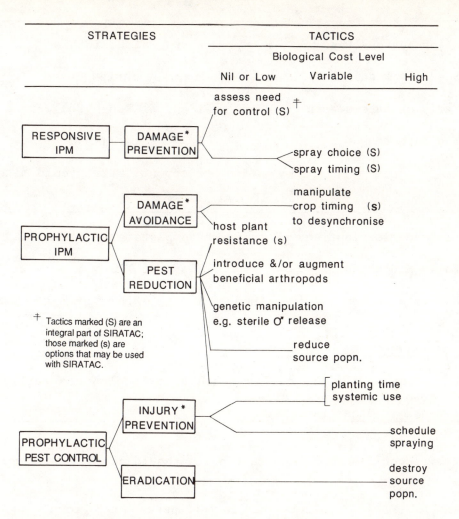

STRATEGIES	TACTICS

Biological Cost Level

| Nil or Low | Variable | High |

RESPONSIVE IPM — DAMAGE* PREVENTION

assess need for control (S) ‡

spray choice (S)
spray timing (S)

PROPHYLACTIC IPM — DAMAGE* AVOIDANCE

manipulate crop timing (s) to desynchronise

host plant resistance (s)

PEST REDUCTION

introduce &/or augment beneficial arthropods

genetic manipulation e.g. sterile ♂ release

reduce source popn.

‡ Tactics marked (S) are an integral part of SIRATAC; those marked (s) are options that may be used with SIRATAC.

planting time systemic use

PROPHYLACTIC PEST CONTROL — INJURY* PREVENTION

schedule spraying

ERADICATION

destroy source popn.

Figure 2. Some of the tactics available for various strategies of integrated pest management and pest control, and their usual level of biological cost. (* Injury and damage as defined by Pedigo et al. (29))

Control actions need be taken only when pests, if uncontrolled, would cause preventable losses in yield value exceeding the cost of control. When that occurs, by definition pest abundance exceeds economic thresholds (36), provided that these economic calculations are made for a single time period only, and a single pest or pest type, e.g., defoliators. Because of the logistic difficulties of doing otherwise, both the interdependence of decisions in different time periods (18), and the indirect and delayed financial costs caused by biological costs, are usually disregarded. How much this affects the appropriateness of the response will vary among crop-pest systems. In the absence of rigorously determined economic thresholds, nominal thresholds (30), also called "action thresholds," are often used. These approximate economic thresholds, but are usually lower and less flexible.

Tactics available for implementing a given IPM strategy may differ widely in their biological costs (Figure 2), but the feasibility of using particular tactics depends on the complex of pests to be managed, and characteristics of the cropping system, its environment and existing farmer practices.

In Australian cotton about 25 species of insects and 2 species of mites may reach pest abundance (see Appendix 1). Of these, 8 species are regular economic pests over most areas, and three others regularly cause economic damage within limited areas (Table 1). The remaining species only sporadically cause economic

253

Table 1. Regular Economic Pests of Cotton in eastern Australia

Major Season-long	2 spp. of Heliothis
Seedling Pests	2 " " Thrips
Early or Late season	1 sp. of Rough Bollworm
Mid Late season	1 " " of Aphids
" " "	2 spp. of Tetranychid Mites

Limited distribution, or range where pests are regular:	
Early Mid season	1 sp. of Tipworm
Mid Late season (v. localised)	1 " " Pink Spotted Bollworm
Early season	2 spp. of Mirids #

* Both aphids & mites may be induced to reach pest levels much earlier by heavy
  early use of broad-spectrum pesticides.
Are sporadic pests in other regions.

damage, but several could become important pests should conditions favor them more
often.

    Cotton and other host plants of cotton pests are grown, and _Heliothis_
disperse and overwinter, over large areas; commercial use of broad spectrum
pesticides is heavy, and adequate information on the effectiveness of the local
beneficial arthropods is lacking. Hence neither genetic nor natural enemy
manipulation would have been feasible as first steps in introducing IPM in
Australian cotton and sufficient reduction of _Heliothis_ source populations (27)
would have been difficult to achieve.

    The diversity of minor pests limits, but does not negate, the effectiveness
of two other prophylactic tactics: host plant resistance, and alteration of crop
timing to reduce coincidence of susceptible crop stages with presence of the most
damaging pests. The relative costs (both biological and financial) of crop timing
alternatives will vary from case to case. Setting the crop early may avoid late
season damage, but it precludes compensation for early season injury. If heavy
spraying is needed to effect early setting, it may induce aphid or mite outbreaks.

    In a responsive system satisfactory and flexible compromises can be achieved
between responses optimal for the most important pests and those for minor pests.
Also a responsive IPM strategy is probably easier to introduce to cotton farmers
unfamiliar with IPM than a prophylactic one, because it shares tools (pesticides)
with preventive schedule spraying, although it uses them more judiciously.

    Cotton growers also make strategic and tactical decisions for crop
management in general (23). SIRATAC was developed as a support system for
tactical decisions within responsive strategies, primarily for pest management,
but now also for irrigation decisions. Compatible prophylactic tactics, such as
resistant varieties, or alteration of crop timing, or both, may be used with
SIRATAC, to avoid some damage.

CONTENT OF THE SIRATAC SYSTEM

Throughout its development, operation of the SIRATAC system has always comprised three steps:
  (i) the user collects field data on the crop and the arthropods in it, and enters them into the computer,
 (ii) the computer processes the field data -- using information from research and previous use of the program, and weather data (in early years entered by the user, but now entered centrally)
(iii) the computer reports to the user with pest management advice, and summaries of pest densities and crop status. The latter was originally only densities of squares (buds), green bolls and open bolls, but now includes a yield estimate, an estimated date for crop maturation, and a plot of crop development.

Because of the overriding importance of the _Heliothis_ spp., they have been the subject of most Australian cotton pest research, and the SIRATAC system, although giving advice for other pests, mostly focuses on managing _Heliothis_.

Field Sampling

Appropriate responsive IPM depends on reliable population estimates. But there is a trade-off between the costs of sampling and of inappropriate decisions. SIRATAC sampling has evolved from very intensive (33), through progressive reductions in time and labor (23, 44).

In the prototype system, fruit and arthropods were sampled daily by visual inspection of whole plants, individually located by a stratified random design. The first time savings, in 1978/79 (23), reduced sampling frequency to 3 times a week, and ceased recording beneficial arthropods (41 categories). The following year the frequency of fruit sampling was reduced to once per week; for arthropods it remained at three times a week, but several changes reduced the time it took (23, 25). Presence/absence sampling was introduced, and once the crop began to produce squares, only the terminals (top 12 cm) of plants were sampled. The proportion of terminals infested was converted to population densities per meter-row by the computer, using relationships derived from the early years' data (43).

To reduce walking while retaining sample representativeness, the single randomized 96-plant sample of the prototype was replaced by 2 or 3 smaller samples, one (1 card) each in widely separated portions of each area for which a single management decision was made. In each sample, clusters of five adjacent plants (terminals) were sampled by a systematic stepwise transect. For 2 seasons the number of plants to be sampled on each card was determined by sequential sampling. But walking was still a major component of total sampling time, and because many categories, some with very low thresholds, were of interest, sequential sampling usually required a sampler to check many plants for at least one category. So the time saved proved to be negligible. A fixed sample of 30 plants/card was therefore adopted, as it requires less education of commercial samplers. The other time saving methods were retained.

It has become necessary to supplement visual sampling sometimes, and also to sample for more of the sporadic and localized pests such as cotton tipworm, mirids, and the pink spotted bollworm (PSBW) (31, 32). Treatment decisions for PSBW are now made on the basis of pheromone trap catches; formerly, samples of bolls were collected to determine percent infested with PSBW. This method is now used for rough bollworm (RBW) when larvae are found in visual samples of terminals. Effective control methods differ for RBW in terminals, and in bolls. With reduced sampling intensity, too few late stage _Heliothis_ larvae were found to give reliable estimates of the proportion in each species needed to guide spray choice. Determination of species composition from 1981 to 1985 was made from pheromone trap catches. Subsequent work by G.Fitt (pers. comm.) has shown pheromone traps are insufficiently reliable for this task. However, the need to distinguish between the species (for spray choice) has diminished due to developments in other aspects of the program. As mirids are regular early season pests in some areas of Queensland (Pyke, Adams & Stone, unpubl., cited in (2)), sweep sampling, undertaken in response only to a "mirid alarm", has been replaced by more efficient shake sheet sampling, which can be done with every visual sample if mirids are a concern (2). The "mirid alarm" is a warning generated by the SIRATAC program when numbers of counted squares do not progress as expected according to climatic and agronomic information.

Except in cool weather, an insect sampling interval of not more than 3 days is still recommended. With longer intervals Heliothis eggs can be laid, hatched, and larvae feeding in the terminals or squares (where they are temporarily protected from sprays) before the eggs have been detected, resulting in serious damage when unexpectedly heavy oviposition occurs.

Computer Processing

Computer processing has evolved considerably from the prototype, but the major principles by which SIRATAC uses field data for decision making are unchanged. It uses both direct predictions of yield loss due to insects (Heliothis only), and comparisons of observed (all pests) and predicted (Heliothis and cotton tipworm only) population estimates with nominal thresholds. When Heliothis must be sprayed, SIRATAC uses a graded spray response. Persistent, broad spectrum, high kill chemicals are applied only when a more short-lived and selective insecticide is not likely to be effective. The approach of spraying only when observed or predicted pests are over threshold, and using a graded spray response, is designed to make maximum use of Heliothis' natural mortality, to thereby reduce the rate of selection for pesticide resistance (14), pesticide effects on non-target organisms and costs. Because less is known about pests other than Heliothis, and a narrower range of pesticides is registered for their control on cotton in Australia, SIRATAC advice for them is less flexible.

THE CROP: PREDICTION AND ASSESSMENT OF TIMELINESS   The cotton plant produces 2 to 3 times more fruit than it can mature (22). SIRATAC takes advantage of this indeterminate fruiting habit. Not only is there no need to prevent damage to fruit that will not mature anyway, but also the plant can compensate to a degree and replace fruit that are lost (3, 22, 28). Because reliance on compensation may delay crop maturation, careful assessment of crop timeliness and the risks associated with delays is necessary.

SIRATAC does this by means of a yield development threshold (YDT) (22). The YDT estimates the number of bolls needed at any date during the season to achieve a specified yield by a specified date. It is defined by three parameters chosen by the crop manager at the start of the season: a target yield, and dates for start and completion of boll setting. These parameters must be set realistically with regard to physiological limits, historical yields for the site, and current agronomic conditions. Fruit counts through the season are used to determine whether the crop is early or late compared with the YDT. If the crop is early, pest control need not be so stringent.

In the prototype, separate YDTs were calculated for squares and bolls, and fruit counts were compared with these directly (22). Subsequently a simple but powerful simulation model of crop development was developed (19, 21), reversing the initial decision not to use crop simulation (22). Initially the model predicted survival of fruit as a function of fruit set, and daily fruit production as a function of both fruit set, and total fruit production. Later it was further refined by making both fruit survival and further fruit production also functions of fruit carrying capacity, itself a function of plant size.

Because fruit are counted at regular intervals, and shedding is predicted, the model can estimate the age distribution of fruit on the crop without the labor of counting them in more age classes. Surviving fruit are also weighted according to the growth rate appropriate to their age class (21). The model is easily recalibrated using fruit count data collected during use of the SIRATAC program.

The model predicts the number of counted squares and bolls likely to contribute to harvest, and compares this with the YDT (23). Later in the season, the model predicts yield, and the date by which 60% of harvestable bolls will be open (the conventional criterion for applying defoliant). It also enables algorithms predicting Heliothis feeding damage (see next section) to be used. Heliothis larvae differentially attack different sized fruit, including developing squares that are not yet large enough to be counted (42). The effects of water and nitrogen stress are currently being incorporated into the model (20), so that it will be able to predict crop development under a wider range of field conditions.

PESTS: POPULATION AND DAMAGE PREDICTION   The SIRATAC program is run whenever insect data are entered. The prototype predicted survival and temperature dependent development of six life stages (2 egg and 4 larval) of Heliothis over the 24 hours following sampling. Since the sampling interval was increased, the prediction interval has been progressively increased to 72 hours, assuming current temperatures continue. Predictions of Heliothis survival on residue-free cotton when unsprayed (24) are influenced by weather and predators, but are still crude pending further analysis. Data on spray residues (41) now allow SIRATAC to estimate their effects on Heliothis survival. The potential for the program to use egg parasitism also exists (31).

Recently a predictive model for cotton tipworm development (G. Hamilton, Univ. of Qld., unpub.) has also been incorporated (2), enabling improved decision making for tipworm control. As yet none of the other pest populations is predicted; their decisions are based on observed abundance only.

Since 1980/81 the program has predicted feeding damage by the counted and predicted Heliothis larvae if they are allowed to survive till the next sampling occasion. Algorithms derived from field studies of larval feeding (42) predict the numbers and age distribution of fruit attacked by the larvae, given the age distribution of fruit on the crop. The crop simulation is run with, then without, this feeding, to determine its effect on yield. Slight feeding damage early in the fruiting cycle can stimulate vegetative growth and branching, resulting in increased boll carrying capacity and yield.

THRESHOLDS AND DECISION MAKING   Any responsive pest management decision potentially has two parts: (i) whether to intervene with control action, e.g., pesticide application, and (ii) which option, e.g., which type of spray, to use, unless only one control option is available. If the control options differ in cost, the economic thresholds (ET's) for using them will differ likewise. Thus if ET's are being used as decision criteria, there is often a range of pest densities for which decisions (i) and (ii) above are interdependent.

In SIRATAC, however, nominal rather than economic thresholds are used, and decisions (i) and (ii) are made separately. Quantitative damage functions are not yet available for Australian cotton pests other than Heliothis, but even in the mid 70's it was known (35) that the plant's susceptibility to damage by a given level of infestation by most pests differs between the three major phases of crop development: pre-squaring (Phase I), squaring and boll setting (Phase II), and boll maturing (Phase III). Action thresholds in the SIRATAC prototype were therefore crop-phase specific "best guesses" (33). Equally nominal thresholds have subsequently been added for additional pests, and all threshold values have been modified in the light of experience or new research data (2, 23, 44).

A decision to spray for _Heliothis_ requires that larvae be over threshold on 2 consecutive days. Thresholds for older _Heliothis_ larvae are lower than those for young larvae and total larvae of all ages (23). Pests other than _Heliothis_ do not have age-specific thresholds, but rough bollworm have separate thresholds for larvae outside the bolls, and percent of sampled bolls infested. For thrips and aphids there are now damage (honeydew contamination of lint for aphids) thresholds, as well as abundance thresholds. Damage may remain after the pest is gone, so treatment is not advised if only the damage threshold is exceeded.

The relatively large number of pests involved, and the need for multiple control interventions in a single crop cycle, mean that Australian cotton pest management ideally should be optimized both over several time periods (i.e., at least over all infestations in the same season) and across pests. In such a situation, the increased precision of even comprehensive, let alone simple, ET's (30) over nominal thresholds, would not equivalently increase the appropriateness of the decision advice. But even with economic optimization over the whole season, a subjective element would still remain in the decision, introduced either by omission of, or in the choice of, values for the biological costs.

SIRATAC currently makes no attempt to optimize over all possible interventions in a season, as adequate predictions of _Heliothis_ population changes are not available. SIRATAC's approach to optimizing across pests is via spray type advice for simultaneous pests and, more recently, pre-emptive thresholds for aphids and mites. Data collected using the SIRATAC program showed that above a certain density, the probability that these pest populations would continue to increase and exceed their action thresholds within a few days, was very high. The program makes a decision for _Heliothis_ before considering other pests; if a spray is advised for _Heliothis_, aphid and mite infestations are tested against lower, pre-emptive thresholds, rather than the full action threshold. A chemical effective against both _Heliothis_ and aphids (or mites) is selected when necessary, enabling fewer applications and hence lower costs.

In the prototype all thresholds were fixed, but since 1981/82 progressively more of them have been made dynamic. Their values depend primarily on crop timeliness, which influenced pest management through spray type advice in the prototype, but now does so through thresholds. Thresholds for all pests except aphids are now dynamic in Phase I (2, 6). They depend on seasonal conditions, the manager's expectation of crop development, and the chosen crop timing strategy expressed through the YDT (6).

Dynamic thresholds for _Heliothis_ were initially restricted to the period from first square till first flower (i.e., the start of boll setting), when the crop is most able to compensate. Since seasonal conditions in Phase I have been used to predict first flower (6), and experience has shown that in Phase II the feeding predictions can be relied on to lower thresholds when damage would occur, thresholds have been made dynamic throughout Phases I and II. Terminal damage in Phase I will delay cotton about 7 days, so predicted first flower must be 7 days ahead of the planned date before the thresholds can be raised. The thresholds of both phases are continuous bounded functions of the number of days by which the crop is earlier than planned (Figure 3); the earlier the crop, the higher the thresholds, up to the upper bound. When the crop is on time or later than the YDT in Phase II, or less than 7 days early in Phase I, the thresholds equal the lower bound. Thus a user can obtain lower thresholds by setting the YDT earlier. If feeding is predicted to cause a yield loss, the thresholds drop to the base even if the crop is early, to avoid eroding a yield potential that is above target; in practice this currently means that thresholds drop to the lower bound ceasing to be dynamic from mid January onwards.

SPRAY CHOICE   The program's decision tree has evolved considerably (19, 23, 33). In the prototype, time since last spray, type of spray used last, crop timeliness

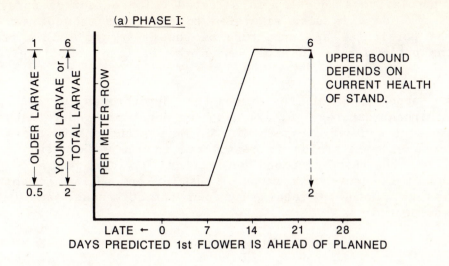

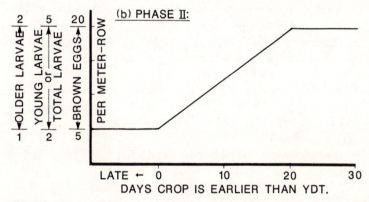

Figure 3. Dynamic thresholds for (a) <u>Heliothis</u> larvae in Phase I (presquaring) and (b) <u>Heliothis</u> larvae and brown (older) eggs in Phase II (squaring and boll setting), as functions of crop earliness with respect to planned crop timing, expressed by the YDT.

and species composition of the <u>Heliothis</u> population all influenced spray choice, according to the principles of the graded spray response. The decision tree then asked about other pests, to decide whether the <u>Heliothis</u> advice needed to be changed to control other pests as well.

By 1981/82 both the <u>Heliothis</u> and other pest decision trees had become intricate and inconvenient to update. So for 1982/83, the process of choosing the spray type for <u>Heliothis</u> was totally reorganized. A "select and try" procedure was developed, to build in flexibility for updating, while expressing quantitatively the principles and opinions that had guided spray choice hitherto. Insecticides (including acaricides) with short persistence, a narrow spectrum of toxicity, or a fairly low level of toxicity to susceptible species and in particular to beneficial arthropods, are classed as "soft". The softest sprays are a virus and bacterium. However, the virus never achieved commercial acceptance because of poor results, lack of technical support by the marketing company and lack of local research. Those toxic to an intermediate number of species, with high or intermediate toxicity to those species, but short persistence, are classed "intermediate". Highly persistent, broad spectrum insecticides, with high or intermediate toxicity, are classed as "hard". In the new procedure for spray choice, sprays are ranked in order of "softness", starting with the softest. The effect of each spray in turn on observed and predicted <u>Heliothis</u> populations is simulated. Data on the efficacy of each insecticide against the various age classes of <u>Heliothis</u> have been accumulated during the use of SIRATAC. The first and therefore softest insecticide shown by the simulation

259

to reduce the population adequately is selected. The effect of spraying on each of three successive days is also simulated to determine the optimum application time, given the particular age structure of the _Heliothis_. This procedure can also select the optimum rate to apply, when the efficacy of different rates is known.

Since the appearance of resistance to pyrethroids by _H_. _armigera_, a resistance containment strategy (8, 12) has been developed and voluntarily adopted (13) by all users of synthetic pyrethroids in Queensland and northern NSW. Under the strategy, use of pyrethroids is restricted to a 6 week period corresponding approximately to the third (second on cotton) of the 4 or 5 _H_. _armigera_ generations that may occur in a season. Use of endosulfan in cotton is also restricted in time, both to retain its own effectiveness, and because cross resistance is suspected.

IMPLEMENTATION

The first trials used the prototype system in the Namoi Valley, either on experiment stations or on grower's properties underwritten against financial loss. Subsequently, in the Namoi and elsewhere, the system has undergone at least two seasons of implementation trials referred to as Direct Research SIRATAC (Table 2), and sometimes a stage called Indirect Research SIRATAC, before being made available commercially. In these trials commercial feasibility was one of the goals, and farmers cooperated at their own risk. They were thus free to override the advice, or pull out of the trial at any time.

Cotton farmers whose management practices and farm size varied widely were canvassed as potential cooperators in Direct trials; a subset agreed to participate. Direct SIRATAC trials were run by research staff, who collected the data and ran the program. In the Indirect trials, a grower (preferably a former Direct SIRATAC cooperator) used SIRATAC himself, with advice from the Direct SIRATAC personnel, on a larger area of crop than had been used in Direct trials. The grower arranged for the data to be collected and the program to be run, but was not charged for using it.

The demand from growers for a tenfold expansion in the acreage using SIRATAC, led the three organizations that gin and market cotton in N.S.W. and Queensland to form a company, SIRATAC Ltd., in 1981/82. The company now distributes the system commercially, provides technical support, and charges growers for these services.

Organizational Structure and Procedures

RESEARCH TRIALS   In 1979/80, the first year with multiple on-farm trials, a Development Officer (DO) was appointed by CSIRO, to run the trials, interpret the program to the growers (later including Indirect cooperators), and liaise between growers and researchers. As implementation expanded, a DO was hired for each region in N.S.W. where there were on-farm research trials. This role in Queensland was filled by research staff and a private consultant. A SIRATAC Coordinator was appointed from among the researchers to oversee the activities of the DO's, provide a channel of communication between them and researchers, and organize the updating of the SIRATAC program. He also decided, after consulting with the whole team, which aspects of the program or procedures should be changed.

The Coordinator instituted weekly 1-2 hour workshops during the growing season for DO's from nearby regions and interested researchers. DO's from further regions reported to the Coordinator by phone. The week's points of interest or anomalous decisions were discussed. If necessary, the Coordinator authorized minor changes to the program. Major changes were deferred to the end of the season, when separate, longer workshops were held, first for users, and then re-

Table 2. Responsibility for SIRATAC operations at various stages of implementation

Operation.	Research SIRATAC Direct	Indirect	Commercial SIRATAC
Sample crop	scientist (D.O.) or assistant	farmer or commercial scout	farmer or commercial scout (A.S.O.)
Input data to computer	" " "	" " "	" " "
Discuss output with farmer	scientist (D.O.)	scientist (D.O.)	SIRATAC Co. (T.O.)
Supervise, advise and liaise with farmer on SIRATAC operations	" "	" "	" " "
Update computer program	research scientist	research scientist	research scientist

D.O.   = Development Officer
A.S.O. = Approved SIRATAC Operator
T.O.   = Technical Officer

searchers, to meet with the Coordinator and DO's. Doubtful decision advice from the program was discussed more fully than was possible during the season. The Coordinator evaluated participants' suggestions for improving the program or procedures, and planned their implementation for next season. Queensland research and development personnel also attended end of season workshops at Narrabri, and at least once each season, members of each state's research teams visited SIRATAC trial sites in the other state.

COMMERCIAL SIRATAC   SIRATAC Ltd. is a private, non-profit company whose members are cotton growers, chemical company personnel, private consultants, and others (5). Initially, the ACGRA lent the Company money to buy a computer, and growers voted support for two years with a levy of 25c per bale of cotton ginned. Most of the company's subsequent income is from charges for service.

The Company employs Technical Officers (TO's) who fill the role played by Development Officers in Indirect Research SIRATAC. The Technical Manager relates to growers and TO's, as did the SIRATAC Coordinator. These support staff ensure that standards of data collection are maintained, reassure users about the program's intentions, and help them use it. The Company also offers very limited Direct SIRATAC service, which includes scouting.

Under a contract with the CSIRO Division of Plant Industry and the NSW Dept. of Agriculture, SIRATAC Ltd. has sole rights in Australia to market SIRATAC services, i.e. use of the computer program and databases, with accompanying

261

technical support. The research organizations retain ownership of the software, and are responsible for maintenance and updating. By 1984/85 there were 5 TO's and 2 trainees; growers paid $9.26/ha to use the program and support services, and $27.50/ha if scouting was included (used on about 3% of the area). In central Queensland, where no Company staff, and hence no support services were yet available, access to the system was free of charge.

A cotton grower who wishes to use SIRATAC must be, or obtain the services of, an Approved SIRATAC Operator (ASO). To become an ASO, originally a person had to attend a training day conducted by research staff each year, and use the system responsibly. ASO training is now done by Company personnel on a one-to-one basis. ASO's include cotton growers, independent cotton production consultants, agri-business company agronomists and, in recent years, pesticide company personnel. An ASO can serve one or more growers. The ASO or casual labor under his/her supervision samples the cotton fields and runs the program. The ASO then consults with the grower about the output, and orders a spray if necessary. When problems arise, prompt support is available by telephone or electronic mail. The Company's computer system manager trouble-shoots computing problems. Serious problems are reported to the SIRATAC research team at the weekly workshop unless urgent.

The weekly and season end workshops instituted by the SIRATAC Coordinator have continued, but the weekly meetings now also involve the SIRATAC Ltd. Technical Manager and company TO's, and the end of season meetings with growers are organized by the Company. Research personnel are still involved in the growers meetings, however, and the Coordinator still decides on program changes and organizes updating.

Practical Aspects of Implementation

MANAGEMENT UNITS AND LARGE SCALE SAMPLING GUIDELINES   At the start of the season, a grower and his ASO divide his cotton into management units (MU's), each requiring a separate tactical decision. An MU consists of one or more fields which are expected to be relatively homogeneous. MU's range in size from about 100 to 1000 hectares, but can be subdivided during the season if necessary. In each MU, a sampler must complete at least two sampling cards for the first 100 hectares and at least one more card for each further 100 hectares.

COMPUTER RELATED ASPECTS   It was decided to locate the program on a central mini-computer, rather than distributing it to growers on their own micro-computers. A centralized system has several advantages: any adjustments made to the program are immediately available to all users, which facilitates the frequent updating integral to SIRATAC; building and managing an integrated database for updating the program, from data collected on-farm by SIRATAC, is easier; and users are not responsible for computer maintenance. This last is especially important for users in remote locations.

The feasibility of centralizing an on-line pest management system, however, depends on both the efficacy and cost of data transmission. All data from the early SIRATAC on-farm trials was brought back to Narrabri for entry into the computer. A few attempts to use on-farm terminals with acoustic couplers on the existing telephone system have been successful, but others suffered corruption of data in transmission, because telephone lines were old and substandard. Most problems were solved by installation of hard-wired modems. At one site, data is even successfully transmitted through a manual telephone exchange, with the educated cooperation of exchange operators. Given the large area over which the system is now distributed, the cost of long distance telephone connection to the computer could deter some growers from using it. Cheaper alternatives for distant growers are now available: a batch version of SIRATAC, and a distance independent data transmission service for low volume traffic using packet switching (AUSTPAC), offered by the national telephone authority. Batch SIRATAC, in which the user can enter the data, disconnect, and call back later to receive the output, would be cheapest with an intelligent terminal that allows entry and storage of data before the phone connection is made to send it.

Each MU has a separate computer account, with its own password, known only to the ASO. After the season the data becomes available anonymously for research, but remains confidential to SIRATAC Ltd. and the cooperating research organizations. No user can access another user's data, nor is it available to any other government agency (5).

The SIRATAC program has been made progressively more unambiguous and convenient to use. Data is entered as answers to a series of questions. Wherever possible, the program checks the answer given; if the answer is impossible or unlikely the question is repeated. An experienced user may use a version with abbreviated questions; the questions for novices are more explicit. The output is split into different logical groupings. If busy, a user can select vital parts of it and receive the rest later. Spray advice is always given.

For a system such as SIRATAC, with users, Company support personnel, and even some researchers, not involved in development of the program, four levels of documentation are necessary. For users, a manual; for Company technical staff, scientists collaborating in program development and outside scientists and administrators, a jargon-free overview of the program, showing the modular structure and role of each subroutine; and for a newcomer to work on the coding, both a manual describing how each subroutine works, and adequate commenting in the code. The last three levels are currently inadequate in SIRATAC.

SYSTEM EVALUATION

Acceptance By Growers

Starting in the Namoi Valley, NSW, the SIRATAC system has now spread to all commercial cotton growing areas of Australia (Table 3). By the 1984/85 season it was used on 260 management units constituting 44,500 hectares, or 27.5% of all Australian cotton acreage. The MU's were on 138 farms; this was 27% of all cotton farms, suggesting that average farm size in SIRATAC was similar to the overall average. The level of acceptance of SIRATAC reflects both growers' attitudes to innovation, and receptiveness to a pest management approach, and varies widely between cotton growing regions.

But even growers who pay for the right to use SIRATAC vary in their willingness to follow its advice (see next section). However, in 1982/83 resistance to synthetic pyrethroids by H. armigera caused field failures of control in the Emerald region. Resistant individuals were subsequently found at low frequencies in several other regions where control was still satisfactory (17). Some larvae tested showed resistance to the full range of synthetic pyrethroids currently marketed in Australia. This has renewed awareness among growers of the risks of indiscriminate pesticide use, and hence somewhat increased receptiveness to SIRATAC.

Progress Toward Achievement of Goals

Direct Research SIRATAC usually involved paired block comparisons. Two blocks of cotton on the same farm, chosen to be as similar as possible in agronomy and yield expectation, were used. SIRATAC managed pests on one block; the farmer used his own criteria to manage pests on the other, and was responsible for all other operations on both blocks. The Development Officer and his assistants sampled both blocks, but gave the farmer information only about the SIRATAC block, not the one he was managing, lest it influence his control decisions.

Table 3. Expansion of SIRATAC through nine seasons, starting with the first trial with the prototype in 1976/77.

Year	Total Hectares of Cotton in Australia	SIRATAC	%	SIRATAC used in # Regions	# M.U.'s	# Cotton Farms in Aust.	SIRATAC
76/77	32000	R. 10		1 (N)	1	N.A.	1
77/78	38000	" 9		1 (N)	1	"	1
78/79	45000	" 14		1 (N)	1	"	1
79/80	71414	" 410	0.6	1 (N)	5	"	5
80/81	83800	" 1143	1.4	4 (2N,2Q)	9 (6N,3Q)	"	11
81/82		R. 436		6 (1N,5Q)	12 (3N,9Q)		
		C. 8292		2 (both N)	49		
	103800	T. 8718	8.4	8 (3N,5Q)	60	276	N.A.
82/83		R. 2829		7 (1N,6Q)	33 (9N,24Q)		
		C. 10065		2 (both N)	87		
	95870	T. 12894	13.5	9	120	388	N.A.
83/84	138000	T. 32000	23.5	9 (3N,6Q)	230	465	N.A.
84/85	162000	T. 44500	27.5	9 (3N,6Q)	260	500	138

R. = Research SIRATAC; C. = Commercial SIRATAC; T. = Total.
N = in New South Wales, Q = in Queensland.
M.U. = management unit
N.A. = data not available

Results of N.S.W. and Queensland trials are shown separately (Table 4), as climate, soils, and pest complex, all of which influence potential yields, differ more between states than between sites within states. Comparisons invalidated by the two blocks being differentially damaged by such things as hail, herbicides, etc., are omitted. Mean yields and quality of SIRATAC and commercially managed comparison blocks were very similar over nearly all years in both states (2, 25, 33). An exception occurred in 1981/82, before SIRATAC had the mirid alarm. A mirid outbreak caught cotton farmers unawares, but other pest numbers were low so SIRATAC cotton had received few sprays. So where mirids were abundant, it was more delayed, and hence lost more quality, than cotton which had been receiving prophylactic sprays.

The average number of sprays on SIRATAC blocks showed no trend over the six seasons of N.S.W. trials, although there were year to year differences reflecting variation in pest pressure. Over this period, however, there was a decrease in the average number of sprays applied both to the commercial comparison blocks and on cotton farms not using SIRATAC. This reflected a shift from prophylactic to responsive spraying, as more farmers used the services of independent consultants or hired scouts. These trends appear to have been encouraged by the on-farm SIRATAC trials (7,9). SIRATAC thus influenced even farmers who were neither cooperators nor users and hastened improvement of cotton pest management in general. However, this reduced the contrast between SIRATAC and commercial practice.

Either spray timing or spray type often differed between the SIRATAC and commercial comparison block, even when both had the same total number of sprays. Except in some early Queensland trials, SIRATAC management usually tolerated more

Table 4. Results of Direct SIRATAC paired comparison trials.

| Season | n | Ave. # Sprays | | Cost Saved | Ave. Yield | | Increase in net return |
		S	C	S-C ($/ha.)	S	C	S-C ($/ha.)
New South Wales trials:							
76/77	1	8	11	-19	5.0	4.9	21*
78/79	1	11	16	125	7.1	7.4	17*
79/80	5	6.6	10.6	57	6.5	6.5	38
80/81	4	8.3	8.8	24	5.5	5.5	16
81/82	3	8.3	8.7	28	6.6	6.5	48
Queensland trials:							
80/81	3	10.0	8.0	26	5.7	5.3	85
81/82	9	9.7	9.8	13	4.8	4.8	26

n = number of trials   * scouting costs not included at this stage
S = SIRATAC fields
C = commercial comparison fields

Heliothis larvae than commercial management. Differences in spray type mainly reflected the SIRATAC philosophy of grading the spray response to Heliothis from "soft" to "hard" according to need, but sometimes also reflected less need to spray the SIRATAC block for induced aphids and mites. Fewer of the sprays applied to SIRATAC blocks were "hard" chemicals; more of them were "soft" than on the commercial blocks, and often more were "intermediate" type (Figure 4).

In 6 of the 7 within-state comparisons (Table 4), spray costs were lower under SIRATAC than commercial management. Average net returns on the SIRATAC blocks were higher than on the comparison blocks in all 7 trials, by $16 to $85 per hectare. Although only a 2% increase on average (range 0.7 to 4.5%) in net returns, this was worth thousands of dollars in the larger, later trials; e.g. in 1981/82 net returns on the 436 ha. in Direct SIRATAC trials exceeded those on the comparison blocks by $11,000.

Thus in the paired comparison trials SIRATAC showed real progress towards its goals of reducing biological costs while maintaining profitability of cotton production. To maintain this progress since commercial release has been more difficult. In the first year of commercial use (1981/82), only 34% of the sprays applied by users followed SIRATAC advice: 41% were applied when SIRATAC advised no spray, the other 26% used a spray type different from that advised by SIRATAC. Even when the number of sprays applied was similar to the number advised, often sprays were applied when not recommended, but were balanced by recommended sprays that were not applied. This discrepancy between advice and practice has continued but varies between regions. By 1984/85 the number of sprays applied ranged among regions from 10% less to 75% more than recommended, while spray type advice was disregarded in 15 to 52% of cases. The lower figures were from the region where SIRATAC had been used longest suggesting increased acceptance with experience and education.

Not all the sprays applied when not advised can be regarded as unnecessary because often such a spray would have pre-empted a subsequent recommendation so the discrepancy became one of timing. Nevertheless the question arises: why do growers pay to use SIRATAC and then often disregard its advice? Growers' comments in conversation and at meetings suggest they are attracted by the prospect of in-

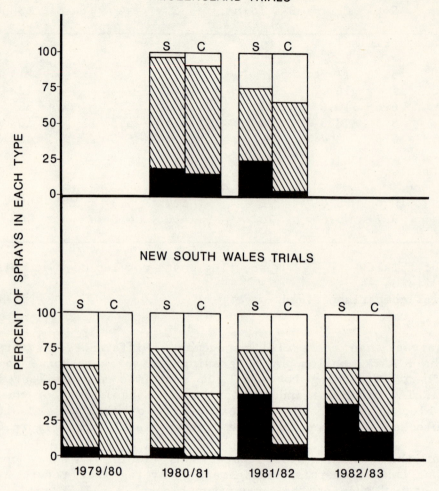

Figure 4. Percent of sprays that were "soft" (black), "intermediate" (hatched) and "hard" (white) (as defined in text) on SIRATAC (S) and commercial comparison (C) blocks in paired comparison Direct SIRATAC trials (data unavailable for some years and some farms).

creased net returns, and by the crop information, and convenient fruit, pest, and management records. But many prefer to make their own final decision, and use SIRATAC as an objective yardstick to make them consider it more thoroughly.

There appeared to be two main reasons users chose to spray more often than advised. Firstly, cotton growers are risk averse, and tend to apply pesticides when faced with uncertainties such as the initially unrecognized mirid outbreak in 1981/82, or the threat in 1982/83 that water shortage would reduce the crop's capacity to compensate for early damage. Secondly, growers tend to pool heterogeneous management units and so spray MU's that don't need it, with ones that do.

SIRATAC often advises inaction, which is psychologically unpalatable, and growers need constant assurance of the program's intent. The need for experienced staff to maintain frequent contact with users was initially underestimated, and it has taken time to build a team of trained TOs, especially as none of the TO's have been former DO's. The requirement for annual registration of ASO's has helped by eliminating those most likely to abuse the system, but much more than annual training is necessary for SIRATAC's full potential to be realized.

CONCLUDING REMARKS

SIRATAC still needs refining, its scientific base needs strengthening so it gives better advice, and the level of discriminating compliance by users must be increased. Nevertheless it has had more commercial success, i.e., more of an impact on pest management practices in the industry, than have some other pest management systems. It is worth considering the features not shared by such systems, that appear to have been advantageous to SIRATAC:

(i)   Models in SIRATAC have been kept simple so they require as few input variables as possible, and so the program is quick to run. They are only made more complex if decision making is improved (itself often a subjective decision!). Changes are evaluated initially by test runs with previous seasons' data and subsequently by performance in the field.

(ii) Development personnel have been essential. The traditional roles of research and extension are adapted to fine-tuning a relatively stable technology that is well understood by both groups. They are not well suited to the application of computer technology, which is more like a quantum jump in traditionally less quantitative areas such as farming.

(iii) Continuing service-type support from research staff, i.e., addressing grower meetings, and continuing to educate and trouble-shoot for Company personnel, has been needed. Again, this conflicts with traditional roles.

(iv) Because of the high investment in and returns from cotton growing, growers tend to be more technologically aware and innovative than many other farmers. Also, the cotton industry in Australia is cohesive, with relatively few growers, mainly aggregated in areas where irrigation water is available. Good communication among growers, and between growers and researchers results.

(v)   Cooperation has outweighed competition, both between and within the research groups working to develop pest management for Australian cotton.

Despite these advantages, SIRATAC's ability to achieve its goals of long term economic and ecological viability for the cotton industry will ultimately depend on a high proportion of growers sharing these goals.

ACKNOWLEDGEMENTS

SIRATAC is the result of team work and in this article we have had the privilege of reporting the work of many people. We warmly acknowledge the contributions to SIRATAC of: N. Ashburner, K.D. Brook, P.O. Cull, N.W. Forrester, D. Murray, W.A. Palmer, B. Pyke, P.M. Room, G. da Roza, N.J. Thomson, L. Tuart, P.H. Twine, L.T. Wilson and A.G.L. Wilson. We also acknowledge the enthusiastic support of cotton growers led by Mr. R.A. Williams and Mr. R.S. Browne and the directors and staff of SIRATAC Ltd. We also thank D.A. Andow for reading and constructively questioning an earlier draft of the manuscript.

LITERATURE CITED

1. Anon. 1984. Cotton: world statistics. Q. Bull of the International Cotton Advisory committee. 38:2 & 3.
2. Ashburner, N.A., Brook, K.D., da Roza, G.D. 1984. SIRATAC for all regions. Aust. Cotton Grow. Res. Conf. 1984. Proc. 207-215.
3. Blood, P.R.B., Wilson, L.T. 1978. Field validation of a crop/pest management descriptive model. In Proceedings SIMSIG Conference, Canberra, ACT, Australia. pp. 91-94.
4. Bottrell, D.G., Adkisson, P.L. 1977. Cotton insect pest management. Ann. Rev. Entomol. 22:451-481.
5. Brook, K.D., Hearn, A.B. 1983. Development and Implementation of SIRATAC: a computer based cotton management system. In Computers in Agriculture, Proc. 1st Nat. Conf. Univ. West. Australia. 222-241.

6. Brook, K.D., Hearn, A.B. 1984. A comparison of different SIRATAC systems of pest management -- initial results. Aust. Cotton Grow. Res. Conf. 1984. Proc. 167-177.

7. Browne, R. 1981. SIRATAC -- a grower's experience. Aust. Cotton Grower 2(4):7.

8. Colton, R.T., Cutting, F.W. 1983. Cotton Pesticides Guide 1983-84. NSW Dept. of Agric. Agdex #151/680. 26 pp.

9. Dowling, D., Cull, P.O. 1982. SIRATAC -- a summary of the 1980/81 season. Aust. Cotton Grower. 3(1):18-20.

10. Eichers, T.R. 1981. Farm pesticide economic evaluation 1981. USDA Econ. Res. Serv. Agric. Econ. Rep. 464. 21 pp. cited by H.T. Reynolds et al., Chapt. 11. In Introduction to Insect Pest Management, ed. R.L. Metcalf, W.H. Luckman. 2nd ed. pp. 69-91.

11. Eichers, T., Andrilenas, P., Blake, H., Jenkins, R., Fox, A. 1970. Quantities of pesticides used by farmers in 1966. USDA Econ. Res. Serv. Agric. Econ. Rep. 464. 21 pp. cited by H.T. Reynolds et al., Chapt. 11. In Introduction to Insect Pest Management, ed. R.L. Metcalf, W.H. Luckman. 2nd ed. pp. 69-91.

12. Forrester, N.W., Cahill, M. 1984a. Field evaluation of an insecticide management strategy for pyrethroid resistant *Heliothis armigera* (Hubner) in Australia. 17th Internat. Congr. Entomol, Hamburg. Abstract # S17.1. 13.

13. Forrester, N.W., Cahill, M. 1984b. Field evaluation of an insecticide management strategy for the control of pyrethroid resistant *Heliothis armiger*. Aust. Cotton Grow. Res. Conf. 1984, Proceedings, pp. 88-96.

14. Georghiou, G.P. 1983. Management of resistance in arthropods. In Pest Resistance to Pesticides, ed. G.P. Georghiou, T. Saito. Plenum, NY. 809 pp.

15. Goodyer, G.J., Greenup, L.R. 1980. A survey of insecticide resistance in the cotton bollworm *H. armigera* (Hubner) (Lepidoptera: Noctuidae) in New South Wales. Gen. Appl. Entomol. 12:38-40.

16. Goodyer, G.J., Wilson, A.G.L., Attia, F.I., Clift, A.D. 1975. Insecticide resistance in *Heliothis armigera* (Hubner) (Lepidoptera: Noctuidae) in the Namoi Valley of New South Wales, Australia. J. Aust. Entomol. Soc. 14:171-173.

17. Gunning, R.V., Easton, C.S., Greenup, L.R., Edge, V.E. 1984. Pyrethroid resistance in *Heliothis armigera* (Hubner) (Lepidoptera: Noctuidae) in Australia. J. Econ. Entomol. 77:1283-1287.

18. Headley, J.C. 1982. The economics of pest management. In Introduction to Insect Pest Management, ed. R.L. Metcalf, W.H. Luckman, pp. 69-91. 2nd Edition. 577 pp.

19. Hearn A.B., Brook, K.D. 1983. SIRATAC -- a case study in pest management of cotton. In New Technology in Field Crop Production. Refresher Training Course, Aust. Inst. Agric. Sci. Gatton, Queensland.

20. Hearn A.B., da Roza, G.D. 1984. Incorporating irrigation and nitrogen management in pest management decision making. Aust. Cotton Grow. Res. Conf. 1984. Proc. 238-249.

21. Hearn A.B., da Roza., G.D. 1985. A simple cotton fruiting model for crop management application. Field Crops Res. (in press).

22. Hearn A.B., Room, P.M. 1979. Analysis of crop development for cotton pest management. Prot. Ecol. 1:265-277.

23. Hearn A.B., Ives, P.M., Room, P.M., Thomson, N.J., Wilson, L.T. 1981. Computer-based cotton pest management in Australia. Field Crops Res. 4:321-332.

24. Ives, P.M. 1982. Natural mortality of *Heliothis* eggs and larvae and its utilization in pest management. Aust. Cotton. Grow. Res. Conf. 1982. Proc. 91-97.

25. Ives, P.M., Wilson, L.T., Cull, P.O., Palmer, W.A., Haywood, C. Thompson, N.J., Hearn, A.B., Wilson, A.G.L. 1984. Field use of SIRATAC: an Australian computer-based pest management system for cotton. Prot. Ecol. 6:1-21.

26. Kay, I.R. 1977. Insecticide resistance in *Heliothis armigera* (Hubner) (Lepidoptera: Noctuidae) in areas of Queensland, Australia. J. Aust. Entomol. Soc. 16:44-45.

27. Knipling, E.F., Stadelbacher, E.A. 1983. The rationale for areawide management of *Heliothis* (Lepidoptera: Noctuidae) populations. Bull. Entomol. Soc. Am. 28: 29-38.

28. Morton, N. 1979. Time related factors in *Heliothis* control on cotton. Pesticide Sci. 10:254-270.

29. Pedigo, L.P., Hutchins, S.H., Highley, L.G. 1986. Economic injury levels in theory and practice. Ann. Rev. Entomol. 31:342-388.

30. Poston, F.L., Pedigo, L.P., Welch, S.M. 1983. Economic injury levels: reality and practicality. Bull. Entomol. Soc. Am. 29: 49-53.

31. Pyke, B., Twine, P. 1982. SIRATAC in Queensland. Aust. Cotton Grower 3(1):17.

32. Pyke, B.A., Stone, M.E., Struss, S. 1984. SIRATAC development trials in the Callide and Dawson Valleys. Aust. Cotton Grow. Res. Conf. 1984 Proc. 250-258.

33. Room, P.M. 1979. A prototype 'on-line' system for management of cotton pests in the Namoi Valley, New South Wales. Prot. Ecol. 1:245-264.

34. Smith, R.F. 1971. Economic aspects of pest control. Proc. Tall Timbers Conf. on Ecol. Anim. Control by Habitat Manage. Tallahassee, Fla. 3:53-83.

35. Sterling, W.L. 1976. Sequential decision plans for management of cotton arthropods in South-east Queensland. Aust. J. Ecol. 1:265-274.

36. Stern, V.M., Smith, R.F., van den Bosch, R.H., Hagen, K.S. 1959. The integrated control concept. Hilgardia 29:81-101.

37. Thomson, N.J. 1979. Cotton. In Australian Field Crops, ed. J.V. Lovett, A. Lazenby, pp. 113-136. Vol. 2. London: Angus and Robertson. 328 pp.

38. Vandermeer, J., Andow, D.A. 1986. Prophylactic and responsive components of an integrated pest management program. J. Econ. Entomol. 79:299-302.

39. Wilson, A.G.L. 1974. Resistance of *Heliothis armigera* to insecticides in the Ord Irrigation Area, North Western Australia. J. Econ. Entomol. 67:256-258.

40. Wilson, A.G.L., Greenup, L.R. 1977. The relative injuriousness of insect pests of cotton in the Namoi Valley, New South Wales. Aust. J. Ecol. 2(3):319-328.

41. Wilson, A.G.L., Desmarchelier, J.M., Malafant, K. 1983. Persistence on cotton foliage of insecticide residues toxic to *Heliothis* larvae. Pestic. Sci. 14:623-633.

42. Wilson, L.T., Waite, G.K. 1982. Feeding patterns of Australian *Heliothis* on cotton. Environ. Entomol. 11:297-300.

43. Wilson, L.T., Room, P.M. 1983. Clumping patterns of fruit and arthropods in cotton, with implications for binomial sampling. Environ. Entomol. 12:50-54.

44. Wilson, L.T., Hearn, A.B., Ives, P.M., Thomson, N.J. 1983. Integrated pest control for cotton in Australia. In Guidelines for integrated control of cotton pests, ed. R.E. Frisbie. FAO Plant Production and Protection Paper 48. FAO, Rome. pp. 134-143.

Appendix 1.   Arthropod species that injure, and may damage Australian cotton.

Order	Family	Genus	Species	Common Name
CLASS INSECTA:				
Lepidoptera	Noctuidae	Heliothis	armigera	(introduced) bollworm
	"	Heliothis	punctigera	native budworm
	"	Earias	huegeli	rough bollworm
	"	Anomis	flava	cotton looper
	"	Chrysodeixis	argentfera	tobacco looper
	"	Spodoptera	litura	cluster caterpillar
	"	Spodoptera	exigua	lesser armyworm
	"	Pseudaletia	convecta	common armyworm
	"	Persectania	dyscrita	inland armyworm
	"	Agrotis	infusa	Bogong moth (a cutworm)
	Gelechiidae	Pectinophora	scutigera	pink spotted bollworm
	Tortricidae	Crocidosema	plebeiana	cotton tipworm
	Pyralidae	Loxostege	affinitalis	cotton web spinner
		Dichocrocis	punctiferalis	yellow peach moth
	Lyonetiidae	Bucculatrix	gossypii	cotton leaf perforator
Thysanoptera	Thripidae	Thrips	imaginis	plague thrips
		Thrips	tabaci	onion thrips
Homoptera	Aphididae	Aphis	gossypii	cotton aphid (=melon aphid)
	Cicadellidae	Austroasca	viridigrisea	vegetable jassid
		Austroasca	terraereginae	cotton jassid Hemiptera
	Miridae	Creontiades(=Megacoelum)	dilutus	green mirid
		Campylomma	livida #	apple dimpling bug
		Taylorilygus	pallidus *#	broken backed bug
	Pentatomidae	Nezara	viridula	green vegetable bug
Orthoptera	Acrididae	Austracris	guttulosa	spur throated locust
Coleoptera	Chrysomelidae	Monolepta	australis	red shouldered leaf beetle

---

CLASS ARACHNIDA:				
Acarina	Tetranychidae	Tetranychus	urticae	two spotted mite
		Tetranychus	ludeni	bean spider mite

---

* Not proven to be damaging, but if sufficiently abundant its injury probably
  would cause damage.
Predators as well as plant feeders.
